Communications in Computer and Information Science   2743

Series Editors

Gang Li , *School of Information Technology, Deakin University, Burwood, VIC, Australia*
Joaquim Filipe, *Polytechnic Institute of Setúbal, Setúbal, Portugal*
Zhiwei Xu, *Chinese Academy of Sciences, Beijing, China*

Rationale

The CCIS series is devoted to the publication of proceedings of computer science conferences. Its aim is to efficiently disseminate original research results in informatics in printed and electronic form. While the focus is on publication of peer-reviewed full papers presenting mature work, inclusion of reviewed short papers reporting on work in progress is welcome, too. Besides globally relevant meetings with internationally representative program committees guaranteeing a strict peer-reviewing and paper selection process, conferences run by societies or of high regional or national relevance are also considered for publication.

Topics

The topical scope of CCIS spans the entire spectrum of informatics ranging from foundational topics in the theory of computing to information and communications science and technology and a broad variety of interdisciplinary application fields.

Information for Volume Editors and Authors

Publication in CCIS is free of charge. No royalties are paid, however, we offer registered conference participants temporary free access to the online version of the conference proceedings on SpringerLink (http://link.springer.com) by means of an http referrer from the conference website and/or a number of complimentary printed copies, as specified in the official acceptance email of the event.

CCIS proceedings can be published in time for distribution at conferences or as post-proceedings, and delivered in the form of printed books and/or electronically as USBs and/or e-content licenses for accessing proceedings at SpringerLink. Furthermore, CCIS proceedings are included in the CCIS electronic book series hosted in the SpringerLink digital library at http://link.springer.com/bookseries/7899. Conferences publishing in CCIS are allowed to use our online conference service (Meteor) for managing the whole proceedings lifecycle (from submission and reviewing to preparing for publication) free of charge.

Publication process

The language of publication is exclusively English. Authors publishing in CCIS have to sign the Springer CCIS copyright transfer form, however, they are free to use their material published in CCIS for substantially changed, more elaborate subsequent publications elsewhere. For the preparation of the camera-ready papers/files, authors have to strictly adhere to the Springer CCIS Authors' Instructions and are strongly encouraged to use the CCIS LaTeX style files or templates.

Abstracting/Indexing

CCIS is abstracted/indexed in DBLP, Google Scholar, EI-Compendex, Mathematical Reviews, SCImago, Scopus. CCIS volumes are also submitted for the inclusion in ISI Proceedings.

How to start

To start the evaluation of your proposal for inclusion in the CCIS series, please send an e-mail to ccis@springer.com

Frédéric Barbaresco · Gerin François
Editors

Quantum Engineering Sciences and Technologies for Industry and Services

First International Conference, QUEST-IS 2025
Paris, France, December 1–4, 2025
Proceedings, Part I

 Springer

Editors
Frédéric Barbaresco
Thales
Vélizy-Villacoublay, France

Gerin François
SEE
Paris, France

ISSN 1865-0929   ISSN 1865-0937 (electronic)
Communications in Computer and Information Science
ISBN 978-3-032-13851-4   ISBN 978-3-032-13852-1 (eBook)
https://doi.org/10.1007/978-3-032-13852-1

This Springer imprint is published by the registered company Springer Nature Switzerland AG
The registered company address is: Gewerbestrasse 11, 6330 Cham, Switzerland

If disposing of this product, please recycle the paper.

Preface

From Quantum Engineering to Applications for Citizens

At the heart of French deep tech, academic and industrial, QUEST-IS 2025, the 1st International Quantum Engineering conference and exhibition, gathered many companies working in the quantum ecosystem from various countries.

It thus addressed technical matters such as quantum computing and algorithms, quantum sensors, quantum communications, crypto and the Internet, including enabling technologies, and obviously quantum engineering targeting applications for citizens.

This multidisciplinary event let attendees from various backgrounds discuss the technical challenges of applied quantum technologies and business opportunities.

We present in this proceedings peer-reviewed papers addressing challenges on quantum engineering.

Quantum Engineering Challenges and Issues for Quantum Computing, Quantum Sensors and Quantum Communication

Quantum engineering denotes the systematic endeavour to transform the abstract formalism of quantum mechanics into practical technologies. It occupies a unique position at the confluence of physics, materials science, control theory, computer science, and systems engineering. The field encompasses three principal technological domains, quantum computing, quantum sensing, and quantum communications, each facing distinct, though deeply interconnected, challenges. Despite their differences, these disciplines share a common difficulty: the engineering of systems that rely upon fragile quantum states, whose coherence can be destroyed by the slightest interaction with their environment.

Quantum computing aspires to create fault-tolerant logical qubits but is constrained by decoherence and the immense overheads of error correction. Quantum sensing seeks ultimate precision, yet must balance coherence preservation with environmental coupling. Quantum communications endeavour to distribute entanglement securely over global distances, but remain limited by photon loss, imperfect memories, and the absence of scalable network architectures.

The future of quantum engineering lies in the emergence of a new discipline, quantum systems engineering, which formalises the design, verification, and integration of quantum devices in the same way that aerospace or microelectronics engineering did for their respective domains. The goal is not merely to understand quantum mechanics but to harness it reliably, reproducibly, and at scale, thereby ushering in the true technological era of the quantum world.

Quantum Computing

In quantum computing, the overarching objective is to construct machines capable of performing computations by manipulating coherent quantum states, superpositions and entanglements, across large numbers of qubits. The central challenge lies in the extreme delicacy of such states. Decoherence, resulting from unwanted coupling between a quantum system and its surroundings, rapidly erases the information encoded in superposition and entanglement. Extending coherence times whilst maintaining precise operational control constitutes one of the most formidable undertakings in modern engineering. Achieving this requires ultra-pure materials, cryogenic operation at temperatures frequently below tens of millikelvin, and stringent isolation from electromagnetic and vibrational disturbances. Even the smallest imperfections in fabrication or fluctuations in temperature and electromagnetic fields may induce decoherence and computational error.

Given that all physical qubits are inherently noisy, large-scale computation depends critically upon quantum error correction (QEC). Unlike in classical systems, redundancy cannot be implemented by straightforward duplication, since quantum states cannot be cloned. Instead, information must be encoded across entangled collections of physical qubits to form logical qubits that are resilient to noise. Implementing QEC on this scale is immensely demanding: thousands of physical qubits may be required to realise a single fault-tolerant logical qubit. Quantum dissipation engineering is emerging based on quantum feedback for bosonic qubits, to reduce drastically the number of physical qubits and by regulating qubit errors through thermodynamics dissipation (the Lindblad equation). Each qubit must be precisely calibrated, and errors must be identified and corrected in real time, often within microseconds. This imposes stringent requirements upon low-latency control electronics and cryogenic data processing.

A further source of difficulty arises from qubit control and crosstalk. Quantum gates must be executed with extremely high fidelity, typically exceeding 99.9%, if fault-tolerance is to be achieved. The generation of microwave or optical control pulses must therefore be exact, and must not disturb neighbouring qubits. As systems grow in size, the problems of frequency crowding, wiring density, and thermal load increase substantially. One promising avenue involves the development of cryogenic CMOS electronics placed in close proximity to the qubits, thereby reducing latency and electrical noise.

The issue of scalability is equally fundamental. Moving from devices containing a few dozen qubits to those comprising millions necessitates entirely new architectural paradigms. The wiring, signal routing, and thermal management of large cryogenic systems present severe engineering constraints. Modular architectures, wherein several small processors are interconnected by photonic links, may offer a practical path to scale. Another promising approach lies in the integration of heterogeneous qubit technologies, such as superconducting circuits, trapped ions, spin qubits, and photonic systems, into hybrid platforms. Ultimately, progress in this field depends upon a holistic approach known as quantum systems co-design, in which hardware, control electronics, and software are developed in concert, optimising the entire computational stack from materials to algorithms.

Quantum Sensing

Quantum sensing exploits the sensitivity of quantum systems to external perturbations in order to achieve measurements of unprecedented precision. These sensors can detect magnetic and electric fields, gravitational gradients, accelerations, or time intervals with sensitivities far beyond classical limits. Yet the very sensitivity that grants them such power also renders them vulnerable to environmental noise. The fundamental challenge is to preserve coherence long enough to perform a measurement, despite constant interaction with the environment.

Quantum sensors often employ nitrogen–vacancy centres in diamond, trapped ions, or ensembles of cold atoms in interferometric configurations. In every case, the sensor must be sufficiently coupled to the quantity of interest to register a measurable signal, while simultaneously being isolated from unwanted noise. Achieving this delicate equilibrium requires careful design of electromagnetic shielding, active feedback, and control of the working point so that the sensor operates at regions of reduced susceptibility to fluctuations.

As these technologies mature, miniaturisation and integration have become pressing concerns. Laboratory systems, typically large and fragile, must be transformed into compact, robust devices suitable for deployment in the field. This entails the integration of optical excitation, detection, and control subsystems within constrained physical volumes, as well as effective thermal management. Cold-atom systems, in particular, must balance vacuum integrity, temperature stability, and portability. Long-term calibration and drift correction present further engineering challenges, since even small instabilities in control parameters may degrade sensitivity over time. Autonomous recalibration and machine-learning-based feedback loops are therefore being investigated to ensure sustained performance.

Quantum Communications

Quantum communications seek to employ the principles of quantum mechanics, particularly entanglement and the no-cloning theorem, to enable the secure transmission of information. The best-known application, quantum key distribution (QKD), allows two parties to share cryptographic keys with absolute theoretical security. Beyond this, researchers envisage a quantum internet, capable of distributing entanglement between distant nodes for both secure communication and distributed computation.

The primary obstacle to such a network is photon loss and decoherence during transmission. Photons propagating through optical fibres are absorbed and scattered, restricting the direct distribution of entanglement to a few hundred kilometres. Free-space and satellite channels mitigate this limitation but suffer from atmospheric turbulence, beam divergence, and pointing errors. To overcome these difficulties, the community is developing quantum repeaters, which rely on intermediate nodes equipped with quantum memories to store, purify, and retransmit entanglement. This, however, requires quantum memories with long coherence times, high efficiency, and rapid read–write cycles, capabilities that remain technically elusive.

The development of reliable quantum memories is thus one of the foremost engineering challenges in the field. Candidate systems include ensembles of cold atoms, rare-earth-doped crystals, and spin systems in solid-state matrices. Each exhibits trade-offs between coherence time, storage efficiency, and compatibility with telecommunication wavelengths. The interface between photonic and matter qubits is similarly critical: efficient, low-noise conversion between optical photons and stationary qubits is essential for scalable networking. Approaches such as cavity quantum electrodynamics, optomechanical transduction, and electro-optic conversion are under active investigation, yet none has achieved the required combination of efficiency, fidelity, and scalability.

At a higher level, network synchronisation and control present formidable obstacles. Entanglement distribution and quantum teleportation protocols require extremely precise timing and phase stability across multiple nodes, often at the nanosecond scale. Large-scale quantum networks must integrate quantum channels with classical communication layers for coordination, routing, and error management. The absence of universally accepted standards and protocols compounds these difficulties. Different experimental platforms employ incompatible encodings, operating wavelengths, and modulation schemes. Achieving interoperability across heterogeneous systems will be crucial to realising a practical quantum internet.

Enabling Technologies and System Integration

Across quantum computing, sensing, and communications, certain engineering challenges are universal. Materials science remains a decisive factor: impurities, defects, and surface roughness all shorten coherence times and reduce gate fidelity. Cryogenics presents another formidable constraint; maintaining stable temperatures at the millikelvin level demands complex and expensive refrigeration systems that are difficult to scale industrially. Quantum control theory is assuming an increasingly central role, providing the mathematical framework for designing optimal pulse sequences and feedback algorithms that stabilise quantum dynamics in the presence of noise.

System integration poses further difficulties, since quantum subsystems must interact seamlessly with classical electronics responsible for control, measurement, and feedback. Timing, thermal management, and signal integrity must all be maintained under extreme conditions. Even the task of verification and validation is non-trivial, for measurement inevitably disturbs the quantum state being tested. Benchmarking and certification of quantum devices thus require sophisticated statistical and tomographic techniques. Lastly, the issue of scalability and manufacturability looms large. Many existing devices are artisanal laboratory constructs, far removed from the robust, reproducible systems required for industrial deployment. The transition to mass-manufacturable quantum technologies will necessitate advances in fabrication processes, design automation, and quality assurance.

Reviewing, Support

Again, we present in this proceedings peer-reviewed papers addressing challenges on quantum engineering. We thank the 102 reviewers from 17 countries who provided an average of 3.5 reviews per paper for the submitted 120 reviewed papers, using a single-blind process. Of the reviewed papers, 68 were selected papers for inclusion in these proceedings.

Support for the event came from Belgium, Canada, Denmark, Finland, France, Germany, Greece, Italy, Japan, The Netherlands, Saudi Arabia, Singapore, South Korea, Spain, Sweden, Switzerland, Taiwan, UAE, and the UK, also from the EU, CERN, and Quantum Flagship, and from our Platinum sponsors, Agence Innovation Défense, EDF, IBM, and Thales, and our Gold sponsors, Laboratoire National de Métrologie et d'Essais, CEA, NATO, Alice & Bob, and Quandela.

October 2025   François Gerin
                                                     Frédéric Barbaresco

Organization

General Chairs

Frédéric Barbaresco   Thales, France
François Gerin   SEE, France

Local Organizing Chair

Loïc Cantu   SEE, France

Program Committee Chairs

Frédéric Barbaresco   Thales, France
François Gerin   SEE, France

Program Committee

Neil Abroug	Inria, France
Khulud Almutairi	King Saud University, Saudi Arabia
Yasutaka Amagai	AIST, Japan
Nina Amini	CentraleSupélec, France
Dimitris Angelakis	TU Crete and AngelQ Quantum, Greece and CQT, Singapore
Thomas Antoni	CentraleSupélec, France
Sofiane Bahbah	SEDI.ATI, France
Bhashyam Balaji	DRDC Ottawa, Canada
Quentin Barbeau	Danish Embassy, Paris, France
Pascale Bendotti	EDF R&D, France
Mathieu Bertrand	Thales Alenia Space, France
Quentin Bodart	Crystal Q. Computing, France
Nadia Boutabba	IAT, Abu Dhabi, UAE
Philippe Bouyer	Quantum Delta, Netherlands
Harry Buhram	Quantinuum, UK
Andrea Busch	Quobly, France
Marie-Elisabeth Campo	EM Armée air & espace, France

Joan Camps	Riverlane, UK
Victor Canivell	Qilimanjaro, Spain
Patrick Carribault	CEA DAM hpc, France
Adam Connolly	Quantinuum, UK
Franck Correia	DGA, France
Giacomo Corrielli	Ephos/IFN-CNR, Italy
Stefan Creemers	UC Louvain, Belgium
Simon Crispel	GIFAS ALAT, France
Nicolas Daval	Quobly, France
Merouane Debbah	Khalifa University, Abu Dhabi, UAE
Fabrice Debbasch	Sorbonne Université, France
Ivo Pietro Degiovanni	INRIM, Italy
Guillaume De Giovanni	Viqthor, France
Philippe Deniel	CEA, France
Amer Delilbasic	FZ Jülich, Germany
Matthieu Desjardins	C12, France
Alain Dessertaine	La Poste, France
Amanda Diez	CERN, Switzerland
Daniel Dolfi	Thales, France
Christophe Domain	EDF R&D, France
Jens Eisert	Freie Universität Berlin, Germany
Andreas Elben	PSI, Switzerland
Jaap Essing	TNO, Netherlands
Olivier Ezratty	Independent Researcher, France
Nicolas Fabre	Télécom Paris, IPP, QuantEduFrance, France
Thomas Fauvel	Choose Paris Region, France
Benjamin Frisch	CERN, Switzerland
Jacques-Henri Gagnon	Canadian Embassy, Paris, France
Luc Gérardin	University of Sussex, UK
James A. Grieve	TII Abu Dhabi, UAE
Fabrice Guillemin	Orange Labs, France
Stefan Haeussler	Hensoldt, Germany
Pascal Halffmann	ITWM Fraunhofer, Germany
Mélanie Hardman	NPL, UK
David Harvey	Thales, France
Winfried Hensiger	University of Sussex, UK
Mohamed Hibti	EDF R&D, France
Israel Hinostroza	CentraleSupélec/SONDRA, France
Magnus Hoijer	FOI, Sweden
Min-Hsiu Hsieh	Foxconn, Australia
Daniel Huerga	ONERA, France
Toshiyasu Ichioka	RIKEN, Japan

Samy Jousset	Région Ile de France, France
Mathieu Juan	Université de Sherbrooke, Canada
Marc Kaplan	Veriqloud, France
Jian Feng Kong	PIC IHPC, Singapore
Naoya Kono	Secretariat of Science, Technology and Innovation Policy, Japan
Romain Kukla	Naval Group, France
Philippe Lacomme	Université Clermont-Auvergne, France
Jacques-Charles Lafoucrière	CEA, France
Catherine Lambert	Académie des Technologies, France
Aolita Leandro	6G RC Abu Dhabi, UAE
Marc Leconte	F2S, France
Megan Lee	Quantum City, University of Calgary, Canada
Jeremy Leow Kwang Siong	FSTD, Singapore
François-Marie Le Régent	Pasqal, France
Tobias Lindstrom	NPL, UK
Craig Lloyd	NPL, UK
Tony Maindron	CEA, France
Joseph Mikael	EDF R&D, France
Swapan Mandal	Visva Bharati Institute, India
Sabrina Maniscalco	Algorithmiq, Finland
Ulrich Mans	Quantum Delta, Netherlands
Luigi Martiradonna	Riverlane, UK
Francesco Mauro	Sannio University, Italy
Mathias Möller	TU Delft, Netherlands
Mikko Möttönen	Aalto University, Finland
Stéphanie Molin	Thales, France
Liran Naaman	Quantum Delta, Netherlands
Yuichi Nakamura	NEC Corporation, Japan
Hiro Nakata	Jij, Japan
Frank Nielsen	Sony Computer Science Laboratories, Japan
Pierre-Elie Normand	Dassault Aviation, France
Myriam Nouvel	Thales, France
Yasser Omar	PQI (Portuguese Quantum Institute), Portugal
Dimitrios Papadimitriou	Université libre de Bruxelles, Belgium
Cécile Perrault	Alice & Bob, France
Frank Phillipson	TNO NL, OTAN Group, Netherlands
Nico Piatkowski	IAIS Fraunhofer, Germany
Jonathan Pisane	Thales, France
Mathilde Portais	Naval Group, France
Tony Quertier	Orange, France
Setra Rakotomavo	La Poste, France

Supporting Countries

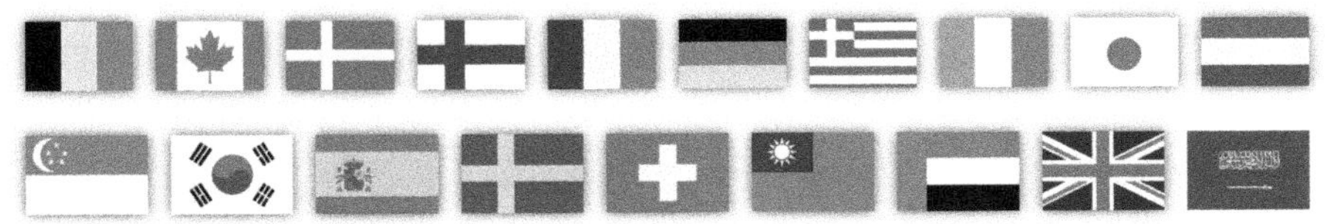

Sponsors

Keynote Talks

The Two Quantum Revolutions: From Concepts to Applications

Alain Aspect

Institut d'Optique - Université Paris-Saclay, France

The first quantum revolution, based on wave–particle duality, led to the society of information and communication. The second quantum revolution is based on entanglement. Will its applications lead to a new societal revolution?

Quantum Error Correction and Feedback

Pierre Rouchon

Center Automatic and Systems, Mines-Paris, University PSL, Member of Académie des Sciences, France

Quantum error correction relies on a feedback loop. This feedback generally corresponds to a classical controller. Quantum error correction can also exploit the dissipation associated with the phenomenon of decoherence. Called autonomous correction by physicists, it then uses feedback where the controller is a dissipative quantum auxiliary system. This talk focuses on the development of such quantum controllers to stabilize logical qubits encoded in harmonic oscillators (bosonic code). Two types of encoding will be considered: cat-qubit encoded in two coherent states of opposite phases for which bit-flip errors induced by usual noises can be experimentally almost suppressed; and GKP-qubit encoded in finite energy grid-states approximating position/impulsion Dirac combs where, in principle, both bit-flips and phase-flips could be almost suppressed.

The Interplay Between Quantum Engineering and Quantum Science

Olivier Ezratty

Freelance quantum engineer, mostly known for "Understanding Quantum Technologies"

Quantum engineering is a relatively new discipline that takes shape as quantum technologies are maturing and turning into commercial products. But what is it exactly? How are science and engineering intermingled in this innovation process? Is the science done, and we are just left with engineering and technology development? What is the engineering scope required for the development of complex quantum systems, particularly fault-tolerant quantum computers? How does quantum engineering connect the dots between the software and hardware stacks? Is the environmental footprint of these emerging technologies integrated in vendors' engineering goals? Are there quantum engineers? How are they trained and how will they be trained?

Quantum Enabling Technologies from Science to Engineering

Richard Versluis

Quantum Enabling Technologies Engineering, TNO/TU Delft, Netherlands

In this talk Richard will share an overview of the most critical engineering challenges in Quantum Computing that we will face in the upcoming years. He will delve into the transition from academic research to practical engineering, emphasising the complexities and hurdles that need to be overcome. Aspects like scalability, reliability and modularity will be quantified and related to practical design choices and design requirements for future quantum computers and their constituent components.

Experimental Quantum Photonics: From Testing Foundations of Quantum Mechanics to Building Practical Quantum Networks

Djeylan Aktas

Experimental Quantum Communications, Slovak Academy of Sciences, Slovakia

In this talk we will explore a few seminal experiments in which quantum photonics made it possible to interrogate nature itself by studying the very foundations of quantum theory. From there, we will see how technological progress allowed for scientists to "surf the wave" of the second quantum revolution to actually create practical quantum technologies that can be useful for society by bringing some quantum advantages like enhanced security for modern communication networks, or better sensing capabilities in various domains of applications.

The European Commission's Vision for Quantum Engineering: Challenges and Opportunities in EU-Funded Projects

Oscar Diez

European Commission Representative, DG CNECT

In this keynote, I will present the European Commission's strategy for quantum technologies and its role in building a globally competitive and resilient European quantum ecosystem. The talk will highlight how EU programmes, including Horizon Europe, the Quantum Flagship, the Chips JU and the EuroHPC Joint Undertaking, support the full quantum value chain, from foundational research and engineering to infrastructure, skills and industrial deployment. It will also address current challenges, such as scaling up quantum platforms, strengthening supply chains and fostering innovation across member states, and will outline upcoming opportunities for industry and research communities to engage with EU initiatives in quantum technologies.

Quantum Sensors

Marco Genovese

Director, Quantum Optics Research, INRIM, Italy

Quantum state coherence and entanglement are very sensitive to interaction with the environment. On the one hand, this represents a problem for developing quantum technologies such as quantum computation or quantum computation, but, on the other hand, it allows for a very high sensitivity to various parameters, leading to the possibility of developing quantum sensors largely surpassing traditional classical sensors. In this talk, I will present this exciting new field, discussing the potentialities of some of the most interesting techniques.

Fair Benchmarking of Quantum Optimisation Applications

Frank Phillipson

Quantum Computing Applications, TNO, Netherlands

Quantum computing is advancing rapidly, and optimisation is one of its most promising application areas. But measuring progress is not as simple as comparing runtimes or claiming "quantum advantage". In this keynote, I will explore why traditional benchmarking approaches – developed for CPUs and GPUs – fall short when applied to quantum and hybrid systems, and how misleading comparisons risk slowing genuine progress. Building on lessons from classical supercomputing, I will introduce principles for fair benchmarking that emphasise transparency, end-to-end workflows, solution quality, and reproducibility. I will also highlight recent initiatives, from Q-Score to TAQOS, and show how energy use and application-driven metrics are shaping the next generation of evaluation standards. The goal is to provide the community with a practical framework for assessing quantum optimisation responsibly – one that enables researchers, industry, and policymakers to interpret results with confidence and set realistic expectations for the future.

Telecom Integrated QKD Networks: The Madrid QCI Example

Vicente Martin

Quantum Engineering, CSS/Univ. Politécnica de Madrid, Spain

In the talk I will revise the types of QKD networks deployed in the field, the problems and different choices, concentrating on the solutions used in the Madrid network. This is a multidomain and highly heterogeneous network currently spanning more than 700 km, with 29 nodes hosting 23 QKD pairs from 10 different vendors. Finally several of the use cases implemented will be shown.

Contents

Session 2 Quantum Sensors – Quantum sensors in detection and sensing

Session 1 Quantum Communications – Test platforms for QKD

Session 2 Quantum Communications – Quantum Key Distribution

Session 3 Quantum Communications – Towards a Quantum Internet Infrastructure

**Session 1 Quantum Algorithms, Computing and Simulation –
Benchmarking A: Methods and Applications**

**Session 2 Quantum Algorithms, Computing and Simulation –
Benchmarking B: Methods and Applications**

**Session 3 Quantum Algorithms, Computing and Simulation – Error
Mitigation and Fidelity**

**Session 4 Quantum Algorithms, Computing and Simulation –
Quantum Algorithms for Finance and Industry**

Session Enabling Technologies

Thales' Roadmap and Product Innovations in Cryogenics for Quantum Applications

Christophe Vasse[1][(✉)], Garmt de Jonge[2], Emilien Durupt[1], Florian Dupe[1], and Daniel Willems[2]

[1] Thales LAS France, 26 Av. Jean François Champollion, 31100 Toulouse, France
christophe.vasse@fr.thalesgroup.com
[2] Thales Cryogenics BV, Hooge Zijde 14, 5626 DC Eindhoven, The Netherlands

Abstract. As the quantum technology landscape continues to evolve, the need for advanced cryogenic solutions becomes increasingly crucial. This paper presents Thales innovative products and strategic roadmap aimed at enhancing the capabilities of quantum applications through cryogenics on a range of temperatures (between 2 K to 80 K). We detail Thales state-of-the-art cryogenic technologies, including systems designed for cooling quantum processors, sensors, and other critical components, demonstrating their potential to optimize performance and stability in quantum systems. Furthermore, we outline our vision for the future of cryogenics within quantum technology, highlighting ongoing research initiatives, collaboration efforts, and upcoming product developments. Through a comprehensive analysis of market needs and technological advancements, we emphasize Thales commitment to addressing the challenges of cooling and maintaining quantum coherence.

Keywords: Cryogenics · Sensing · Cryocooler · SWaP · Thales

1 Introduction

In a rapidly evolving technological landscape, quantum systems are positioned as potential catalysts for transforming the way we process information, ensure the security of communications, and develop innovative solutions across various sectors. Cryogenics plays a vital role in optimizing the performance of quantum technologies by enabling the extremely low temperatures necessary for the manipulation and measurement of quantum sensors or quantum communication. This aspect is clearly identified in several publications [1–3].

Thales, a global leader in defense, security, and aerospace technologies, is actively engaged in contributing to this quantum revolution. This document presents Thales' roadmap for cryogenics, detailing our vision, objectives, and the necessary steps to develop cutting-edge cryogenic solutions. By aligning our research and development efforts with the needs of quantum applications, we aim to explore the frontiers of innovation while addressing the complex challenges facing the industry.

© The Author(s), under exclusive license to Springer Nature Switzerland AG 2026
F. Barbaresco and G. François (Eds.): QUEST-IS 2025, CCIS 2743, pp. 3–9, 2026.
https://doi.org/10.1007/978-3-032-13852-1_1

This roadmap will highlight the approaches and technologies that Thales intends to adopt to create compact, efficient and reliable cryogenic systems for applications between 2 K and 80 K. Through this initiative, Thales aims not only to strengthen its position in the cryogenics domain but also to actively contribute to the emergence of a robust and sustainable quantum ecosystem.

2 Cryogenics and Quantum Applications

2.1 Why Cryogenics for Quantum Applications?

Cryogenics is essential for quantum applications primarily due to the need for maintaining extremely low temperatures to enhance the sensitivity and stability of quantum sensors. Quantum systems are inherently delicate, and their operational stability can be compromised by environmental factors such as temperature fluctuations. At cryogenic temperatures, the thermal energy is significantly reduced, which allows qubits—whether they are based on superconducting circuits or trapped ions—to operate with greater fidelity and reduced error rates. Furthermore, many quantum systems leverage superconductivity, where materials exhibit zero electrical resistance and the expulsion of magnetic fields at low temperatures, enabling more efficient quantum computations and information transfer. Therefore, developing advanced cryogenic technologies is crucial for unlocking the potential of quantum computing, enabling breakthroughs in fields such as cryptography, materials science, and complex system simulations. Without effective cryogenic solutions, the performance and scalability of quantum devices would be severely hindered.

At the same time, cryogenics could have a big impact on system size, power consumption, thermal management and maintenance/reliability, etc.... Thus cryogenics solutions shall be developed with needs of specific applications in mind.

2.2 Cryogenics for Quantum Applications

Several applications are targeted for Thales cryocoolers. These applications are mainly linked to quantum sensing and quantum communication. The next table gives examples of applications with their cold temperature range (Table 1).

Table 1. Quantum applications with their cold temperature range.

Superconducting Quantum Interference Filters (SQIF)	60–80 K
RF superconducting filters	20–80 K
RF wide range signal treatment with doped crystal	2–3 K
Entangled Photon	2–3 K
Single Photon counter	2–3 K

These applications are described in several reports [2, 3].

These examples address both civil or defense applications. They can be embedded in aircraft or aboard ships as well as be ground based. We can summarize some of the main requirements for cryogenic coolers:

- Low cold temperature and large range of cold temperature (2 K = > 80 K)
- Compactness
- Efficiency/Low power consumption
- High availability (24/7 applications)
- No perturbation to the sensor (vibration, noise, electromagnetic compatibility,…)
- Can be easily embedded in aircraft, ships or satellites
- Low constraints regarding export control (dual use or not controlled)

3 Thales Roadmap

Thales Cryogenics today is an important supplier of cryocoolers for infrared sensor cooling and applications that require extremely reliable, long life coolers. Such products are ideally suited to be developed into coolers that can be used for quantum sensor cooling. Thales has started a program to develop cryocoolers for quantum sensors along three main axes:

- Optimization of existing cryocoolers regarding efficiency, compactness, and compatibility with sensors,
- Improvement of dewar maturity for quantum sensors,
- Extending our cooling range to 2–4 K.

3.1 40 K–80 K

For cold temperature between 40 K and 80 K, Thales Cryogenics intends to use existing COTS Stirling coolers with optimizations/actions (Fig. 1):

- Perform Cooler efficiency improvements in order to reach at least 40 K and maybe lower, without a big impact on the SWAP (Size, Weight, and Power),
- Analysis of potential impact of the cooler on quantum sensors (Electro-Magnetic Compliancy or induced vibrations),
- Dewar development for quantum sensors compliant with sensors constraints.

Rotary Stirling cryocooler

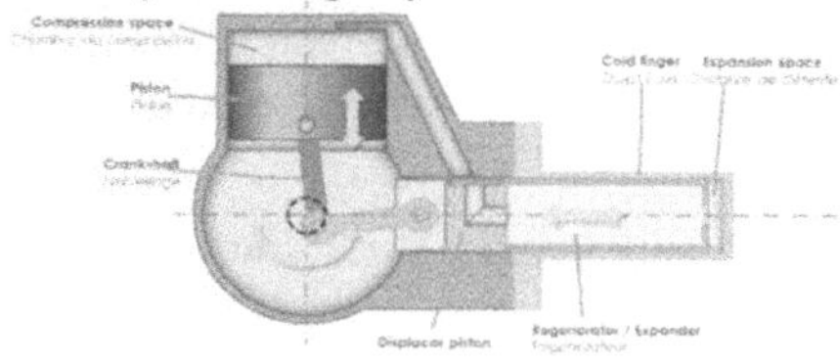

Linear Stirling cryocooler

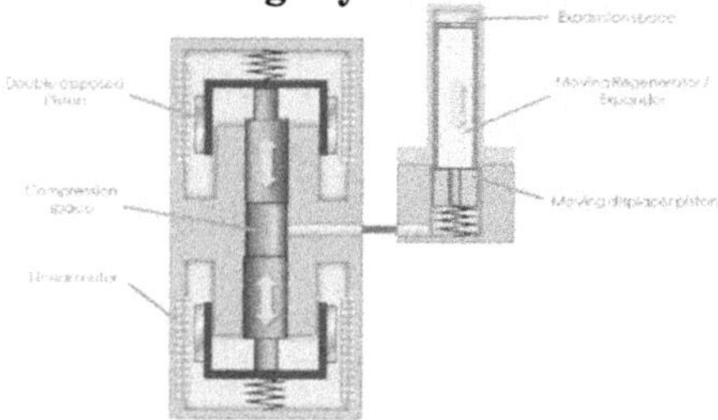

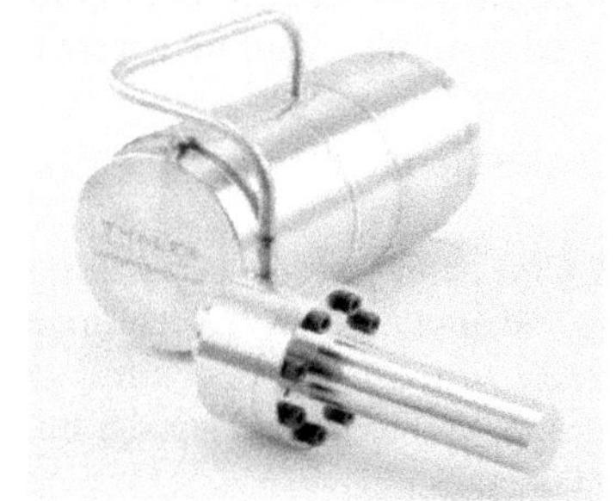

Typical characteristics:
- 60 K ➔ 150 K
- Weight: < 1 kg / Size: < 1 L
- Cooling power: < 1 W @ 80 K
- Electrical consumption: < 6W @ 80 K
- Reliability: > 40 000 h

Typical characteristics:
- 50 K ➔ 150 K
- Weight: < 2.5 kg / Size: < 2 L
- Cooling power: < 2 W @ 80 K
- Electrical consumption: < 50W @ 80 K
- Availability: > 98% after 5 years continuous use

Fig. 1. 40 K–80 K Stirling cryocooler technologies.

3.2 20 K–40 K

For the range 20 K–40 K, Thales Cryogenics will build on its heritage from Space coolers that cover this range, more specific two-stage pulse-tube coolers, by developing non-space /COTS variants.

The next picture shows an example of a two-stage pulse-tube cryocooler for space developed in collaboration with CEA [4] (Fig. 2).

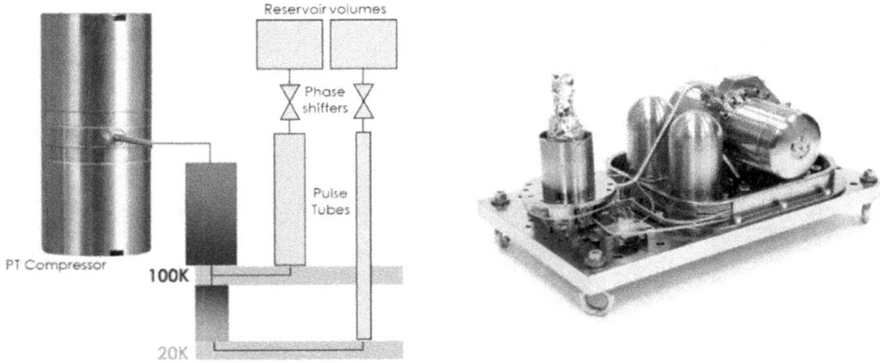

Fig. 2. 20 K–40 K pulse-tube cryocooler technologies.

Typical characteristics could be (development targets):

- 20 K → 80 K
- Electrical consumption: < 200 W
- Weight: < 15 kg/Size: < 10 L
- Availability: > 10 years
- Cooling power: > 3 W @ 80 K
    > 200 mW @ 20K

These pulse-tube coolers are so-called 'Stirling-type' pulse-tube coolers (as opposed to 'GM-Type pulse-tybe coolers). Stirling-type pulse tube coolers are driven at relatively high frequency by direct compression and expansion of the working fluid. There is no external compressor and no gas lines. Stirling-type pulse-tube coolers are maintenance free, extremely reliable, up to twice as efficient as a comparable GM-type pulse-pulse tube and significantly more compact.

3.3 2 K–4 K

For this lowest temperature range, a combination of several cooler technologies is needed. We aim to combine our strengths in pulse-tube cooler and compressor development and manufacturing with complementary technologies from selected partners. A suitable combination could be the combination of the Thales two-stage pulse-tube cooler mentioned earlier with a closed loop Joule Thomson cooler. We aim to have a portable, rack mountable solution that is highly reliable, scalable and cost effective.

The next schematic explains how a cooler solution could be envisaged (Fig. 3).

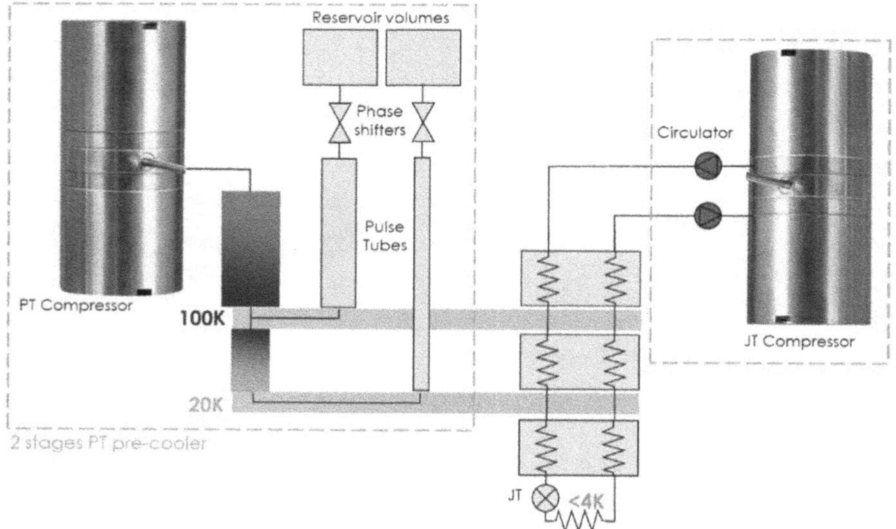

Fig. 3. 2K - 4K cryocooler schematic

Typical characteristics we aim based on feasibility studies are:

- $2\,K \rightarrow 4\,K$
- Weight: $< 20\,kg$
- 19 inches, rack mountable
- Cooling power: $< 3\,W$ of cold @ 80K
    $< 10\text{--}15\,mW$ of cold @ 2 K
- Electrical consumption: $< 300\,W$
- Availability: > 10 years

4 Conclusion

This paper has outlined Thales' strategic roadmap for advancing cryogenic technologies tailored to the demanding requirements of quantum applications. By leveraging our expertise in low-temperature engineering, thermal management, and integration of scalable cryogenic systems, with our capability to manufacture coolers in large numbers, we aim to enable compact, robust, efficient, and reliable quantum devices operating in the 2 K–80 K temperature range. Our roadmap addresses critical challenges such as high availability, manufacturability and cost, efficiency, minimizing perturbations, designing dewars, and ensuring system scalability and maintainability, all of which are essential for the transition from laboratory prototypes to industrial quantum solutions. Moving forward, Thales is committed to pushing the boundaries of cryogenics to support the emerging quantum ecosystem, fostering innovation across communication, sensing, and computation domains.

Thales Cryogenics is part of CryoNext, which is the cryogenics part of the French quantum plan.

References

1. Stratégie nationale sur les technologies quantiques – France relance (2021)
2. Rapport sur les technologies quantiques pour les applications Aéronautique-Spatial-Défense, GIFAS (2024)
3. Ezratty, O.: Understanding quantum technologies (2024)
4. Leenders, H., et al.: Development of a 30–50 K dual-stage pulse tube space cooler Cryogenics **88**, 91–100 (2017)

Preliminary Demonstration of Spin Qubit Control Using the Quantum Instrumentation Control Kit (QICK)

Dana El Hajj[1]([✉]), Vivien Schmitt[2], Xavier Jehl[2], Benjamin Criton[3], Guillaume De Giovanni[1], Jérémie Theze[1], and Arnaud Perrin[1]

[1] Viqthor, 78100 Saint-Germain-en-Laye, France
dana@viqthor.com
[2] Univ. Grenoble Alpes, CEA, Grenoble INP, IRIG, PHELIQS, 38000 Grenoble, France
[3] Université Paris-Saclay, CEA, IRFU, 91191 Gif-Sur-Yvette, France

Abstract. In this paper, we present initial results demonstrating the use of the Quantum Instrumentation Control Kit (QICK) to perform measurements on a spin Qubit. We used the RFSoC to perform energy selective spin readout through a resonator coupled to the qubit. Thanks to of a frequency up-converter we also drive the spin at the Larmor frequency beyond the RFSoC range, typically 15 GHz. These early-stage measurements provide evidence that QICK can be adapted for spin qubit control. This work opens a path towards using an Open-Source Hardware platform (QICK) to characterize and control Spin Qubits.

Keywords: QICK · RFSoC · Spin qubits · control electronics

1 Introduction

Silicon spin qubits are among the leading platforms for scalable quantum information processing, offering long coherence times, high-fidelity operations, and compatibility with industrial fabrication processes [1, 2]. Hole spin qubits, in particular, benefit from strong spin–orbit coupling, allowing for all-electrical qubit control via electric-dipole spin resonance (EDSR) [2]. However, implementing such control presents a significant challenge because of the requirement for precise, flexible, and low-latency electronics. In this work, we demonstrate the use of an FPGA-based solution for qubit control and readout using the Quantum Instrumentation Control Kit (QICK) [4].

Originally designed for superconducting qubits, we adapted QICK for the control of a single-hole spin qubit in a quantum dot formed in a natural silicon nanowire. This system performs resonator-based readout, qubit manipulation, and data acquisition. We validated this platform with a sequence of typical spin-qubit experiments performed on a characterized qubit [9] running in a dry dilation refrigerator at 100 mK.

F. Barbaresco and G. François (Eds.): QUEST-IS 2025, CCIS 2743, pp. 10–18, 2026.
https://doi.org/10.1007/978-3-032-13852-1_2

2 Background and Related Work

2.1 Spin Qubits

Hole spin qubits in silicon nanostructures offer fast and electrically driven spin control due to the strong spin-orbit coupling in the valence band [5]. Recent advances have demonstrated high-fidelity [6] and long coherence in nanowire geometries and SiGe/Ge/SiGe heterostructures [3].

2.2 Quantum Instrumentation Control Kit (QICK)

The Quantum Instrumentation Control Kit (QICK) [4] is an open-source qubit control platform built around a Xilinx RFSoC. The RFSoC integrates digital-to-analog converters (DACs), analog-to-digital converters (ADCs), and programmable logic within a single device, making it an ideal platform for waveform generation, acquisition, and feedback.

The use of an RFSoC significantly reduces latency and allows real-time control logic; this makes it an ideal platform for quantum experiments which require fast and deterministic control. In addition to the architectural advantages of the RFSoC, QICK is mainly made up of a timed processor and a combination of different signal generation and readout blocks, all of which can be modified by the user and tailored to his specific needs. This makes QICK extremely customizable.

Not to mention that QICK's software layer is based on Python, a very popular programming language, making it accessible to many users. Not to mention that QICK's software layer can be used by anyone as there are high-level classes and functions provided for ease of us, while also providing the option to write the codes on a register level or through assembly if needed. Moreover, it is possible to integrate QICK with QCoDeS, a widely used Python-based framework for instrument control in quantum research. This makes it straightforward to incorporate QICK into automated measurement routines and data acquisition workflows, which is crucial for reproducible and scalable quantum experiments.

In summary, the QICK platform enables high-resolution pulse generation and digitization on a single FPGA chip with real-time processing. It has been used in superconducting systems [4], but applications to spin qubits remain underexplored.

3 Experimental Setup

3.1 Control Hardware

The required pulses were generated using VIQTHOR's MIMOTHOR platform (Fig. 1) which features an RFSoC with QICK installed on it. Through MIMOTHOR, the user can output RF or DC signals, both of which are needed to perform spin qubit readout and control experiments. Furthermore, the qubit resonance frequency of the device tested was in the 12–18 GHz range which is beyond the output range of the RFSoC. To overcome this limitation and reach the qubit resonance frequency, we relied on VIQTHOR's up converter module (Fig. 1). The QICK FPGA generated intermediate-frequency (IF) signals between 1-2 GHz, which were mixed with an LO tone to produce the high-frequency pulses needed.

Fig. 1. VIQTHOR's MIMOTHOR Platform (left) and VIQTHOR's Up Converter Module (right) which takes a signal between 1–2 GHz and outputs a signal between 2–18 GHz

3.2 Spin Qubit

The tests were performed at CEA Grenoble on a single-hole spin qubit in a quantum dot formed in a natural silicon nanowire. The qubit has been thoroughly characterized as reported in [9]. Furthermore, a high-resolution, high-stability iTest DAC system was used to generate static bias voltages.

4 Measurements Performed

4.1 Resonator Spectroscopy

We start by characterizing the RF resonator response by sweeping the probe frequency and measuring the reflected signal. Figure 2 shows the reflection spectrum in amplitude, where a clear resonance dip is observed. This resonance shifted in response to charge transitions in the sensing dot. This frequency shift was used to determine the optimal operating point for subsequent readout measurements. Although amplitude is shown here, in practice we primarily use the phase response for high-sensitivity readout measurements, because the phase shows a sharp transition near the resonance.

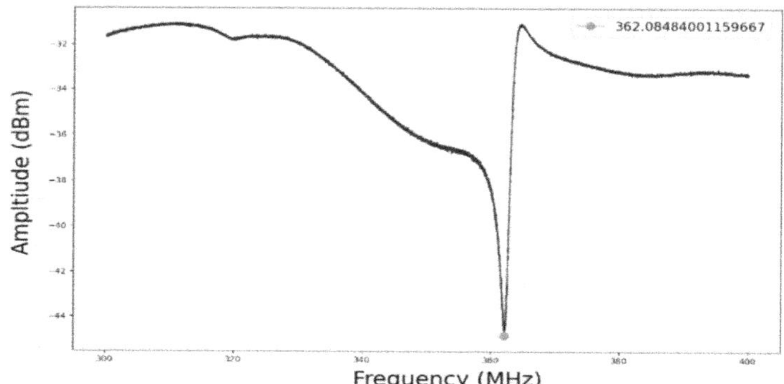

Fig. 2. Resonator Spectroscopy. The sharp dip indicates the resonance frequency of the coupled readout resonator, measured at approximately 362.08 MHz, which is used for subsequent qubit state readout

4.2 Energy-Selective Spin Readout

The so-called"Elzerman" spin readout scheme relies on connecting the quantum dot to a reservoir and adjust their relative position in energy. When the Fermi level in the contact lies in between the spin up and spin down levels, a spike in current can be observed if the spin was initially up, assuming the tunnel rates are appropriate and the detection fast enough. In contrast if the spin was down, no current'blip' is observed [7]. Therefore this readout scheme requires acquiring the resonator's response for a relatively long time (hundreds of microseconds), keep each trace (see Fig. 3a) and accumulate statistics to obtain the spin up probability. This long readout is performed after a series of DC pulses allowing to place the system in any region where we want to manipulate the spin before bringing it back to the readout position (see details in Fig. 1 of supplementary materials of [8]).

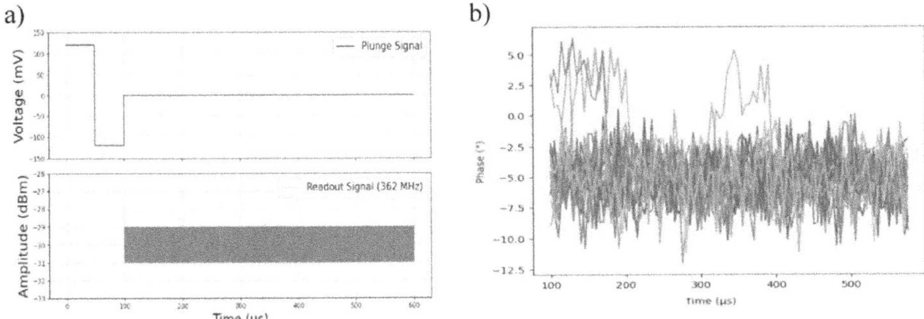

Fig. 3. a) The Plunge and Readout signals needed to perform the Elzerman Readout. b) 20 Elzerman Readout traces, some of them featuring a'blip' indicating a spin up

Figure 3b shows all the individual traces obtained after sending the pulses mentioned above twenty times. Out of the twenty experiments sent, only some of them will exhibit a phase shift, often called a'blip' (i.e., a phase value greater than 0). Performing the experiment several times allows us to compute the probability of observing a phase shift, i.e. the probability of having a spin up. Once we are able to detect a phase shift using Elzerman, we can move on to other tests.

4.3 Spin Tail

As shown in Fig. 4, we perform a sweep of gate voltage values on T3 (B2 is adjusted accordingly) while measuring the reflected phase signal over 500μs following the initialization (plunge) pulse. The resulting phase map reveals a "tail"-like structure corresponding to the spin-dependent tunneling dynamics. This feature helps us tune the gate voltages for optimal spin readout contrast.

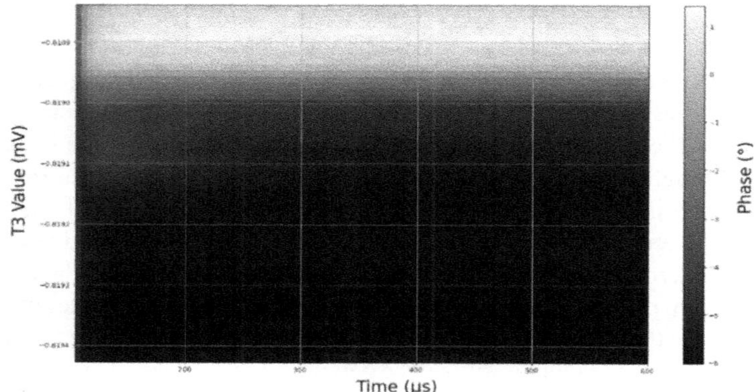

Fig. 4. Phase Reflectometry measurement for various gate voltage values on T3.

The spin tail measurement is critical in order to validate control fidelity and optimizing readout performance when controlling a qubit.

4.4 Qubit Spectroscopy

A qubit spectroscopy is performed to find the qubit's frequency frequency which is determined by the Landé factor g, the Bohr magneton μ_B and the applied static magnetic field which lifts the degeneracy between spins up and down and therefore creates the qubit's 2-level basis: $f = g \times \mu_B \times B$. Since the frequency of a spin qubit is between 12–18 GHz, we utilized VIQTHOR's up converter module which takes an input signals between 1 GHz and 2 GHz and shifts them to a frequency within a 2–18 GHz range while maintaining consistent output power and minimal spectral distortion across the full bandwidth. Therefore in our measurements, we use MIMOTHOR to generate an input frequency between 1–2 GHz, which will be sent to the up converter module who will output the signal at the user specified frequency. Figure 5a shows the signals needed to perform the qubit spectroscopy.

Sinusoidal Microwave signals (the drive pulse) are sent to the qubit for a duration of 2 μs. The drive frequency is varied. If the drive matches the resonance frequency of the qubit, the hole escapes the dot and triggers a noticeable blip in the reflectometry signal. However, if the drive frequency deviates from the resonance frequency, no changes can be seen in the reflectometry signal as the hole spin remains in its ground state. This means that the resonance frequency is the frequency where we have the highest probability of observing blips, i.e. the highest probability of spin up.

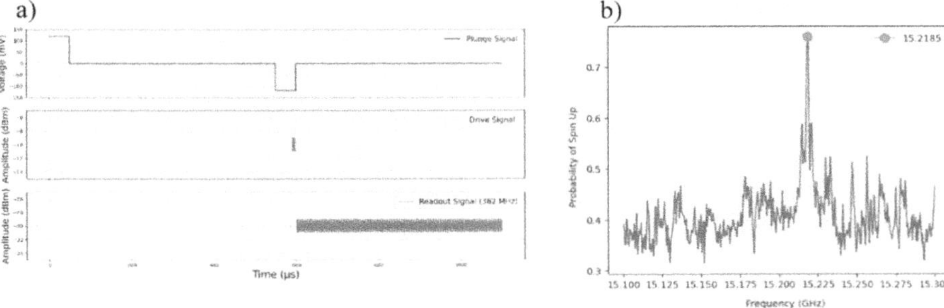

Fig. 5. a) The Plunge, Drive and Readout signals needed to perform Qubit Spectroscopy. b) Output of the Qubit Spectroscopy. The plot shows the probability of spin up as a function of frequency, with a clear resonance peak observed at approximately 15.2185 GHz.

To precisely determine the spin qubit resonance frequency, frequency sweeps with fine step resolution are required. However, when the qubit frequency is initially unknown, performing a high-resolution sweep over a large frequency bandwidth can be time-consuming. To overcome this, we could drive the qubit with a chirp signal. The Chirp signal enables us to rapidly scan a wide frequency range which allows us to identify the approximate resonance region. Once the resonance region is found, we can perform a frequency sweep with fine steps.

4.5 Rabi Oscillations

To characterize the coherent control of the spin qubit, we performed a Rabi oscillation experiment by applying resonant microwave pulses of varying duration. By increasing the pulse length, we notice oscillations in the qubit state population (Fig. 6), which correspond to coherent rotations of the spin state on the Bloch sphere. Rabi Oscillations measurements confirm our ability to drive the qubit.

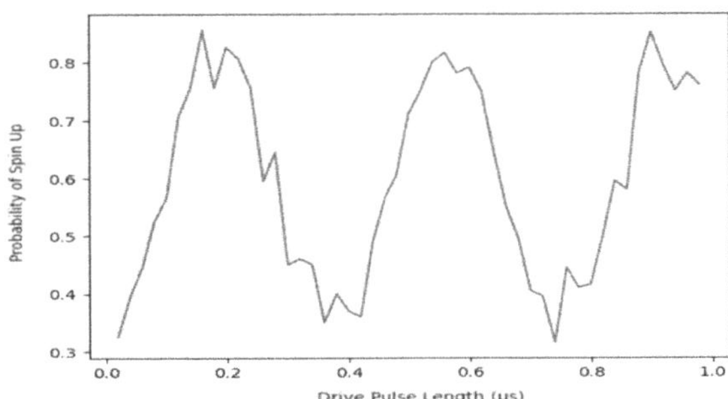

Fig. 6. Rabi Oscillations. The plot shows the probability of spin up as a function of the drive pulse length.

4.6 Rabi Chevron

The Rabi Chevron experiment is done by sweeping both the drive frequency and pulse duration in order to map the coherent Rabi oscillations of the hole spin qubit versus frequency detuning. The resulting 2D plot (Fig. 7) displays a chevron-shaped interference pattern, which indicates coherent oscillations occurring near the resonance frequency. The chevron peak corresponds to the qubit's resonance frequency, and the fringe spacing reflects the Rabi frequency of 2.5 MHz as a function of detuning.

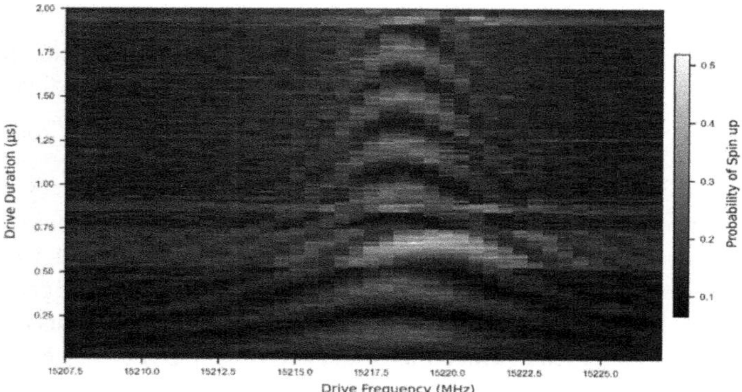

Fig. 7. Chevron plot showing qubit excitation probability as a function of drive frequency and pulse duration. The resonance condition is marked by the peak of the chevron. A shift in the qubit frequency over time is visible.

During this measurement, we observed a gradual shift in the resonance frequency, visible as a horizontal drift in the chevron peak over time. This shift is likely due to environmental or device-level factors such as gate voltage drift due to prolonged operation or thermal cycling caused by repeated measurements.

4.7 Resonance Frequency Drift

The Ramsey Experiment consist of two $\pi/2$ pulses separated by a variable delay time. As previously mentioned, during the Rabi Chevron measurement, we observed a shift in the qubit resonance frequency of approximately 1 MHz. Therefore, the subsequent Ramsey Experiment was performed using the pre-drift resonance frequency which resulted in reduced fringe contrast and rapid dephasing, as seen in Fig. 8.

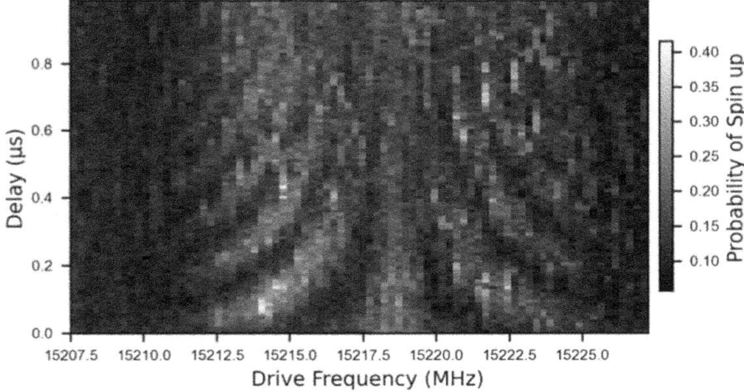

Fig. 8. Ramsey measurement taken after a qubit resonance frequency shift observed during the preceding chevron scan. The drive frequency was not updated to the new resonance after the jump observed in Fig. 7, leading to fast decoherence and reduced fringe visibility

To maintain an accurate control over the qubit, resonance calibration must be done frequently in order to account for the shifts in the qubit frequency. The Ramsey experiment's result highlights the necessity of incorporating automated resonance tracking or real-time frequency tuning in future works.

5 Conclusion

In this work, we demonstrated control and readout of a single hole spin qubit in a silicon nanowire using VIQTHOR's MIMOTHOR platform which is based on the QICK platform, as well as VIQTHOR's up converter module. Through a series of characterization measurements which include: resonator spectroscopy, Elzerman readout, spin tail, qubit spectroscopy, Rabi, Rabi Chevron and Ramsey; we were able to validate the effectiveness of the control hardware in generating and delivering the needed pulses accurately.

Additionally, the Ramsey experiment revealed a rapid decay of coherence due to a frequency drift observed during the rabi chevron experiment. This highlights the sensitivity of spin qubit control to environmental or device-level variations thus emphasizing the need for frequent resonance frequency recalibration.

The results presented in this work show promise of using QICK as a platform for spin qubit control. This work validated the system's ability to generate the needed pulses for all the experiments performed and thus establishes a foundation for scalable and low-cost control architectures for spin qubits.

This work was supported (in part) by the French National program "Programme d'investissement d'avenir, IRT Nanoelec, n° ANR-10-AIRT-05 (Q-Loop)" as well as the CEA PTC program (ERDAQS).

References

1. Stano, P., et al.: Review of performance metrics of spin qubits in gated semiconducting nanostructures. Nat. Rev. Phys. **4**(10), 672–688 (2022). https://doi.org/10.1038/s42254-022-00484-w
2. Burkard, G., et al.: Semiconductor spin qubits. Rev. Mod. Phys. **95**(2), 025003 (2023). https://doi.org/10.1103/RevModPhys.95.025003
3. Maurand, R., et al.: A CMOS silicon spin qubit. Nat. Commun. **7**(1), 13575 (2016). https://doi.org/10.1038/ncomms13575
4. Stefanazzi, L., et al.: The Qick (quantum instrumentation control kit): Readout and control for qubits and detectors. Rev. Sci. Instruments **93**(4) (2022). https://doi.org/10.1063/5.0076249. arXiv:2111.04249
5. Voisin, B., et al.: Electrical control of g-factor in a few-hole silicon nanowire MOSFET. Nano Lett. **16**(1), 88–92 (2015). https://doi.org/10.1021/acs.nanolett.5b02920
6. Wang, C.-A., et al.: Operating semiconductor quantum processors with hopping spins. Science **385**(6707), 447–452 (2024). https://doi.org/10.1126/science.ado5915
7. Elzerman, J.M., et al.: Single-shot read-out of an individual electron spin in a quantum dot. Nature (2004). https://www.nature.com/articles/nature02693
8. Piot, B.A., et al.: A CMOS silicon spin qubit demonstrating fast control. Nat. Nanotechnol. **17**, 1072–1077 (2022). https://doi.org/10.1038/s41565-022-01196-z.11
9. Bassi, M.: Tunable resilience to charge noise of a hole spin. Physics [physics]. Universit´e Grenoble Alpes [2020-..] (2024). English. ⟨NNT : 2024GRALY018⟩. ⟨tel-04718717⟩

Microwave Interconnectivity as a Scalable Solution for Quantum Computing: Engineering, Measurement, and Standardization

Laurent Petit[✉], Evan Ichir, Fabrice Janot, Fleury Grandjean, Guillaume Marion, Claude Brocheton, Thierry Le Nadan, Bastien Huon, Jacques Martinet, and Julien Legrand

Radiall, 642 Rue Emile Romanet, Voreppe, France
`laurent.petit@radiall.com`

Abstract. The transition from prototype to large-scale quantum computers requires a new class of microwave interconnect solutions capable of operating reliably at ultra-low temperatures. This paper presents a modular cryogenic interconnect framework designed for compatibility with superconducting quantum processors, addressing critical requirements for signal fidelity, thermal anchoring, and system scalability. The proposed architecture integrates cryogenic attenuators, filters, and switches optimized for operation in dilution refrigerators down to 10 millikelvin. Multiphysics modeling, including electro-thermal co-simulation and mesoscopic heat transport analysis, guides component design. Measurement validation combines calibrated RF methods with qubit-based performance metrics to assess thermal noise impact and signal integrity. Standardized calibration and benchmarking protocols ensure reproducibility across platforms. This work supports ongoing standardization efforts within CEN/CENELEC and paves the way toward scalable, low-noise quantum infrastructures.

Keywords: Cryogenic Microwave Interconnects · Quantum Hardware Scalability · Thermal and Electromagnetic Co-Simulation · Sub-Kelvin Metrology · Standardized RF Component Integration

1 Introduction

1.1 Microwave Interconnectivity Solution to Scale up Quantum Computer

Quantum computing promises to revolutionize industries ranging from molecular simulation to optimization. However, the practical deployment of such processors demands precise control and readout of qubit states. These processes rely on signal integrity, minimal cross-talk, and thermalization, particularly within cryogenic environments. This paper is structured as follows: Sect. 2 reviews design functions and constraints. Section 3 details key components. Section 4 covers metrology, and Sect. 5 discusses standardization efforts. Microwave interconnectivity plays a central role in ensuring these conditions. This domain intersects materials science, RF and microwave engineering, cryogenics, and systems-level integration.

F. Barbaresco and G. François (Eds.): QUEST-IS 2025, CCIS 2743, pp. 19–25, 2026.
https://doi.org/10.1007/978-3-032-13852-1_3

Components such as attenuators, filters, cables, and connectors must preserve their electrical characteristics while operating at temperatures below 100 mK, with tight constraints on electromagnetic compatibility and thermal management.

2 Cryogenic Microwave Interconnects: Functions and Design Requirements

2.1 Functional Roles in Quantum Architectures

Microwave interconnects serve as conduits for qubit control and readout signals for systems scaling from tens to potentially millions of qubits. These include:

- Drive lines that deliver RF pulses to qubits
- Readout chains capturing quantum states via resonators and amplifiers
- DC biasing lines and shielding connections

Each line is subject to frequency-specific losses, phase noise, thermal noise injection, and electromagnetic crosstalk. Control lines typically span DC to 12 GHz, while readout paths extend beyond 20 GHz in dispersive regimes. And the constraints are rising with the "scale up" from tens [1] to several hundred [2] and even millions of qubits that must be precisely controlled and measured [3].

2.2 Multiphysics Design Constraints

Operating in a dilution refrigerator introduces several interdependent constraints:

- Thermal anchoring at each stage (300 K, 3 K, 800 mK, 100 mK, 10 mK)
- Minimized attenuation at low temperatures, preserving signal amplitude
- Avoidance of thermal shorts and vibration sensitivity
- Material selection: superconducting alloys (NbTi), copper-nickel, stainless steel, alumina substrates

3 Component-Level Developments

3.1 Cryogenic Attenuators

Resistive attenuators function as both thermal noise suppressors and signal conditioners, improving qubit readout fidelity and stability. Their design must reconcile the Wiedemann–Franz law and Kapitza resistance, ensuring heat extraction without excessive RF loss. The attenuator structures are based on resistive layers (e.g., TaN or NiCr), embedded in high thermal conductivity ceramics (e.g., AlN). Designs include cartridge-based modules and ganged arrays for high-density integration.

Modeling includes:

- Heat flow simulation using finite element tools
- Mesoscopic modeling of phonon bottlenecks, characterized by ballistic transmission and Kapitza resistance metrics.
- Coupled electromagnetic-thermal co-simulation

Measurements involve both calibrated S-parameters and indirect thermometry based on qubit decoherence and readout fidelity shifts.

Sub Kelvin Attenuator Design and Characterization
At sub-Kelvin temperatures, attenuator performance must be characterized not only in terms of RF properties (e.g., insertion loss, VSWR). At millikelvin temperatures, attenuators must exhibit predictable thermal dissipation while maintaining low insertion loss and impedance stability. Mesoscopic modeling captures phonon transmission and heat backflow from resistive zones, and is validated by comparing device-integrated thermometry with indirect readout fidelity degradation in qubits exposed to thermal noise. Such combined approaches enable refining the resistive film design, thermal interface resistance, and ceramic substrate layout (Fig. 1).

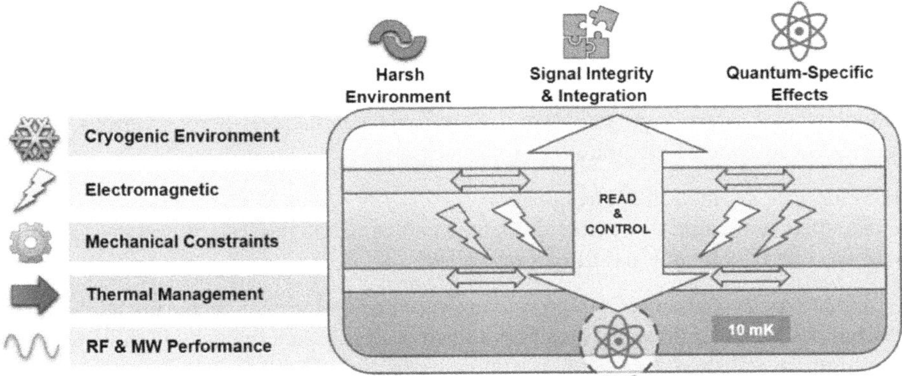

Fig. 1. R&D Constraints for RF&MW Interconnectivity for QT

3.2 Cryo Switches and Multiplexers

Switches are essential for reducing the cable count and enabling reconfigurable experiments. Cryogenic SPDT to SP8T switches are designed to operate with low actuation power (<10 mW), enabling use in 10 mK environments. SMT versions are tested down to 20 mK, showing temperature resilience over multiple actuation cycles.

A key innovation involves packaging with high-conductivity materials, controlled magnetic susceptibility, and hybrid actuation circuits with minimal transient heat load.

3.3 Filters and Infrared Blockers

Filters mitigate spurious modes, reflections, and infrared noise. Their performance in cryogenic conditions depends on:

- Precise impedance matching
- Effective IR absorption (using resin composites or crystalline structures)

- Low-pass filters with cutoff frequencies below 20 GHz

Evaluation involves both network analysis and qubit-based reflectometry techniques monitoring changes in qubit thermal population and parity switching before and after filter integration.

Infrared Filter Thermal Design Validation

Infrared (IR) filters, typically implemented with resin-embedded or multilayer absorbing structures, are essential to block high-energy thermal photons that would otherwise perturb qubit states. Their thermal design effectiveness is indirectly validated by monitoring qubit excited-state populations, parity switching rates, or qubit thermalization dynamics before and after integration of filter modules. These measurements complement electromagnetic simulations and provide system-level evidence of IR filtering efficacy under realistic operating conditions.

4   Integrated Interconnectivity Architectures

Modular sideloaders aggregate over 100 microwave channels across various thermal stages. These systems integrate:

- Solderless, hermetic SMPM clusters
- Bundled semi-rigid or flexible cryogenic cables (CuNi, NbTi)
- Embedded attenuator and filter assemblies

CAD-based mechanical design is complemented by thermal anchoring analysis, and mechanical stress simulations under cooldown cycles. Signal integrity is validated using time-domain reflectometry and noise figure extraction (Fig. 2).

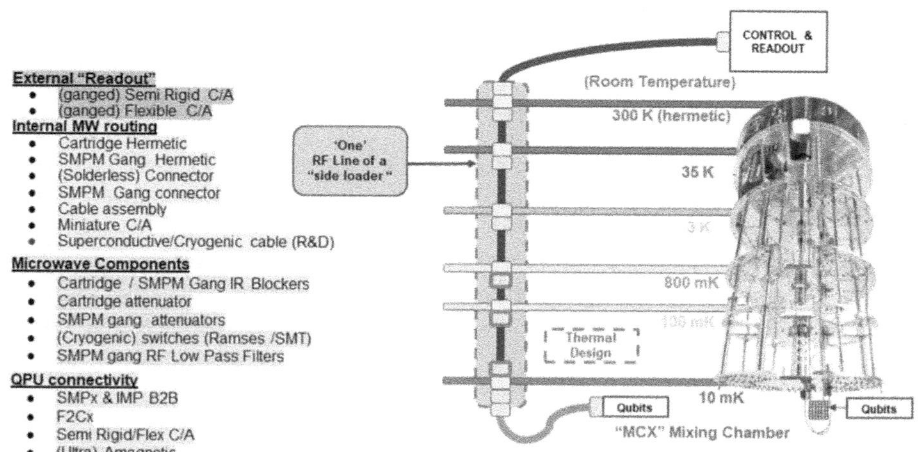

Fig. 2. Integrated Interconnectivity Architectures

5 Cryogenic Metrology and Test Platforms

Precision in RF measurements at cryogenic temperatures is non-trivial. A robust metrological framework includes:

- Cryostat-integrated test benches equipped with a Vector Network Analyzer (VNA)
- TRL (Thru-Reflect-Line), pulse modulation, and arbitrary waveform control
- Temperature-staged calibration using TRL and SOLT methods adapted to cryogenic interfaces
- Local thermometry and thermal imaging for characterizing heat dissipation in components

Both direct methods—such as network analyzer S-parameter measurements, time-domain reflectometry, and resonator Q-factor extraction, and indirect methods such as qubit decoherence signatures, photon statistics, tunneling rates, and parity switching analysis—are employed to assess the behavior and stability of components under cryogenic conditions.

Uncertainty quantification is carried out using both the Guide to the Expression of Uncertainty in Measurement (GUM) and Monte Carlo propagation. The latter estimates uncertainty by simulating numerous outcomes based on probability distributions of input variables, particularly valuable when analytical models are impractical.

In parallel, work is ongoing to improve thermal anchoring reproducibility and to establish benchmarking protocols across laboratory platforms, all coordinated with system-level modeling efforts.

6 Standardization Initiatives

From a measurement and evaluation perspective, standardization aims to:

- Define traceable metrics and test protocols for cryogenic performance
- Harmonize both direct and indirect measurement methods across test sites
- Build consensus on connector and cable specifications to ensure compatibility and integration

These efforts are part of a broader strategy to bridge laboratory R&D and scalable manufacturing by embedding reproducibility, comparability, and traceability into the qualification of quantum interconnects. Standardization is a strategic pillar for the industrialization of quantum technologies. Target parameters include an insertion loss below 3 dB, a VSWR under 1.3, and a heat load below 5 μW per channel at 10 mK.

Radiall coordinates the interconnectivity working group within JTC22 in its scope of "enabling technologies" with the objective of finalization and ongoing adoption within ISO JTC3: Quantum Technologies [5].

7 Modular System Architecture for Cryogenic Quantum Platforms

Building on the above, a comprehensive cryogenic interconnectivity offer has been structured to meet both R&D and industrial scaling needs. This includes:

- Customizable sideloaders with up to 208 coaxial lines using SMPM clusters
- Thermally-optimized attenuator modules for each temperature stage (300 K to 10 mK)
- Filter blocks and IR absorbers designed to minimize thermal radiation and EMI
- High-density harnesses composed of flexible and semi-rigid cryogenic cables based on CuNi or NbTi alloy conductors
- Integration support: mechanical CAD, thermal anchoring, signal routing, and EMI modeling

This modular architecture is validated through mK testbeds and aligned with current measurement practices, offering scalable, low-noise, and reproducible signal paths tested with qubit setups showing $< 2\%$ thermal-induced fidelity loss across 1–20 GHz (Fig. 3)."

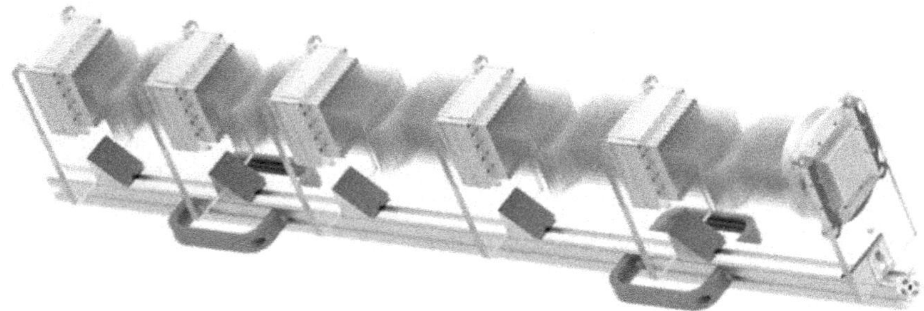

Fig. 3. High Density Sideloader integrating Microwave Cryogenic Interconnectivity [6]

8 Conclusion and Perspective

Microwave cryogenic interconnectivity has emerged as a strategic enabler of scalable quantum computing. Innovations in component design, thermal engineering, and test methodology contribute to reducing noise, improving integration, and enabling high-density scaling.

Future work includes:

- Full qualification of flexible micro-coaxial and FPC-based harnesses
- Deployment of modular sideloaders in production dilution refrigerators
- Expansion of inter-laboratory validation and reference datasets

As quantum processors evolve toward 10k–100k qubit systems, interconnectivity must evolve from a bespoke subsystem into a mature, standardized backbone that integrates RF, thermal, mechanical, and systems engineering into a coherent quantum infrastructure.

References

1. Krinner, S., et al.: Engineering cryogenic setups for 100-qubit scale superconducting circuit systems. Rev. Sci. Instrum. **90**(12), 123528 (2019)

2. Masluk, N.: Cryogenic infrastructure for 400 qubits and beyond. In: Proceedings of Cryogenic Engineering Conference/International Cryogenic Materials Conference (CEC/ICMC), Honolulu, HI, USA (2023)
3. Mohseni, M., et al.: How to build a quantum supercomputer. arXiv preprint arXiv:2411.10406 (2024)
4. CENELEC. Quantum Technologies Committee. CEN-CENELEC (2023). https://www.cencen elec.eu
5. ISO/IEC JTC 1/SC 42. Quantum Technologies. International Organization for Standardization (ISO). https://www.iso.org/committee/10138914.html
6. Radiall. Quantum Technologies—Enabling Scalable Interconnectivity. Radiall Industries (2023). https://www.radiall.com/industries/industrial/quantum-technologies.html?f=0.35382200%201744907020

Scalable Photonic Control System for Large Scale Quantum Computers

Abhilash Amsanpally[1]([✉]), Maeva Franco[2], Arnaud Perrin[1], Jérémie Theze[1], Dana El Hajj[1], and Guillaume De Giovanni[1]

[1] VIQTHOR S.A.S, Saint-Germain-en-Laye, France
abhilash@Viqthor.com
[2] Lab-STICC, Ecole Nationale d'Ingénieurs de Brest, Plouzané, France

Abstract. Scaling quantum computers require control systems that are both high-performance and cryogenically efficient. Traditional microwave-based electronics face limitations in such environments due to heat generation, signal loss, and electromagnetic interference. To address these challenges, we present a fully photonic control architecture based on Viqthor's Menthor, a multi-channel RF-over-Fiber (RFoF) transmitter designed for quantum computing applications. Menthor converts microwave signals up to 18 GHz into optical signals, transmitting them over single-mode fiber to minimize thermal load and signal degradation. Paired with IceQube, a cryo-compatible multi-channel microwave converter, the system enables reliable signal recovery directly within dilution refrigerators at temperatures as low as 55 K. Experimental results demonstrate low relative intensity noise (RIN), high dynamic range, and flat frequency response, ensuring high-fidelity control of quantum operations. This platform establishes a scalable, fiber-based approach for high-density qubit control, representing a significant advancement toward next-generation quantum processors.

Keywords: Photonic control systems · Large scale quantum computers · Cryophotonics · RF-over-Fiber

1 Introduction

Quantum computing represents a groundbreaking technological advancement with the potential to transform industries such as cryptography, artificial intelligence, drug development, and climate modeling. Despite their promise, realizing large-scale, practical quantum computers face major challenges in scalability, reliability, and operational efficiency. Central to these challenges is the control of qubits, the fundamental units of quantum information, which currently relies predominantly on electronic systems. Traditional electronic control is limited by heat generation, bandwidth constraints, and electromagnetic interference, creating bottlenecks for scaling quantum architectures [1]. In modern systems, superconducting qubits are typically controlled and measured using microwave pulses generated at room temperature and delivered into cryogenic environments via long coaxial cables with multiple stages of attenuation to suppress thermal noise. While effective for small-scale systems, this method imposes serious obstacles when scaling to large qubit arrays.

F. Barbaresco and G. François (Eds.): QUEST-IS 2025, CCIS 2743, pp. 26–36, 2026.
https://doi.org/10.1007/978-3-032-13852-1_4

To address these limitations, we propose the development of Photonic Control Systems (PCS) for large-scale quantum computing. By adopting a fully photonic approach, this initiative aims to redefine how qubits are manipulated, controlled, and measured, enabling quantum computers that are scalable, efficient, and reliable. The system leverages the inherent advantages of photonics including high speed, precision, and minimal thermal load to establish a fundamentally new paradigm for quantum control.

2 Advantages of Photonic Control Systems

Conventional quantum computing systems often rely on electronic control mechanisms, which, while effective for small-scale systems, encounter significant limitations as the number of superconducting qubits increases. Heat generation, susceptibility to noise, and spatial constraints imposed by electronic circuitry present major obstacles [3]. In contrast, photonic systems offer a fundamentally different control paradigm. Optical signals can interact with qubits with extraordinary precision, operate at the speed of light, and avoid the heat dissipation issues inherent to electronic components. Additionally, integrating photonics with quantum systems leverages mature technologies from the telecommunications and photonics industries, reducing development timelines and costs associated with scaling [4, 5].

This work presents a photonic control system (PCS) designed to replace traditional electronic control in quantum computing with advanced optical technologies. By utilizing optical signals, waveguides, and integrated photonics, such as silicon photonics, the PCS addresses the thermal, spatial, and bandwidth constraints that limit current architectures. It supports large-scale qubit arrays with enhanced coherence, lower error rates, and improved scalability. While current electronic approaches struggle with heat, interference, and limited bandwidth, particularly in cryogenic environments [6], our optical approach offers a promising path forward.

3 System Architecture Overview

The PCS is built around two core subsystems: Menthor, a room-temperature RF-over-Fiber transmitter, and IceQube, a cryo-compatible opto-electronic receiver. Menthor, illustrated in Fig. 1 with its block diagram in Fig. 2, converts RF signals into optical signals using RF-over-Fiber technology, enabling high-frequency signal delivery into cryogenic environments. Operating at wavelengths in the C-band (1525–1555 nm), it employs low-noise DFB lasers to produce stable optical signals spanning 10 MHz to 18 GHz. With low relative intensity noise (RIN), strong harmonic suppression, and an optical output power of 3–10 dBm, Menthor delivers spectrally pure, distortion-free signals for precise qubit operations. Its flat gain and group delay characteristics ensure signal consistency, while a spurious-free dynamic range (SFDR) of approximately $100 \text{ dB/Hz}^{2/3}$ and a 1 dB compression point at 18 dBm highlight its high-performance capabilities.

Menthor provides three primary advantages for quantum computing. First, it eliminates heat dissipation associated with traditional RF cables by using fiber optics, preserving the cryogenic environment necessary for qubit coherence. Second, its modular, fiber-based design enables effortless scaling from a few to thousands of qubits without compromising signal quality or thermal stability. Third, it delivers ultra-clean, stable RF signals with low noise and precise shaping, enhancing gate fidelity and reducing operational errors. The current implementation is a compact 2U system with eight independent RF-over-Fiber channels, each equipped with its own modulator and laser for fully parallel, scalable operation. Control is provided via an intuitive Ethernet GUI or SCPI commands, supporting both remote and automated operation. Menthor integrates seamlessly with Viqthor's Mimothor platform, providing a complete quantum control solution with remote monitoring and management capabilities.

IceQube, shown in Fig. 3, serves as the modular photonic receiver within the cryogenic environment and is specifically designed for high-density integration. Optical signals generated at room temperature travel through robust, cryo-stable Teflon-jacketed fibers, passing through sealed feedthroughs to maintain vacuum integrity. These microwave-modulated optical signals are delivered to high-speed photodetectors located inside the cryochamber. The fibers, coated with PTFE jackets, are engineered to withstand ultra-low temperatures and minimize outgassing effects [7, 8]. Each IceQube module can accommodate hundreds of photodetectors, aligned with cryo-compatible fiber ribbons containing 12 to 48 fibers. The geometry is carefully optimized, with detector arrays spaced at 250 μm to match standard fiber array connectors. The entire assembly is constructed from cryo-compatible materials, including OFHC copper for thermal anchoring, gold-plated surfaces, and shielded cable interfaces to reduce RF crosstalk and outgassing.

Fig. 1. Menthor RF Transmitter system to control qubits

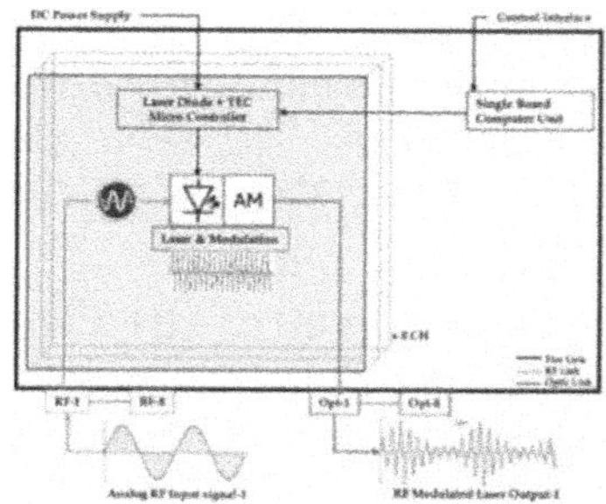

Fig. 2. Block diagram of Menthor

At the cryogenic stage, IceQube's high-speed photodetectors convert the incoming optical signals back into electrical RF, which is routed to the quantum processor through cryo-compatible cabling. This module bridges optical control and quantum processors at millikelvin temperatures using reliable SMP and MMCX interfaces. Its modular design supports flexible stacking inside dilution refrigerators, allowing adaptation to different processor sizes and layouts. By addressing challenges in signal routing and cryogenic stability, IceQube minimizes heat load, eliminates bulky electronics near the qubits, and enhances overall control fidelity. Together with Menthor, IceQube completes the photonic control chain, enabling precise, scalable, and low-noise qubit manipulation.

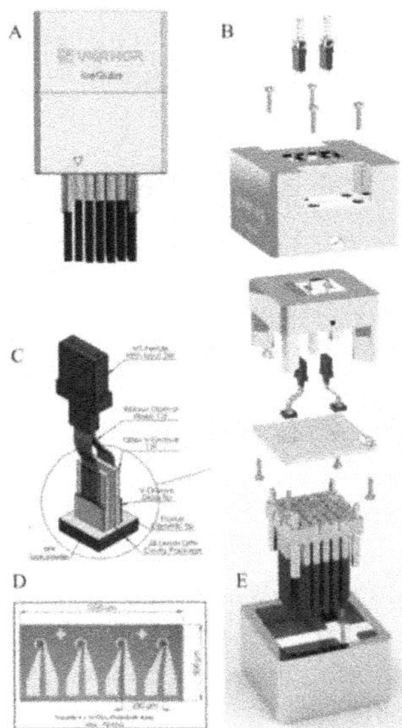

Fig. 3. (A) External and (B) internal views of the IceQube module, its internal (C) fiber array integration with (D) photodiode array & (E) High density SMP matrix array.

4   Procedure of Controlling Qubits with the Proposed PCS

The integration of Menthor with IceQube establishes a scalable architecture for delivering precise microwave control signals to qubits inside dilution refrigerators. In this system, classical control electronics at room temperature generate microwave signal up to 18 GHz, corresponding to qubit operations such as single-qubit rotations or two-qubit gates. These signals are converted directly into optical carriers by Menthor modulated lasers, producing optical signals in the 1525–1560 nm range that are transmitted over single-mode fiber. This approach eliminates bulky coaxial cabling and reduces thermal load on the cryostat. The optical signals are carried into the refrigerator through Teflon-jacketed, cryo-stable fibers, which maintain vacuum integrity and support dense multiplexing with fiber ribbons of up to 48 channels for large-scale qubit control. Inside the cryogenic environment, IceQube's high-speed photodetectors convert the optical modulation back into RF signals with minimal distortion, demonstrating stable, linear performance from room temperature down to 55 K, with modulation bandwidths up to 12 GHz. Bias voltages between 0.25 V and 1.0 V optimize carrier extraction, and while responsivity decreases by ~ 10–14 dB under cryogenic conditions, this can be compensated with bias tuning, increased laser power, or low-noise amplification. The recovered signals are then routed via cryo-compatible coaxial cables to the quantum processor at the millikelvin stage, where they drive transitions and entangling gates with high fidelity. Crucially, the preserved linearity and broad frequency response ensure that control signals arrive undistorted, enabling reliable qubit operations. Finally, the modular IceQube design allows hundreds of photodetectors to be integrated in parallel, providing a pathway to scaling photonic qubit control systems without the thermal and spatial limitations of conventional coaxial wiring.

5 Cryogenic-Photonic Test Results and Analysis

The primary goal of our experimental demonstration was to validate the performance of the Menthor and IceQube system and to determine whether the photonic links remain efficient and stable in the extreme environment of a dilution refrigerator. The experiments were conducted using a comprehensive setup: Menthor was configured to output modulated optical signals in the 1525–1560 nm range, which were transmitted into the cryostat via fiber optics. Inside the cryogenic chamber, IceQube converted the optical signals back into RF and delivered them to measurement equipment, including spectrum analyzers and RF power meters.

Tests were performed at multiple temperature stages, from room temperature down to 55 K, across RF frequencies ranging from 10 MHz to 18 GHz, with input power levels up to $+$ 18 dBm. The resulting data formed the basis for the performance analysis presented in the following sections.

5.1 Photocurrent as a Function Input Laser Current at 8 GHz Modulation.

ZXThis test evaluated the photocurrent behavior of high-speed photodetectors (HSPDs) under varying laser currents and temperatures, serving as a critical validation of the proposed photonic control architecture. Figure 4(A) presents curves showing photocurrent output across a range of laser input currents under room temperature (300 K) and cryogenic (55 K) conditions. The analysis was performed for four photodetector bias voltages: 0.25 V, 0.5 V, 0.75 V, and 1.0 V. The laser modulation frequency was fixed at 8 GHz, using a directly modulated laser (DML) at 1550 nm wavelength, consistent with the internal configuration of Menthor. At room temperature (300 K), the photocurrent exhibits a clear linear relationship with increasing laser input current, confirming that the photodetector operates in a highly responsive and stable regime. This behavior aligns with theoretical expectations, as the responsivity of photodiodes under standard conditions typically follows a near-linear dependence on incident optical power [9].

Under cryogenic conditions (55 K), the photocurrent decreases across all bias voltages. For example, at a laser current of 35 mA, the photocurrent at 55 K is approximately 6 μA lower than at 300 K. Despite this reduction, the linear relationship with laser current is preserved, indicating that the photodetector maintains predictable operation at extremely low temperatures. The reduction in responsivity at cryogenic temperatures can be attributed to intrinsic semiconductor effects, including reduced carrier mobility and changes in the absorption coefficient, which lower the internal quantum efficiency. The retention of linearity is critical, as it ensures that the system can still operate within a predictable regime, allowing calibration and compensation via control or software algorithms. The observed behavior confirms that photonic downlink channels can reliably deliver microwave signals in cryogenic quantum systems. This addresses a key challenge in scaling quantum computers, enabling precise, low-heat, and low-noise control of qubits at millikelvin temperatures. The bias-dependent photocurrent characteristics

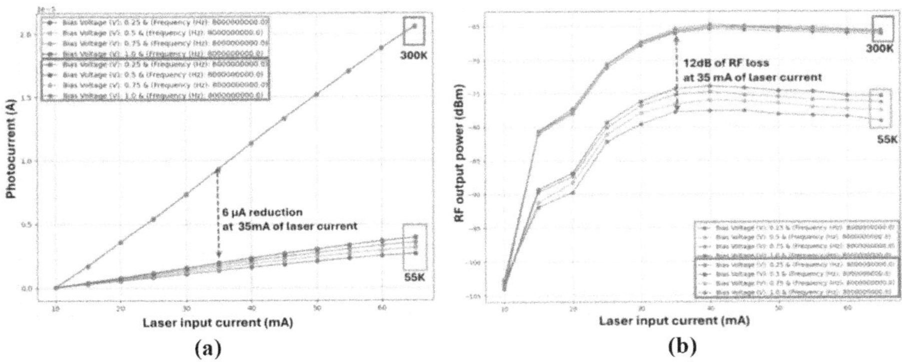

Fig. 4. (A) Photocurrent at 300 K and 55 K at 8 GHz modulation frequency, showing reduced photocurrent of 6micro amps at cryogenic temperatures. (B) RF output power at 300 K and 55 K, highlighting a 12 dB drop under cryogenic conditions.

align with known photodiode performance, demonstrating their suitability for scalable, high-speed quantum control applications.

5.2 RF Output Power as a Function Laser Input Current (300 K vs. 55 K)

The results in Fig. 4(B) compare RF output power as a function of laser input current across bias voltages ranging from 0.25 V to 1.0 V, at a fixed modulation frequency of 8 GHz. Two temperature conditions are shown: room temperature (300 K, red box) and cryogenic temperature (55 K, blue box).

Temperature Impact on RF Output: A significant drop in RF output power is observed at 55 K compared to 300 K. At a laser input current of 35 mA, the RF output power decreases by approximately 12 dB, indicating a strong thermal dependence of the photodetector's conversion efficiency or a potential impedance mismatch under cryogenic conditions. At both temperatures, RF output power increases with laser current up to 35–40 mA, beyond which it begins to saturate or slightly decline. This saturation suggests that the photodetector is approaching its optimal operating range or that thermal and electrical effects start to influence performance.

Bias Voltage Dependence: Across both temperature ranges, higher bias voltages (0.75 V and 1.0 V) consistently produce stronger RF output. This supports the earlier observation that increasing bias improves carrier extraction efficiency, an effect particularly critical in cryogenic operation.

Cryogenic Behavior and Stability: Although overall RF output is lower at 55 K, the response curves remain stable and predictable, confirming that the system continues to function reliably. The observed 12 dB reduction at 35 mA underscores the importance of optimization in cryogenic tuning and impedance matching to maximize system performance.

5.3 Photocurrent Response as a Function Bias Voltage at Fixed Laser Current

The experimental investigation shown in Fig. 5 characterizes the photocurrent behavior of high-speed photodetectors (HSPDs) under varying bias voltages while maintaining a constant laser current of 35 mA. Measurements were performed at two temperatures, 300 K and 55 K, across four modulation frequencies: 6 GHz, 8 GHz, 10 GHz, and 12 GHz. These results provide key insights into the effects of temperature and electric field strength on carrier transport efficiency and device responsivity, both critical parameters for photonic control systems in quantum computing environments.

As shown in Fig. 5(A), at room temperature, the photodetectors exhibit a non-linear increase in photocurrent with applied bias voltage, eventually saturating at higher voltages. This saturation indicates full depletion of the photodiode's intrinsic region and efficient carrier collection. The maximum observed photocurrent under these conditions was approximately 10 μA. In contrast, at 55 K, the photocurrent is significantly lower, plateauing near 2.5 μA. The reduction is especially pronounced at lower bias voltages; for instance, at 0.25 V and 8 GHz, the photocurrent decreases by about 8 μA compared to room temperature.

At cryogenic temperatures, this reduction is attributed to decreased carrier mobility, altered recombination dynamics, and increased activity of deep-level traps, common effects in semiconductors at low thermal energy. These observations underscore the importance of precise bias tuning in cryogenic photonic systems. Despite the reduced responsivity, the photodetectors maintain a flat frequency response from 6 to 12 GHz, demonstrating a high modulation bandwidth suitable for broadband quantum control. ~~Overall, these results indicate that with proper biasing and thermal calibration, photodetectors remain well-suited for high-speed photonic links in scalable, low-temperature quantum computing architectures.~~

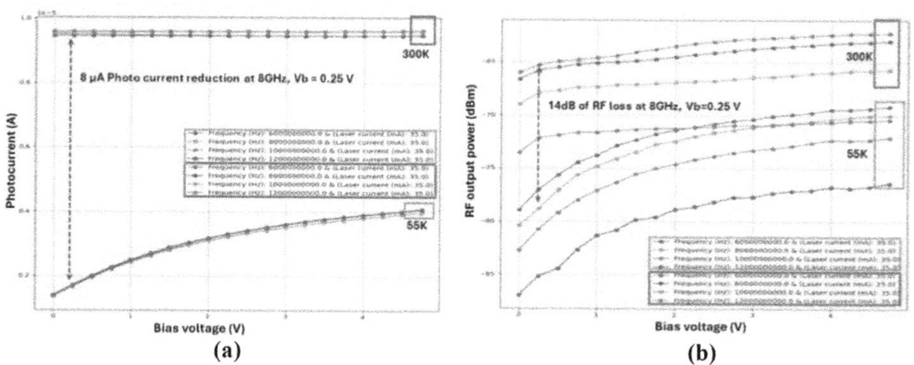

Fig. 5. (A) Photocurrent at 300 K and 55 K across 6–12 GHz, showing reduced responsivity at cryogenic temperatures. (B) RF output power at 300 K and 55 K, highlighting a 14 dB drop under cryogenic conditions.

5.4 RF Output Power as a Function Photodetector Bias Voltage

This test further evaluates photodetector performance by analyzing RF output power as a function of bias voltage under a fixed laser current of 35 mA and a constant RF modulation frequency of 8 GHz. The RF output was recorded at both 300 K and 55 K, providing a comparative assessment of signal demodulation efficiency in room-temperature versus cryogenic conditions.

As shown in Fig. 5(B), at 300 K, the RF output power increases steadily with bias voltage, eventually stabilizing near -65 dBm at higher voltages. This behavior reflects enhanced carrier drift velocities and reduced transit times due to stronger internal electric fields, resulting in efficient optical-to-RF signal conversion. At 55 K, the maximum RF output power drops to approximately -80 dBm, a 14 dB reduction at 0.25 V bias. This attenuation highlights the challenges of signal recovery at cryogenic temperatures, where altered semiconductor properties, such as junction capacitance and carrier dynamics, reduce the photodiode's ability to generate strong RF signals. Despite the lower absolute output, the cryogenic measurements show a consistent, monotonic increase in RF power with bias voltage, mirroring the trend at room temperature. This consistency indicates that the fundamental RF generation mechanisms remain intact, albeit with reduced efficiency, under cryogenic conditions. The maintenance of linearity across the operating range confirms the device's reliability in preserving signal fidelity, which is critical for quantum control operations.

These findings emphasize the importance of electrical bias optimization to mitigate cryogenic-induced performance losses. They also confirm that, while absolute RF power levels are lower at 55 K, the photodetectors retain functional integrity and can operate across the GHz regime with acceptable linearity and frequency response. Collectively, these results validate the feasibility of deploying RF-over-Fiber (RFoF) links in photonic control systems for large-scale quantum computing, where operation inside cryostats at deep cryogenic temperatures is essential. Overall, the data empirically supports the design principles of cryogenic-compatible, high-speed photonic control architectures, demonstrating that strategic bias voltage management and thermal calibration enable photodetectors to meet the stringent performance requirements of next-generation quantum processors.

5.5 RF Output Power as a Function RF Input Power at 8 GHz Modulation

Test results shown in Fig. 6(A) examine the photodetector's linearity and RF gain characteristics by analyzing RF output power as a function of RF input power at a modulation frequency of 8 GHz. The experiment was conducted at varying bias voltages (0.25 V, 0.5 V, 0.75 V, and 1.0 V) and under two thermal conditions, 300 K and 55 K, with the laser current fixed at 35 mA. This test provides insight into the photodetector's behavior under dynamic power conditions, which is critical for evaluating its suitability in scalable and power-efficient quantum control systems.

34   A. Amsanpally et al.

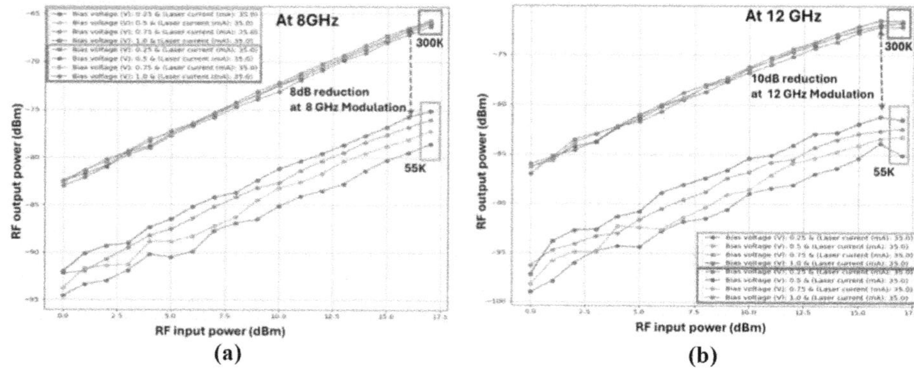

Fig. 6. (A) Linear RF gain across input power range; 8 dB cryogenic loss observed. (B) RF response at 12 GHz with ~ 10 dB loss at 55 K compared to 300 K.

At 300 K, the results show a near-linear increase in RF output power with rising RF input power across all bias voltages. The slope of the curves indicates consistent gain and minimal distortion over the tested dynamic range. As expected, higher bias voltages produce stronger RF output, due to improved carrier sweep-out efficiency and reduced transit-time effects. This linear behavior highlights the device's robustness in transmitting modulated signals without introducing nonlinear artifacts, thereby preserving signal fidelity.

At 55 K, the RF output exhibits an ~ 8 dB reduction compared to room temperature, with losses most pronounced at low input powers, reflecting increased threshold sensitivity and higher minimum power requirements. Despite this attenuation, the output maintains linearity at higher bias voltages, indicating preserved intrinsic gain but reduced efficiency at cryogenic temperatures. This behavior arises from reduced carrier mobility, suppressed recombination, and altered photodiode impedance. While output power decreases, the sustained linear response confirms suitability for cryogenic RF-over-Fiber applications, provided sufficient input levels and optimized biasing, which is critical for deterministic qubit manipulation in photonic quantum computing.

5.6   RF Output Power as a Function of RF Input Power at 12 GHz Modulation.

This section builds upon the findings from Fig. 6(A) by reanalyzing RF gain and linearity at a modulation frequency of 12 GHz. This test is critical for assessing the photodetector's bandwidth limitations and its ability to handle high-frequency modulations required for fast qubit gate operations. As in the previous test, measurements were conducted across multiple bias voltages (0.25 V to 1.0 V), with the laser current held constant at 35 mA, under both 300 K and 55 K conditions. As shown in Fig. 6(B), at 300 K, the results demonstrate strong RF output power scaling with increasing RF input power across the entire tested range. Like the 8 GHz case, the curves maintain excellent linearity, and higher bias voltages yield improved output strength. These results confirm the detector's ability to operate efficiently at frequencies up to 12 GHz without encountering saturation or bandwidth roll-off, which is essential for high-speed quantum control systems.

Under cryogenic operation at 55 K, the output power exhibits a notable reduction, approximately 10 dB below room-temperature performance at equivalent input levels. This reduction is slightly greater than that observed at 8 GHz, indicating modest frequency-dependent sensitivity when the device is cooled. Nevertheless, the linear trend of RF output versus input power is preserved, and higher bias voltages continue to produce superior output levels. This consistent behavior demonstrates that the photodetector remains effective for high-frequency operation in cryogenic conditions, albeit with slightly reduced gain. The observed 10 dB loss highlights the need for thermal compensation techniques, such as increased laser power, optimized biasing, or post-demodulation signal amplification. However, the absence of signal compression or nonlinearity at higher input levels confirms that the photodetector preserves signal integrity, a crucial requirement for precision-driven quantum computing systems.

Overall, the results in Fig. 6 validate the operational linearity and frequency responsiveness of the photodetector across a broad RF input power range and under extreme temperature conditions. These characteristics are instrumental in enabling scalable, low-noise, and high-fidelity photonic control systems for large-scale quantum computing architectures.

6 Conclusion

The experimental results demonstrate how operating parameters influence the performance of our PCS in controlling solid-state qubits. Optimal operation was achieved with a laser current near 60 mA, consistent with Menthor's default configuration, and a photodetector bias voltage around 0.75 V, balancing efficiency and thermal considerations. The system tolerated RF inputs up to $+$ 16 dBm without distortion, highlighting the robustness and linearity of the photonic control platform.

This work validates Viqthor's cutting-edge photonic control system built around Menthor and IceQube, capable of transmitting and demodulating high-frequency RF signals with minimal loss and excellent stability down to cryogenic temperatures as low as 55 K. Beyond simple signal conversion, the PCS provides a modular, scalable, low-noise, and thermally efficient foundation for quantum control. The experimental results establish a practical roadmap for integrating photonic control into next-generation quantum computing systems, representing a significant step toward modular, high-fidelity, and scalable quantum processors.

References

1. Krinner, S., Storz, S., Kurpiers, P., et al.: Engineering cryogenic setups for 100-qubit scale superconducting circuit systems. EPJ Quant. Technol. **6**, 2 (2019)
2. Fowler, A.G., Mariantoni, M., Martinis, J.M., Cleland, A.N.: Surface codes: towards practical large-scale quantum computation. Phys. Rev. A **86**, 32324 (2012)
3. Krantz, P., et al.: A quantum engineer's guide to superconducting qubits. Appl. Phys. Rev. **6**, 021318 (2019)
4. Pintus, P., Soltani, M., Moody, G.: Cryogenic optical data link for superconducting circuits. Nat. Photon. **18**, 306–308 (2024)

5. Joshi, S., Moazeni, S.: Scaling up superconducting quantum computers with cryogenic RF-photonics. J. Lightwave Technol. **42**(1), 166–175 (2024)
6. Lecocq, F., et al.: Control and readout of a superconducting qubit using a photonic link. Nature **591**, 575–579 (2021)
7. Christiansen, A.J.: Characterization of optical fiber at cryogenic temperatures. In: SPIE Proceedings, vol. 12417 (2023)
8. Lu, X.: Impact of the fiber coating on the temperature response of distributed optical fiber sensors at cryogenic ranges. JLT **36**(4), 961–967 (2018)
9. Bardalen, E.: Evaluation of InGaAs/InP photodiode for high-speed operation at 4 K. Int. J. Metrol. Qual. Eng. **9**, 13 (2018)

Session 1 Quantum Sensors – Quantum sensors applications

Design Concept of NV-Center-Based Wideband Current Comparators

Yasutaka Amagai[1,2]($\boxtimes$), Hidekazu Muramatsu[1,2], Yuta Kainuma[1,2], Hiromitsu Kato[1,3], Norihiko Sakamoto[2], Nobu-Hisa Kaneko[1,2], Yuji Hatano[4], Mutsuko Hatano[4], and Takayuki Iwasaki[4]

[1] National Institute of Advanced Industrial Science and Technology (AIST), Global Research and Development Center for Business by Quantum-AI Technology (G-QuAT), 1-1-1 Umezono, Tsukuba 305-8560, Ibaraki, Japan
y-amagai@aist.go.jp
[2] National Institute of Advanced Industrial Science and Technology (AIST), National Metrology Institute of Japan (NMIJ), 1-1-1 Umezono, Tsukuba 305-8563, Ibaraki, Japan
[3] National Institute of Advanced Industrial Science and Technology (AIST), Advanced Power Electronics Research Center (ADPERC), 1-1-1 Umezono, Tsukuba 305-8568, Ibaraki, Japan
[4] School of Engineering, Institute of Science Tokyo, 2-12-1 Ookayama, Meguro 152-8550, Tokyo, Japan

Abstract. Electrical metrology has historically relied on precise current comparators, which play a central role in determining accurate current ratios. Conventional devices, however, depend on inductive pickup coils or cryogenic infrastructures such as SQUIDs, imposing limitations on bandwidth, complexity, and operational environments. To address these challenges, we propose a unified comparator architecture based on nitrogen-vacancy (NV) centers in diamond. The system replaces conventional magnetic pickup or flux feedback mechanisms with a solid-state quantum magnetometer that optically detects magnetic flux in the air-gap of a magnetic core. The NV sensor provides sensitivity to both AC and DC magnetic fields, enabling seamless operation across different metrological regimes, and fully compatible with a modern power electronics requirement. The proposed architecture offers key advantages such as complete electrical isolation, compact system, and potential scalability for portable or embedded applications. Modeling and recent advances in NV magnetometry suggest that the required sensitivity is achievable with current technologies. By addressing both AC and DC current measurement requirements in the field of power industries and fundamental electrical standards, this concept paves the way toward a next-generation current ratio standard.

Keywords: NV center · Diamond · Quantum sensing · metrology · electrical standards · current comparator · current ratio measurement

1 Introduction

The field of electrical metrology has long relied on quantum-based standards such as the Josephson DC voltage standard [1] and the quantum Hall DC resistance standard [2], along with single-electron tunneling pumps [3, 4]. Meanwhile, even with the availability

© The Author(s), under exclusive license to Springer Nature Switzerland AG 2026
F. Barbaresco and G. François (Eds.): QUEST-IS 2025, CCIS 2743, pp. 39–46, 2026.
https://doi.org/10.1007/978-3-032-13852-1_5

of quantum electrical standards, practical current metrology often requires precise comparison between currents rather than absolute current alone. In several applications such as current transformer calibration, bridge circuits for fundamental resistance measurement, current comparators play a crucial role. These devices allow for precision current scaling, filling a gap between quantum standards and practical instruments.

Traditionally, alternating current (AC) measurements have relied on inductive current transformers and their calibration using current comparator systems that operate at power-line frequencies (50/60 Hz). These devices have achieved impressive accuracy on the order of 10^{-6} [5, 6], thereby underpinning much of the infrastructure required for electric power distribution, grid stability, and industrial metering. In parallel, direct current (DC) ratio measurements have been indispensable for resistance bridge circuits, which ensure the traceability of resistance standards derived from the quantum Hall effect. In this context, SQUID-based [7] current comparators have delivered extreme accuracy down to 10^{-9}, but only at the cost of cryogenic operation [8, 9]. Such low-temperature requirements increase complexity and cost, limiting accessibility to a small number of national metrology institutes. This dichotomy between AC and DC comparators illustrates a broader challenge: while existing technologies are highly specialized, they lack a unified platform that can address both regimes with comparable performance. Moreover, emerging applications in modern power electronics, smart grids, and electric mobility demand measurement solutions that combine high accuracy with compactness, portability, and robustness under real-world conditions.

Recent advances in room-temperature quantum magnetometry using nitrogen-vacancy (NV) centers in diamond offer a promising pathway to address these limitations. NV centers allow for the optical detection of magnetic fields with sub-nanotesla sensitivity and a wide dynamic range, simultaneously covering both AC and DC regimes [10–13]. Unlike conventional pickup coils or SQUIDs, NV sensors operate at ambient conditions, provide inherent electrical isolation, and can be engineered into miniaturized and scalable devices. These features open the possibility of developing a unified current comparator architecture—one that seamlessly supports both AC and DC current ratio measurements within a single, solid-state sensing platform.

In this paper, we present the design concept of such a comparator system based on NV-diamond magnetometers. The proposed approach eliminates the need for complex windings or cryogenic environments, while enabling broadband magnetic field detection through an optically read-out sensor placed in the air-gap of a magnetic core. We focus on the design strategy, sensor requirements, and theoretical feasibility, supported by recent modeling efforts and advances in high-purity diamond engineering. Experimental implementation is currently underway, and results will be reported in a separate publication [14]. By bridging practical metrological requirements with cutting-edge quantum sensing technology, this work aims to contribute to the next generation of electrical standards and to provide a scalable route toward international standardization.

2 Measurement Principles

2.1 Quantum Magnetometry Using NV Centers in Diamond

Nitrogen-vacancy (NV) centers in diamond have gained significant attention as versatile quantum magnetometers and represent one of the most mature solid-state quantum sensing platforms available today. An NV center consists of a substitutional nitrogen atom adjacent to a vacancy in the diamond lattice, forming a point defect with an electronic spin triplet ground state ($S = 1$). The spin sublevels ($m_s = 0$ and $m_s = \pm 1$) experience Zeeman splitting in the presence of an external magnetic field, with the splitting magnitude being directly proportional to the field strength [10].

Optically detected magnetic resonance (ODMR) provides a powerful mechanism to initialize, manipulate, and read out these spin states. When the NV center is illuminated with green laser light, it exhibits spin-dependent fluorescence in the red spectral region. The application of a resonant microwave field modulates the fluorescence intensity, giving rise to characteristic ODMR spectra. These spectra display sharp resonance dips that shift linearly with magnetic field, thereby enabling precise and quantitative magnetic field measurements. By carefully monitoring these shifts, one can resolve both AC and DC magnetic components with remarkable sensitivity.

NV centers offer several advantages compared with other magnetometer technologies. They function robustly at room temperature, and coherent spin control has been demonstrated even near 1000 K [11], highlighting their remarkable thermal stability. They achieve sensitivity levels down to the pico-tesla (10^{-9} T) regime for both AC and DC fields [12, 13], and due to their crystallographic orientation, they can provide vector magnetic field information [15], which is difficult to obtain with many other sensors. Moreover, the solid-state nature of diamond allows compact device integration, long-term stability, and the potential for large-scale sensor arrays, making NV centers scalable for both laboratory and industrial applications [16].

Recent engineering advances have further pushed the performance limits. High-purity, isotopically engineered diamond samples combined with optimized optical and microwave control have demonstrated sensitivities below 10 pT/Hz$^{1/2}$ [17, 18]. Such performance enables the detection of extremely weak residual magnetic fields, including those encountered in precision metrology devices, resistance bridge circuits, and current comparators. Taken together, these attributes position NV centers as one of the most promising platforms for bridging the gap between quantum physics and practical measurement standards.

2.2 Conceptual Integration of NV Sensors into Current Ratio Measurement

The proposed architecture leverages NV-diamond magnetometry to overcome the inherent limitations of conventional current comparators and to provide a unified platform for precision AC and DC measurements. In this scheme, the magnetic core of the comparator is intentionally designed with an air-gap, into which the NV sensor is either embedded or externally aligned in close proximity. This configuration enables the sensor to optically detect the residual magnetic flux arising from even very small current imbalances. The principle is thus conceptually simple yet powerful: when the primary and secondary currents are perfectly balanced, the net flux through the core vanishes, whereas any deviation produces a detectable magnetic leakage field at the air-gap.

For AC applications, the system functions in a manner closely analogous to traditional inductive comparators. The NV sensor is able to track oscillating magnetic fields at standard power-line frequencies, thereby supporting feedback control schemes that null the flux in real time. The resulting balance condition provides the desired current ratio, while maintaining full electrical isolation between the sensing unit and the current-carrying windings. Importantly, this approach extends naturally beyond 50/60 Hz operation, suggesting applicability to a broader frequency range relevant to modern power electronics.

In the case of DC measurements, where conventional comparators encounter fundamental challenges due to the limited sensitivity of pick-up coils, the NV sensor introduces a qualitatively new detection principle. By directly measuring the static magnetic field generated by current imbalance, the system establishes a null-detection method for resistance bridge circuits and DC ratio standards without recourse to inductive components or cryogenic devices such as SQUIDs. This eliminates many of the operational constraints that have historically restricted DC comparator deployment to specialized laboratories.

Beyond the immediate metrological benefits, the use of NV centers also simplifies both mechanical and magnetic design. The absence of pick-up coils reduces the complexity of winding arrangements and allows the magnetic core to be made smaller, lighter, and more compatible with modular instrumentation. Optical read-out further enhances robustness against electromagnetic interference, while enabling straightforward electrical isolation and safe operation in complex environments such as high-voltage laboratories or industrial substations. These characteristics naturally support miniaturization, portability, and potential integration into embedded systems.

3 Sensor Design and Feasibility

3.1 Conceptual Architecture

The proposed current comparator architecture is centered around a magnetic core equipped with multiple windings and an intentionally introduced air-gap, within which an NV center–based diamond magnetometer is carefully positioned. This structural arrangement enables the sensor to directly probe the local magnetic field in the vicinity of the air-gap, allowing for the detection of the net magnetic flux that arises from even a slight imbalance between the currents flowing through the windings. By converting this residual flux into an optically detectable signal, the NV sensor serves as the key element that bridges conventional electromagnetic comparator principles with modern quantum magnetometry.

Figure 1 illustrates the simplified structure of the system. The primary and secondary currents, driven through separate windings, produce opposing magnetomotive forces in a high-permeability core. Under ideal balanced conditions, the net magnetic flux through the core is zero. Any residual flux arising from a mismatch between the currents escapes through the air-gap, where it is detected by the NV sensor. The NV sensor is optically interrogated using a green laser and microwave excitation, with red fluorescence intensity being monitored to extract the local magnetic field via optically detected magnetic resonance (ODMR). This optical readout provides complete electrical isolation from the current-carrying paths, enhancing measurement safety and robustness against electromagnetic interference. To evaluate the feasibility of the proposed approach, we estimate the magnetic field that will be detected by the NV sensor in a typical comparator configuration. Finite element simulation of magnetic flux distribution in a toroidal core with and without an air gap. The flux is significantly decreased due to high magnetic reluctance of the air gap. The presence of the gap causes local flux leakage into the surrounding space, enabling detection by a magnetic sensor such as the NV center. These sensors are thus well positioned to achieve the necessary performance for detecting small residual flux levels in both AC and DC conditions. Moreover, the absence of detection windings significantly simplifies the magnetic design, allowing for reduction in core size and may faster dynamic response. This is especially advantageous for time-sensitive calibration systems. It is also worth noting that even if the sensitivity of the NV magnetometer does not meet the required threshold, longer integration times can reduce random noise. This trade-off between measurement time and precision may allow practical use of metrology.

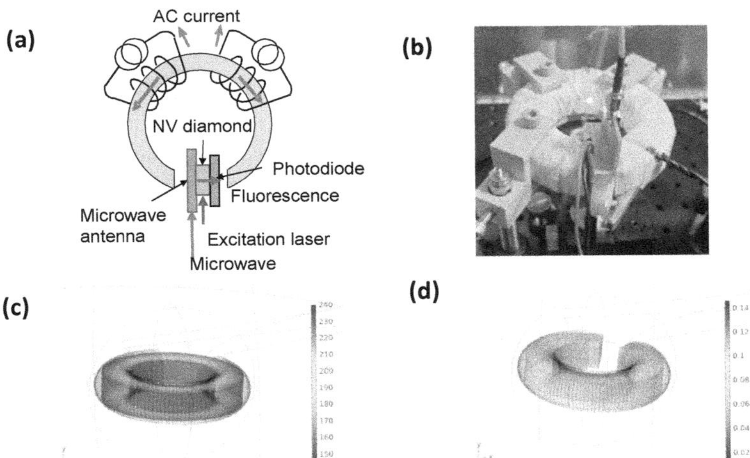

Fig. 1. (a) Schematic drawing of the proposed current comparator architecture using NV centers in diamond as a magnetic flux detector. (b) A photograph of the proposed current comparator architecture. (c) Simulated magnetic flux density in a toroidal core without any gap. The magnetic field remains uniformly confined within the core material, resulting in negligible external flux leakage. (d) Simulated magnetic flux distribution in the same toroidal core after introducing an air gap. The presence of the gap causes local flux leakage into the surrounding space, enabling detection by a magnetic sensor such as the NV center.

3.2 Optical and Electromagnetic Considerations

The NV magnetometer requires direct, line-of-sight optical access to the sensing volume in order to achieve efficient excitation and readout of the spin states. In our design, laser light is introduced laterally through the side of the diamond chip or coupled via an optical fiber, providing flexibility in integration and allowing compact packaging of the sensor head. The resulting red fluorescence is collected either along the same axis or through an orthogonal optical path, then spectrally filtered before reaching a photodiode or avalanche photodiode, ensuring high signal-to-noise detection. The microwave field necessary for spin manipulation is delivered by a coplanar waveguide patterned in close proximity to the sensing region, which ensures homogeneous excitation and minimizes unwanted reflections.

To further enhance sensitivity, the air-gap in the magnetic core is deliberately designed to be narrow and positioned precisely along the anticipated flux leakage path, thereby maximizing the magnetic field concentration at the diamond sensor. Careful attention is paid to the geometry of the sensor head, as a compact form factor is essential to minimize magnetic perturbations and to preserve the intended flux distribution. In addition, multilayer magnetic shielding surrounding the sensor head is considered to suppress environmental magnetic noise and to reduce the influence of mechanical vibrations. Such shielding not only improves measurement stability in laboratory conditions but also enhances robustness for potential deployment in industrial environments, where electromagnetic interference is more severe. Collectively, these optical, microwave, and shielding strategies provide a practical foundation for achieving high-fidelity NV-based current comparator measurements.

3.3 Limitations and Ongoing Work

While initial feasibility studies are promising, several engineering challenges remain to be addressed before the NV-based comparator can be realized as a practical standard. In particular, DC operation may be influenced by leakage flux arising from geometric asymmetries in the magnetic core or from magnetization hysteresis associated with the presence of the air gap. These effects, although less prominent in inductive or SQUID-based comparator architectures, can degrade measurement accuracy if not carefully modeled and compensated. Ongoing efforts therefore focus on accurate finite-element simulations and the development of active compensation techniques to suppress such parasitic flux contributions.

Another critical requirement is the use of homogeneous NV-diamond sensors with minimal device-to-device variation. Inhomogeneities in crystal quality, defect density, or orientation distribution can lead to variability in sensor performance, complicating calibration and long-term reproducibility. Equally important are the engineering aspects of optical alignment, thermal management, and microwave field homogeneity, all of which directly impact the achievable sensitivity and stability. For example, slight drifts in laser coupling or temperature fluctuations can introduce baseline shifts over long measurement times, an issue of particular relevance to metrological applications that demand repeatability at the 10^{-8} level or below.

In addition, practical DC operation often requires the use of auxiliary coils wound around the magnetic core to provide controlled DC offsets for balancing purposes. While such coils are indispensable for fine adjustment, they may themselves act as sources of electromagnetic noise. Careful design of the coil geometry, shielding, and filtering circuitry is therefore essential to suppress these noise contributions and to prevent them from limiting the ultimate sensitivity of the NV read-out.

4 Conclusions

This work presents a conceptual architecture for a novel current comparator system based on nitrogen-vacancy (NV) centers in diamond. The proposed design offers several attractive features, including room-temperature operation, complete electrical isolation, compactness, and compatibility with modern measurement requirements such as portability and broadband functionality demanded by power and energy industries. By eliminating the dependence on bulky pickup coils and cryogenic components, the NV-based approach also simplifies the overall system design, paving the way for instruments that are easier to manufacture, deploy, and maintain. Although the proposed comparator has not yet been fully implemented experimentally, the preliminary design considerations and feasibility analysis strongly suggest that its realization is within reach of currently available NV sensing technologies. Continued improvements in diamond material quality, sensor integration, and optical read-out techniques are expected to further enhance sensitivity and stability, thereby supporting long-term reproducibility essential for metrological applications. Moreover, the architecture is inherently scalable, making it adaptable not only for laboratory-grade standards but also for embedded systems in industrial or grid-level monitoring environments.

Taken together, these advantages indicate that the NV-based current comparator concept represents a promising path toward next-generation current ratio standards. In addition to advancing measurement science, this work may contribute to the broader field of quantum sensing by demonstrating a concrete, application-driven implementation. Experimental validation is currently underway, and results will be presented during the conference presentation.

Acknowledgements. This work was supported by the MEXT Quantum Leap Flagship Program (MEXT Q-LEAP) Grant Number JPMXS0118067395, the CSTI, the Cross-ministerial Strategic Innovation Promotion Program (SIP), "Promoting the application of advanced quantum technology platforms to social issues," and Bridging the gap between R&D and the IDeal society (society 5.0) (BRIDGE) organized by the Council for Science, Technology Innovation (CSTI) (Funding agency: QST). It was also partially supported by the JST ASPIRE programGrant number JPMJAP24C1.

References

1. Kohlmann, J., Behr, R., Funkc, T.: Josephson voltage standards. Meas. Sci. Technol. **14**, 1216 (2003). https://doi.org/10.1088/0957-0233/14/8/305
2. Poirier, W., Schopfer, F.: Resistance metrology based on the quantum hall effect. Eur. Phys. J. Spec. Top. **172**, 207–245 (2009). https://doi.org/10.1140/epjst/e2009-01051-5
3. Pekola, J.P., Saira, O.-P., Maisi, V.F., et al.: Single-electron current sources: toward a refined definition of the ampere. Rev. Mod. Phys. **85**, 1421 (2013). https://doi.org/10.1103/RevMod Phys.85.1421
4. Fujiwara, A., Yamahata, G., Johnson, N., Nakamura, S., Kaneko, N.-H.: Silicon quantum dot single-electron pumps for the closure of the quantum metrology triangle. ECS Trans. **112**, 119 (2023). https://doi.org/10.1149/11201.0119ecst
5. Moore, W.J.M., Miljanic, P.N.: The Current Comparator. Peter Peregrinus Ltd., London (1988)

6. Kusters, N.L., Moore, W.J.M.: The compensated current comparator: a new reference standard for current-transformer calibrations in industry. IEEE Trans. Instrum. Meas. IM **13**, 107–114 (1964). /https://doi.org/10.1109/TIM.1964.4313383

7. Greenberg, Y.S.: Application of superconducting quantum interference devices to nuclear magnetic resonance. Rev. Mod. Phys. **70**(1), 175–222 (1998). https://doi.org/10.1103/Rev ModPhys.70.175

8. Drung, D., et al.: Improving the stability of cryogenic current comparator setups. Supercond. Sci. Technol. **22**, 114004 (2009). https://doi.org/10.1088/0953-2048/22/11/114004

9. Seppä, H., et al.: Thin-film cryogenic current comparator. IEEE Trans. Instrum. Meas. **48**, 365–369 (1999). https://doi.org/10.1109/19.769602

10. Jelezko, F., Wrachtrup, J.: Single defect centres in diamond: a review. Phys. Status Solidi A **203**(13), 3207–3225 (2006). https://doi.org/10.1002/pssa.200671403

11. Liu, G.Q., Feng, X., Wang, N., et al.: Coherent quantum control of nitrogen-vacancy center spins near 1000 kelvin. Nat. Commun. **10**, 1344 (2019). https://doi.org/10.1038/s41467-019-09327-2

12. Doherty, M.W., et al.: The nitrogen-vacancy color center in diamond. Phys. Rep. **528**(1), 1–45 (2013). https://doi.org/10.1016/j.physrep.2013.02.001

13. Taylor, J.M., et al.: High-sensitivity diamond magnetometer with nanoscale resolution. Nat. Phys. **4**, 810–816 (2008). https://doi.org/10.1038/nphys1075

14. Muramatsu, H., et al.: Current comparator for both AC and DC ratio measurements with 10^{-8}-level accuracy. arXiv preprint arXiv:2508.05140 (2025)

15. Maertz, B.J., Wijnheijmer, A.P., et al.: Vector magnetic field microscopy using nitrogen vacancy centers in diamond. Appl. Phys. Lett. **96**(9), 092504 (2010). https://doi.org/10.1063/1.3364135

16. Pogorzelski, J., Horsthemk, L., et al.: Fully integrated LED quantum sensor based on NV centers in diamond. Sensors **24**, 743 (2024). https://doi.org/10.3390/s24030743

17. Sekiguchi, N., Fushimi, M., Yoshimura, A., et al.: Diamond quantum magnetometer with dc sensitivity of sub-10 pT/Hz$^{1/2}$ toward measurement of biomagnetic field. Phys. Rev. Appl. **21**, 064010 (2024). https://doi.org/10.1103/PhysRevApplied.21.064010

18. Sekiguchi, N., et al.: Performance evaluation of a diamond quantum magnetometer for biomagnetic sensing: a phantom study. Appl. Phys. Lett. **126**, 194001 (2005). https://doi.org/10.1063/5.0254828

Detection of Grinding Burns with NV Center in Diamond Magnetometer

Baptiste Vindolet[1], Guillaume Bourcin[1]([✉]), Benjamin Ducharne[2,3],
Hoai Nam Nguyen[1], Xavier Mougenot[4], Christophe Gallais[4], and Thomas Hingant[1]

[1] KWAN-TEK, 1 Rue Galilée Espace Innova, 56270 Ploemeur, France
g.bourcin@kwan-tek.com
[2] LGEF, INSA-Lyon, UR682, Université Lyon, 8 Rue de La Physique, 69621 Villeurbanne,
France
[3] ELyTMaX IRL3757, CNRS, Univ Lyon, INSA, Lyon, Central Lyon, Université Claude
Bernard Lyon 1, Tohoku Université, Sendai, Japan
[4] Audit Et R&T Industriels, Safran Transmission Systems, 18 Bd Louis Seguin, 92700
Colombes, France

Abstract. Grinding burns alter the hardness and mechanical properties of steel components and can result in cracks or failure. Nital etching is widely used in the industry to detect grinding burns, but it poses environmental and safety risks and requires visual detection. Alternative techniques such as magnetic Barkhausen noise inspection have been developed, but automation is hindered by calibration drifts and by the sensor dimensions. Here, we introduce a novel non-destructive approach based on nitrogen-vacancy center in diamond to detect leakage magnetic field from grinding burns. This technology enables quantitative vector magnetic field measurement, is contactless and does not require calibration. Grinding burns with different sizes were successfully detected with a high spatial resolution and sensitivity.

Keywords: NV Diamond Quantum Magnetometry · Non-Destructive Testing · High-Resolution Sensing

1 Introduction

Some essential metallurgic elements such as gears or bearings are machined in high quality steel to resist extreme operational requirements [1–3]. To maintain expected tolerances and high surface quality, a rectification stage is performed to guarantee performance, durability and optimal security [4]. This processing step is the trickiest one, potentially leading to a heating of the material. Excessive heat can alter the material microstructure and bring reduced hardness and intrinsic strain formation [5, 6]. With time, such defects, called grinding burns (GB), can result in cracks or failures in the material [7].

Detection methods exist to detect GB. These include chemical detection or magnetic non-destructive testing (NDT). Because most ground steel possess ferromagnetic phases,

© The Author(s), under exclusive license to Springer Nature Switzerland AG 2026
F. Barbaresco and G. François (Eds.): QUEST-IS 2025, CCIS 2743, pp. 47–54, 2026.
https://doi.org/10.1007/978-3-032-13852-1_6

magnetic NDT is well-suited for detecting defects. Destructive methods also exist offering high precision, but present excessive cost, and are not suitable on production lines due to the invasive inspection [8].

Among chemical methods, nital etching is the most common in industry [9–11]. This method consists of applying a chemical etching component on the sample surface. Material grain (crystalline structure), phase (ferrite, martensite, …), and defect on surface are made visible to the naked eye, becoming darker as usual. While this process is effective and inexpensive, it usually requires an operator to visualize defects, therefore it is difficult to fully automate [12, 13] Moreover, it requires the use of toxic chemical products.

Magnetic NDT has advantages of being non-invasive, inexpensive, and make possible integration and automation in a product line, conversely to chemical etching. Among them, the magnetic Barkhausen noise (MBN) analysis can detect grain, texture, phase, or impurities in the material [14–16]. The difficulty is to differentiate all these magnetic phenomena. These can be differentiated by combining MBN with other magnetic methods but require a lot of data for a specific products line and is not easily transposable for another one [17, 18].

In this study, a novel approach of magnetic NDT is proposed. This method, based on Nitrogen-Vacancy (NV) center technology, exploits the quantum properties of NV color centers in diamonds to achieve precise and reliable magnetic field measurements [19–21]. Due to its quantum nature, this technology is highly sensitive and does not require calibration, while providing digital measurement with a high resolution. Additionally, the probe is small enough (sub-mm) and is suitable to access difficult geometries, such as gear teeth for instance.

2   Implementation of a Stray Field Measurement Setup

The Nitrogen-Vacancy (NV) center is a defect in the crystalline lattice of a diamond, where two adjacent carbon atoms are substituted by a nitrogen atom and a vacant site respectively. Due to its configuration, the NV center diamond absorbs green light, and reemits red fluorescence. The negatively charge NV^- center presents an electronic spin triplet characterized by the projection number $m_s = 0, \pm 1$. Due to spin-spin interaction, the singulet state $m_s = 0$ and doublet states $m_s = \pm 1$ are energetically separated of $D \simeq$ 2.87 GHz. Under green light excitation, the NV is pumped into the $m_s = 0$ and the red fluorescence rate from the $m_s = 0$ state is higher than from the $m_s = \pm 1$ states, enabling optically detected magnetic resonance (ODMR), by collecting photoluminescence (PL) over a MW frequency sweep [22]. Furthermore, with the Zeeman effect, the degeneracy of the $m_s = \pm 1$ states is lifted when a magnetic field is applied, and the transition frequencies between the ground state and both excited states are given by [21]:

$$\nu_{\pm 1} = D \pm \frac{\gamma_{NV}}{2\pi} \boldsymbol{B} \cdot \hat{\boldsymbol{u}}_{NV}, \tag{1}$$

where $\hat{\boldsymbol{u}}_{NV}$ is the NV center axis, and γ_{NV} is the gyromagnetic ratio of the NV center, with $\frac{\gamma_{NV}}{2\pi} = 28.035 \text{ MHz.mT}^{-1}$.

Because of the crystal structure, there are four possible orientations of the NV center, allowing vectorial measurements of the magnetic field. These exceptional spin properties have attracted much interest for the NV center in diamond as a quantum sensor for magnetometry but also electric field sensing and thermometry [19–21].

Inhomogeneous magnetization in ferromagnetic materials results in leakage magnetic field outside the sample. In a magnetized material, these inhomogeneities can occur at cracks, inclusions, corrosions, or grinding burns for instance. The principle of the NV magnetometer measurement is to provide stray field cartography at the sample surface, as depicted in Fig. 1 (a).

The NV diamond is integrated into an endoscopic sensor, glued at the tip of an optical fiber. The diamond is of 500 μm cubic shape, and highly doped with NV center (~ 0.3 ppm). Additionally, a microwave (MW) antenna is localized near to the diamond, to excite the NV magnetic resonances. The antenna is a thin copper coil, connected to an amplified MW synthesizer via an SMA connection. A photograph of the tip of the.

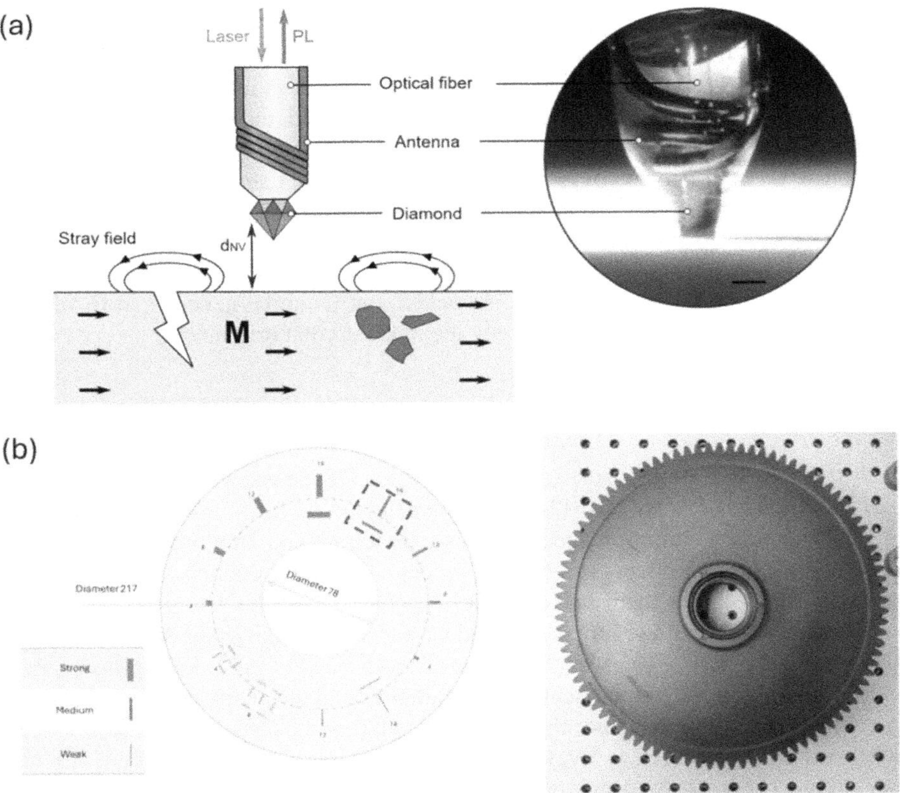

Fig. 1. (a) Schematic representation of inhomogeneous magnetization due to cracks or phase changing in a steel sample resulting in leakage magnetic field. The measurement is performed with a diamond NV center-based endoscope, with a photograph of the tip on the right-hand side. (b) Schematic representation of the gearwheel with grinding burns of different lengths, widths, and depths. Different cartographies of defects presented in this study are emphasized in dashed square regions. A photograph of the gearwheel is provided at the right-hand side.

endoscope is given in Fig. 1(a). The green laser of 520 nm wavelength couples into the optical fiber from free space. The red PL goes back into the fiber and is separated from the green one thanks to a dichroic mirror. Then, the red light is collected on a photodiode which converts the signal into voltage.

In addition to the endoscope, a static magnetic field is applied - using a permanent magnet - to split the ODMR resonances. The frequency position of one of these resonances is tracked [23]. When an unknown magnetic field occurs from the sample, its projection along the NV center axis shifts the resonance. The tracking is performed using the modulation and demodulation method. The applied MW is typically modulated at the kilohertz range around the center frequency of the resonance ν_0. Hence, the PL signal is also modulated at the MW modulation frequency. After demodulation using a lock-in amplifier, the output signal obtained is similar to a derivative of the resonance curve. This output signal presents a linear region around the center frequency. In this region, the amplitude is directly proportional to the MW frequency, hence to the magnetic field from the defect and cancels at the resonance.

To track the resonance, a proportional-integral-derivative (PID) feedback controller is implemented. The PID adjusts the synthesizer frequency to maintain the demodulated signal at zero. To obtain a cartography of the leakage field, the sample is mounted on a XY translation stage, and the endoscope is fixed above the sample, with a lift-off distance of about 100 μm to prevent physical contact and potential damage. A two-dimensional map is obtained by storing the instantaneous resonance frequency at each pixel. Using the transition frequencies formula given in Eq. (1), the leakage magnetic field is retrieved from the measured instantaneous resonance frequency at each pixel.

In this study, only one resonance is tracked, but it can be generalized to vectorial measurement by tracking simultaneously the four different resonances.

3 Experimental Results

In this study, the NV magnetometer is tested on GB detection. For this purpose, a gearwheel made of 16NiCrMo13 with a diameter of 20 cm is used, as illustrated in Fig. 1(b). Burns were reproduced with laser-treatment and resulting defects are like those due to thermal effects of grinding process. A temper etching process allows to make defects visible to the naked eye.

The defects were previously calibrated with a twin sample using destructive metallographic analysis. The gearwheel presents 15 laser burns of different lengths (4, 8, 12, and 16 mm) along and perpendicular to the radial axis of the gearwheel. The defects are categorized according to either their depths or their widths: the defect depth is either of 60 μm and considered as weak, or of 180 μm and considered as strong; similarly, the width is either of 580 μm (weak) or of 1050 μm (strong). Note that detecting weak burns is not straightforward with standard visual inspection [24].

A ferromagnet has a magnetic history, and the magnetization homogeneity around a defect is not guaranteed. In the case of inhomogeneous magnetization, the leakage magnetic field related to the laser burn would be quasi-inexistant. To overcome this issue, material magnetization preprocessing is necessary. Before each measurement campaign, the sample is magnetized around the defect under test with a high amplitude in-plane magnetic field using an electromagnetic yoke. Poles of the yoke were separated of 7 cm and placed at each side of the defect, applying a magnetic field up to 30 mT.

3.1 Influence of the Magnetization Direction

To detect the defect, the pre-processing step of magnetization needs to be performed properly. Indeed, as shown in the following of this study, the defect detection is directly linked to the magnetization direction. Figure 2(a) illustrates a cartography of the magnetic field probed with the NV diamond-based endoscope. The range of cartography is 30 mm in X and Y direction, and the pixel size is 200 μm. These directions are quasi equivalent to the azimuthal and radial directions respectively. The cartography area is emphasized by the square red region in Fig. 1(b), which includes two medium defects with a length of 16 mm in the azimuthal (X-axis) and radial (Y-axis) directions. Additionally, the magnetic field is applied along the X-axis, therefore perpendicular to defect along the Y-axis. A smooth magnetic field gradient of about 15 mT originates from sample localization beneath yoke poles throughout the magnetization processing. The defect along the Y-axis is visible with an amplitude contrast of about 65 μT whilst the defect along the X-axis is not. In Fig. 2(b) the Laplacian of the magnetic field is computed, given by:

$$\Delta B_{\mathrm{NV}}(x, y) = \frac{\partial^2 B_{\mathrm{NV}}(x, y)}{\partial x^2} + \frac{\partial^2 B_{\mathrm{NV}}(x, y)}{\partial y^2} \tag{2}$$

The Laplacian cartography reveals fast spatial magnetic variation, i.e. the default, and suppresses slow variations, i.e. the slow magnetic gradient between the yoke's poles.

In Fig. 2(c) and (d) the map from the magnetic field and from the corresponding Laplacian are represented, with a magnetization orientation at 45° from the X-axis direction. With this magnetization orientation, both the defects in the azimuthal and radial directions are visible. In the perpendicular direction of a defect, the magnetization divergence occurs in a shorter distance as in its parallel direction. Therefore, the resulting leakage magnetic field is higher when the sample is magnetized along the perpendicular direction to the defect than along its parallel direction.

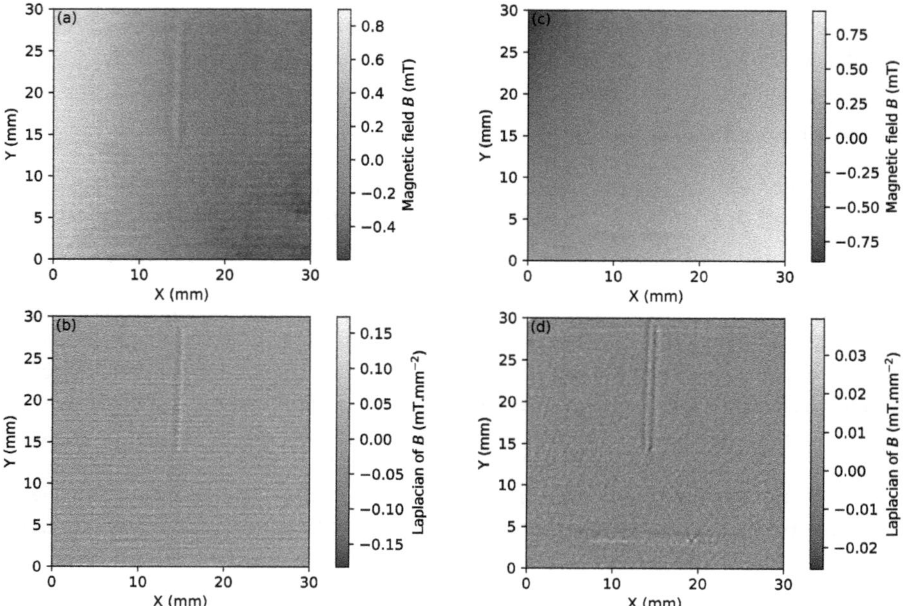

Fig. 2. Measured magnetic field along X and Y axis with a magnetization along X-axis in (a), and at 45° from X-axis in (c). Mapping around two medium defects with a length of 16 mm. In (b) and (d) are depicted the Laplacian of the magnetic field of respectively (a) and (c).

3.2 Weak Grinding Burns Detection

In the previous section, we demonstrated detection with NV magnetometry on medium-intensity GB. In this section, we tried to assess the limit of GB detection with this technology. After performing a setup's parameters optimization with strongest defects to maximize the signal-to-noise ratio, i.e. laser and microwave power, PID loop settings, frequency modulation, scan parameters, and magnetization amplitude, cartographies of weak burns were conducted. Figure 3(a) and (b) show the Laplacian map for two different weak burns of respective lengths 4 mm and 8 mm. The pixel size is 125 μm, and the cartography range of 15 and 10 mm, respectively. As mentioned earlier, the magnetization has been performed along the perpendicular direction of the defect to better visualize it. As observed, the NV magnetometer has successfully detected all the weak burns. Due to their dimensions, the weak burns are barely visible to the naked eye.

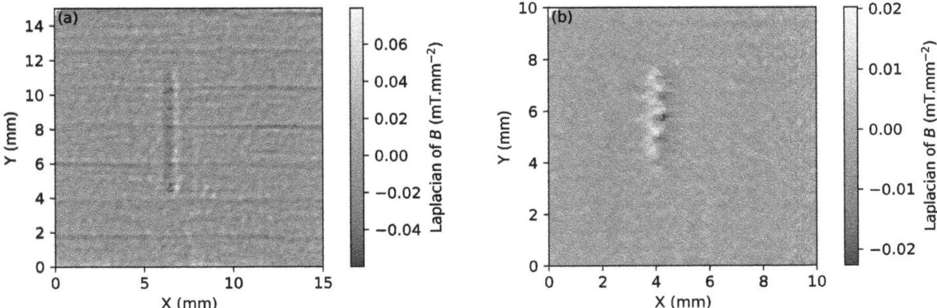

Fig. 3. Measured magnetic field along X and Y axis with a magnetization along X-axis. Maps obtained around a weak defect with a length of 8 mm in (a), and of 4 mm in (b).

4 Conclusion

In this study, we proved the ability of the NV-based magnetometer to detect successfully grinding burns on steel surfaces. With preliminary applying a weak magnetic field of about 30 mT along the perpendicular direction of the defects, all the defects were observed, even the weakest ones. This study demonstrated quantitative measurements of the magnetic flux leakage, with high sensitivity and resolution. More generally, it illustrates the potential interest of diamond quantum sensors for NDT, as a technique combining high spatial resolution, sensitivity and quantitative measurements.

Beyond grinding burns in steel, NV diamond sensors improve existing magnetic NDT, either by increasing resolution or by providing quantitative measurements that do not suffer from calibration drifts.

References

1. M'Saoubi, R., et al.: High performance cutting of advanced aerospace alloys and composite materials. CIRP Ann. **64**, 557 (2015)
2. Tönshoff, H.K., Arendt, C., Amor, R.B.: Cutting of hardened steel. CIRP Ann. **49**, 547 (2000)
3. Zaretsky, E.V.: Rolling bearing steels – a technical and historical perspective. Mater. Sci. Technol. **28**, 58 (2012)
4. Wegener, K., Bleicher, F., Krajnik, P., Hoffmeister, H.-W., Brecher, C.: Recent developments in grinding machines. CIRP Ann. **66**, 779 (2017)
5. Malkin, S., Guo, C.: Thermal analysis of grinding. CIRP Ann. **56**, 760 (2007)
6. King, R.I., Hahn, R.S.: Handbook of Modern Grinding Technology. Springer, Heidelberg (2012)
7. Del Re, F., Dix, M., Tagliaferri, F.: Grinding burn on hardened steel: characterization of onset mechanisms by design of experiments. Int. J. Adv. Manuf. Technol. **101**, 2889 (2019)
8. Wang, B., Zhong, S., Lee, T.-L., Fancey, K.S., Mi, J.: Non-destructive testing and evaluation of composite materials/structures: a state-of-the-art review (2020)
9. He, B., Wei, C., Ding, S., Shi, Z.: A survey of methods for detecting metallic grinding burn. Measurement **134**, 426 (2019)
10. Mayer, J.E., Price, A.H., Purushothaman, G.K., Dhayalan, A.K., Pepi, M.S.: Specific grinding energy causing thermal damage in helicopter gear steel. J. Manuf. Process. **4**, 142 (2002)

11. Sauter, E., Winter, M., Wegener, K.: Analysis of robustness and transferability in feature-based grinding burn detection. Int. J. Adv. Manuf. Technol. **120**, 2587 (2022)
12. E04 Committee, Test Method for Macroetching Metals and Alloys (2015)
13. Seidel, M.W., Zösch, A., Härtel, K.: Grinding burn inspection: tools for supervising and objectifying of the testing process. Forsch. Im Ingenieurwesen **82**, 253 (2018)
14. Sorsa, A., Ruusunen, M., Santa-aho, S., Vippola, M.: Sub-surface analysis of grinding burns with barkhausen noise measurements. Materials **16**, 159 (2023)
15. Thanedar, A., Dongre, G.G., Singh, R., Joshi, S.S.: Surface integrity investigation including grinding burns using barkhausen noise (BNA). J. Manuf. Process. **30**, 226 (2017)
16. Ducharne, B.: Non-destructive testing of ferromagnetic steel components based on their magnetic response. In: Otsuki, A., Jose, S., Mohan, M., Thomas, S. (eds.) Non-Destructive Materials Characteristics Methods, pp. 707–725. Elsevier (2024)
17. Withers, P.J., Turski, M., Edwards, L., Bouchard, P.J., Buttle, D.J.: Recent advances in residual stress measurement. Int. J. Press. Vessels Pip. **85**, 118 (2008)
18. Kern, R., Wolter, B., Altpeter, I., Dobmann, G.: Industrial Applications of 3MA ? Micromagnetic Multiparameter Microstructure and Stress Analysis, (n.d.)
19. Vindolet, B., Ducharne, B., Nguyen, H.N., Mougenot, X., Gallais, C., Hingant, T.: High-resolution non-destructive detection of grinding burns with NV diamond quantum magnetometer. NDT E Int. **155**, 103439 (2025)
20. Degen, C.L., Reinhard, F., Cappellaro, P.: Quantum sensing. Rev. Mod. Phys. **89**, 035002 (2017)
21. Rondin, L., Tetienne, J.-P., Hingant, T., Roch, J.-F., Maletinsky, P., Jacques, V.: Magnetometry with nitrogen-vacancy defects in diamond. Rep. Prog. Phys. **77**, 056503 (2014)
22. Doherty, M.W., Manson, N.B., Delaney, P., Jelezko, F., Wrachtrup, J., Hollenberg, L.C.L.: The Nitrogen-Vacancy Colour Centre in Diamond (2013)
23. Clevenson, H., Pham, L.M., Teale, C., Johnson, K., Englund, D., Braje, D.: Robust high-dynamic-range vector magnetometry with nitrogen-vacancy centers in diamond. Appl. Phys. Lett. **112**, 252406 (2018)
24. Inês Silva, M., Malitckii, E., Santos, T.G., Vilaça, P.: Review of conventional and advanced non-destructive testing techniques for detection and characterization of small-scale defects. Prog. Mater. Sci. **138**, 101155 (2023)

Uncut Gem - An Open-Source Hackable Quantum Sensor

Mark Carney[1,2]([envelope])[iD] and Victoria Kumaran[1,2]

[1] Quantum Village Inc., Wilmington, DE, USA
{mark,victoria}@quantumvillage.org
[2] Quantum Village Inc., London, UK

Abstract. This work presents an overview of our fully open-source, hackable quantum sensor platform based on nitrogen-vacancy (NV) center diamond magnetometry. This initiative aims to democratize access to quantum sensing by providing a comprehensive, modular, and cost-effective system. The design leverages consumer off-the-shelf (COTS) components in a novel hardware configuration, complemented by open-source firmware written in the Arduino IDE, facilitating portability, ease of customization, and future-proofing the design. By lowering the barriers to entry, our sensor serves as a compact platform for education, research, and innovation in quantum technologies, embodying the ethos of open science and community-driven development.

Keywords: Quantum sensing · Open Source

1 Introduction

Quantum technology needs an 'Apple II' moment. By this, we refer to the importance that the Apple II holds in the history of the desktop/personal computer. From the first spreadsheet, *VisiCalc*, to the first graphical adventure game *Mystery House*, the Apple II was a consumer-grade platform that enabled much of the modern technological world we are now so familiar with [16].

In a similar vein, the release of the Raspberry Pi was viewed as a key enabling technology in the development of the Internet of Things, driving the decentralization of compute in society even further and giving prominence to the Single Board Computer (SBC) [10]. Our work has taken inspiration from this concept and applied this reasoning to Quantum Technologies, and sensing in particular. How can we push the boundary of what is conventionally considered to be possible?

Open Source software has, in the first quarter of the 21st Century, formed the bedrock of much of our technology stacks and infrastructure. Hoffman *et al.* [8] determined that companies would spend around 3.5× more on software annually if open-source options did not exist. With this, the importance of open sourcing the key parts of the quantum technology stack become apparent - it drives adoption upwards whilst driving costs down [8]. This sentiment of increasing

F. Barbaresco and G. François (Eds.): QUEST-IS 2025, CCIS 2743, pp. 55–64, 2026.
https://doi.org/10.1007/978-3-032-13852-1_7

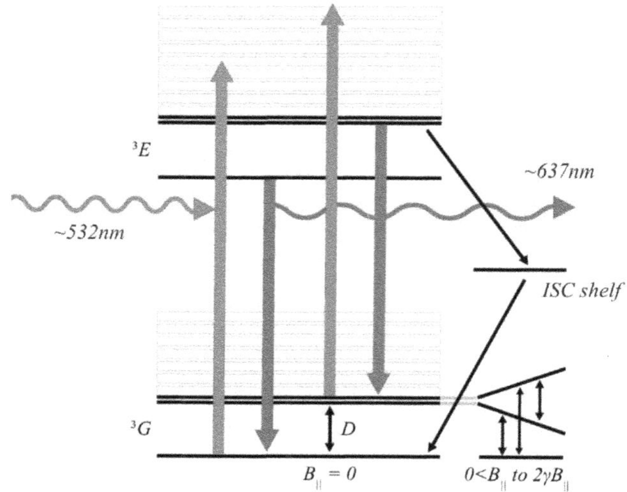

Fig. 1. Energy Level diagram for NV Centre Magnetometry, detailed in Sect. 1.1. **NB** - the intersystem crossing (ISC) shelf involving two singlet states with an infrared (IR) transition occurring at 1042 nm, but this is not relevant to our magnetometry.

openness, interoperability, and institutional momentum is echoed by other open source projects and studies in quantum technology, see [2,19].

Stegemann *et al.* [20] managed to get the overall cost of an NV Centre diamond experimentation platform down to under €200, utilising COTS parts and a 3D printer. This made our challenge apparent - could we reduce this cost down even further? And could we fill in some of the gaps and create a full-stack device that is fully open-source, moving away from the optics table and towards a hand-held form factor?

We first began designing and developing devices for quantum sensing and magnetometry back in 2023, and this project has been the culmination of 2 years of work towards making at least part of the quantum technology space truly open.

1.1   Brief Overview of NVC Magnetometry

The best surveys of the physics behind Nitrogen-Vacancy (NV) centre diamond magnetometry are [18] for detailed theory, and [7] for a high-level overview.

A Nitrogen Vacancy Centre (NV Centre) is the replacement in a diamond lattice of a single carbon (C) atom with a single Nitrogen (N) atom. In lab-grown diamonds this replacement is typically done through electron bombardment [13]. The mismatch in atomic valencies (4 for C to 3 for N) means that a vacancy is introduced. This defect essentially acts as a virtual atom contained within the lattice structure that has a strong zero-phonon line at 1.945 eV ($\lambda_{ZPL} = 637$nm) between the excited states ^3E and ground states ^3G. The hyperfine structures at both these levels yield sub-levels that are split into a lower singlet state $m_s = 0$ and two upper states $m_s = \pm 1$ separated by a difference of $D = 2.87\text{GHz}$ with no magnetic field present [18]. The vacancy also has no Doppler shift relative to any interacting photons, which is why its effects may be utilized at room temperature.

The ground states Hamiltonian of this, ignoring the hyperfine interactions, is given in [20] and [18] as:

$$\mathcal{H} = hDS_z^2 + hE(S_y^2 - S_x^2) + g\mu_B \mathbf{B} \cdot \mathbf{S} \tag{1}$$

where z is the parallel axis to the NV centre, D and E are the Zero-Field Splitting parameters, S_x^2, S_y^2, and S_z^2 are the spin operators, g is the Landé g-factor, and μ_B is the Bohr magneton. D arises via spin-spin interaction between the two uncoupled electrons in the defect, and E is proportional to the internal strain within the lattice. The resulting resonant frequency ν (see [20]) for the applied microwave field at frequency D is given by

$$\nu = D \pm \sqrt{E^2 + (\gamma \cdot \mathbf{B}_{||})^2} \tag{2}$$

with $\gamma = g\mu_B/h \approx 28\text{GHzT}^{-1}$ [18]. The important takeaway here is that under this simplified regime, provided E is much less than $\mathbf{B}_{||}$, the main variable regarding the splitting is the magnetic field $\mathbf{B}_{||}$ and its orientation to the NV centre axis.

The Energy-level scheme found in Fig. 1 summarizes the key physical concepts. We can excite these NV centres with green light $\lambda_E \approx 532\text{nm}$. The NV centres then relax into the upper excited states ^3E, and will become magnetosensitive under the application of microwaves. In practice, we sweep the microwave spectrum surrounding 2.87GHz, utilizing a similar sweep to the sweep in [20] by starting at 2.614GHz and sweeping up to 3.126GHz in discrete steps of 4MHz.

2   Gap Analysis

Many possible design options were reviewed, including the designs and build characteristics from [4,5,9,12,17,20,21]. The most complete guide was from Stegemann *et al.* [20], which provided a basis for our final design.

Upon analysis, however, we found a number of key things were missing from the available designs; The firmware used in NV Center diamond experiments is generally not available. Likewise, a reliable photodetector design schematic is generally not available, with most examples in the literature not always accurate to best electronic design principles. There are few easily configurable diamond mounts, as most require specialist optical mounting apparatus. Lastly, none of the solutions were designed to be 'hackable' for rapid deployment or prototyping, with little to no reconfigurable electronics and software. Similarly, none of the designs were geared to being ruggedized towards real world deployment.

3   Design Overview

In order to remediate these gaps, we came up with an initial design that made the following key decisions:

1. **Use off the shelf components** - the only specialist part should be the diamond material and any custom PCBs.
2. **Make as many things 'software problems' as possible** - by careful hardware design we can relinquish most control to a single software source. This massively improves debugging, maintenance, and updating a device.
3. **Simplify the design as much as possible.**
4. **Make the device 'hackable'** - by this we mean that the device should be fundamentally and easily reconfigurable and/or extendable in both software and hardware.

3.1   Microcontroller and Firmware

We wrote the source code for this device as Arduino IDE code - this taps into a large user base, with lots of help and built-in support for 26 native Arduino boards and hundreds of derivative boards [1]. This does two things; first, it immediately makes the design much more accessible to more people who are familiar with Arduino, over using an IDE or development tool-chain that is manufacturer specific.

Second, it allows the design to be ported to other chips through Arduino's universal hardware abstraction layer. This future-proofs the design, as any issues affecting the availability of one microcontroller means another more available one may be substituted with ease.

To this end, we used an ESP32 microcontroller - this is very well supported by Arduino IDE, has high quality 12-bit ADC inputs, is cheap and widely available, and has built-in WiFi and Bluetooth Low Energy (BLE) for future connectivity.

The default software libraries for driving an ADF4351 chip (see below) use a short calculation loop to generate the input values for the chip based off the input of a desired frequency. However, the sensitivity of the electronics (see below) means that this interferes with measurements significantly. To mitigate this, we pre-computed all of the desired sweep values and store them in firmware. This only consumes around 30 bytes per frequency at most, and practically eliminates the noise from the microcontroller interfering with the measurement operations.

3.2   Electronics

We made use of off-the-shelf operational amplifiers that are in a standard transimpedence configuration. We use the BPW34 photodiode as the main detection part as it has a large detection area - $3 \times 3\,\mathrm{mm}$ - and is widely available. This is then amplified by a TL082 dual op-amp, a common low-noise dual operational amplifier.

To minimize the parts required and to only require one 5 V power input, we utilize a voltage divider formed from two 1,000 Ω resistors in series from 5V to Ground. This gives us a floating 2.5 V which we utilize as a 'fake ground' to which the non-inverting inputs ('+' pins) of the operational amplifiers are tied. This achieves two things - firstly, it means we can do away with additional inverting voltage regulators, and secondly means this guarantees that output will be below the 3 V limit on the ESP32's ADC input, without needing level shifting.

An ADF4351 PLL signal generator was chosen as a microwave generator as it allows control via SPI bus with a range of 35 MHz to 4.4 GHz. The output of this is then amplified by two monolithic microwave RF amplifiers, specifically GALI-84+ (or compatible) amplifiers providing $\approx$ 16 dB amplification each.

The PCBs were designed to use single-sided soldering and assembly of surface mount components, reducing cost and complexity of production. The PCBs also use top and bottom layer ground planes to reduce parasitics and noise.

Additionally, to make the device hackable beyond just giving full software access, we also incorporated a small prototyping area. These generally consist of 1.0mm copper (Cu) plated holes, spaced 2.54 mm apart in a grid array. This allows the incorporation of any device which has output pins with this standard pitch, which is historically derived from the DIN 41612 standard, see [6].

3.3 Light Source

Canonically, 525–532 nm light is sourced from a laboratory laser, however these may sometimes prove to be problematic, especially where there is little to no guarantee of adequate laser safety protocols.

One of the main advantages of lasers is the generation of phase coherent photons. This is something we do not utilize for our magnetometry (see Eq. 1), so it suffices to use any source with the correct frequency output. 'Superbrite' LEDs were sourced that emit the correct frequency, and were found to perform on par with laser sources. This significantly reduces cost and power consumption, whilst also increasing the safety of the device.

3.4 Diamond and Photodiode Encapsulation

We make use of fast-UV curing epoxy to encase the photodetector, a microwave aerial, and NV centre diamond samples. As pointed out in [20], we can omit the need for a dichroic mirror and instead place a square of easily available red lighting filter gel between the photodiode and diamond to act as our primary optical filter.

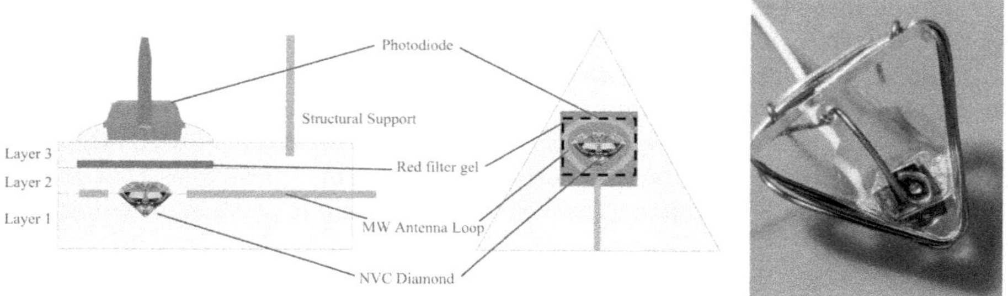

Fig. 2. Left: A diagram depicting the constructions layers of the diamon epoxy prism mount. **Right:** A mounted sample of diamond.

The epoxy prism is created during a 3-step layering process, corresponding to the diagram in Fig. 2; Layer 1 is a base layer of epoxy upon which the diamond and microwave aerial are placed after it is cured. A second layer of epoxy is then applied to fix the diamond and aerial in place. A third layer that fixes in place the filter, photodiode, as well as any mechanical fixing parts.

The density of diamond means that it will naturally sink in most epoxies, hence a layering approach is used. Even for fast UV curing epoxy, the excellent optical properties of clear epoxy are not compromised by the layering technique. It is also easy to reconfigure, as most epoxies only need to be heated to $\sim 200\,^{\circ}\mathrm{C}$ in order to delaminate. The diamond is completely unaffeced by this heating, and so can be retrieved for subsequent encapsulation. In particular, this move away from the optics table towards very low cost epoxy mounting is a key part of the fast iteration and innovation potential for this design.

4   Testing

Although full characterization of the quantum sensor is still yet to be performed, we have managed to get some preliminary checks and estimates of sensitivity. In Fig. 3 we can see the output from the sensor both with and without the presence of a small magnet placed on top of the diamond prism. These are in line with the measurement of the magnetism of a small bar magnet, $\approx 0.005 - 0.01$ Teslas, found in Fig. 7 in [20] and Fig. 3 in [18], by rough approximation.

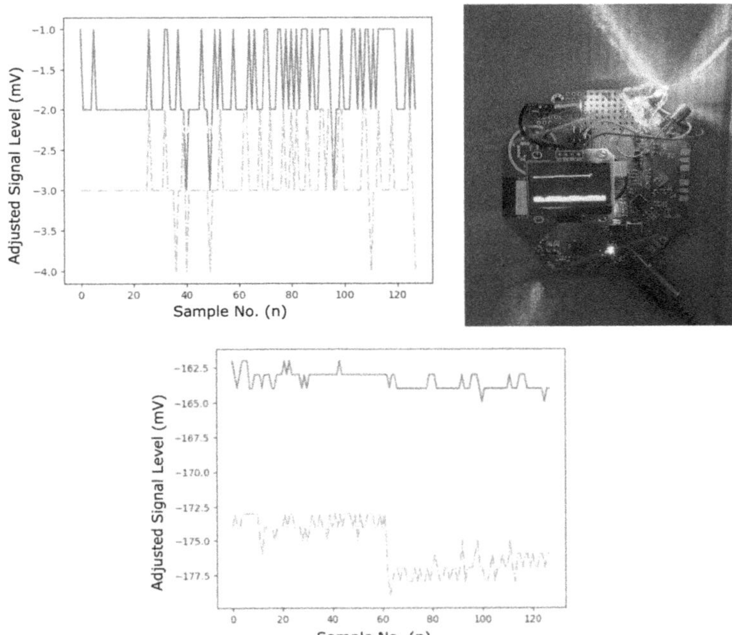

Fig. 3. Upper Left: Plotted readings in noise-adjusted millivolts from the ESP32 via serial over USB. Each point is an average of 6 measurements, with the dashed orange showing measurements with, and blue lines measurements without the presence of a small magnet. The shift in the splitting can be observed both in the drop in levels and in the shift of the valleys moving further apart. The frequency for each measurement is given in MHz from the sample number n by $(2838 + n)$. **Upper Right:** The device during operation. **Lower:** Plotted readings in noise-adjusted millivolts with the sensor in the presence of a large magnet (dashed orange) compared to no magnet (blue). Same scale as other plot, but in a different noise environment, hence lower adjusted readings.(Color figure online)

5   Conclusions and Future Work

On April 14th 2025 we released the full stack as open source on GitHub in [3]. To this end, the schematics, PCBs, parts lists, software code, and build instructions are all fully available, free and open source. We hope to build on this success with future releases and updates to both the hardware and base software to enable more people to learn about and practice with quantum sensing. Table 1 details the Bill of Materials (BoM) for the first release of the device, detailing a total cost that should not exceed £120. In Aug 2025 we released a single-board variant at DEF CON 33 [11, 15] with a lower total price of just under $100. With our subsequent version releases we expect this cost to fall even further.

Table 1. Bill of Materials for the first release version sensor.

Item	Number	Cost
Battery pack	1	Free to £5
OpAmp + electronics	1	up to £5
LED	1	~£2–£5
ADF4350/1 MW Generator	1	£15–25
MW Gain (40dB) + BNC	1	£10
PCB (unit from run)	1	~£5
OLED Screen	1	£5
ESP32 board	1	~£10
NV Center Diamond	1	~£10
Epoxy and Cu Wire	1 ea.	£10–15
	TOTAL	~£115

5.1 Future Work

Firstly, a full and accurate characterization should be performed on multiple diamond samples in order to gather data for both assessing the direct measurement and further analysis. There is plenty of scope for improving the output analysis and measurement estimation using edge machine learning and TinyML, something that Homrighausen *et al.* demonstrated in [9], and which we think would be an important step forward in improving the output of these devices.

We would like to see a space-ready version of the device prepared. This would mean changing the form factor to that of a CubeSat, as well as ensuring that all of the materials, soldering, conformal coating, and mechanical properties are suitable for the stresses and strains of space.

There are many use cases that should also be further explored for their suitability to this kind of low-cost quantum sensing setup, including; medical technology use cases like magnetocardiography (MCG), industrial sensing applications, and low-cost magnetometry for Position, Navication and Timing (PNT) use cases, building on work such as Muradoglu *et al.* [14].

Acknowledgments. The authors are grateful to the following people for their help during development of the device: Prof. Ben Varcoe, Max Shirokawa-Aalto, David Benoit, Matthew Markham, Rick Altherr, David Bengtson, and Brian McDermott.

Disclosure of Interests. The authors have no competing interests to declare that are relevant to the content of this article.

References

1. Banzi, M.: Getting started with arduino. Make: Projects, 2 edn. O'Reilly Media, Boston (2009)
2. Bülau, A., Walter, D., Fritz, K.P.: Components for an inexpensive cw-odmr nv-based magnetometer. Magnetism **5**(3), 18 (2025). https://doi.org/10.3390/magnetism5030018
3. Carney, M., Kumaran, V.: Uncut Gem - A Fully Open-Source, Hackable Quantum Sensor (2025). https://github.com/QuantumVillage/UncutGem
4. Chatzidrosos, G., et al.: Fiberized diamond-based vector magnetometers. Front. Photon. **2** (2021). https://doi.org/10.3389/fphot.2021.732748
5. Dix, S., Lönard, D., Barbosa, I.C., Gutsche, J., Witzenrath, J., Widera, A.: A miniaturized magnetic field sensor based on nitrogen-vacancy centers (2024). https://doi.org/10.48550/ARXIV.2402.19372
6. Fletcher, A.: Connector industry: A profile of the European Connector Industry - market prospects to 1999. Elsevier (2013)
7. Frellsen, L.F., Ahmadi, S.: Sensing magnetic fields with diamonds and green laser light. Quant. Views **2**, 10 (2018). https://doi.org/10.22331/qv-2018-07-17-10
8. Hoffmann, M., Nagle, F., Zhou, Y.: The value of open source software. SSRN Electron. J. (2024). https://doi.org/10.2139/ssrn.4693148
9. Homrighausen, J., Horsthemke, L., Pogorzelski, J., Trinschek, S., Glösekötter, P., Gregor, M.: Edge-machine-learning-assisted robust magnetometer based on randomly oriented nv-ensembles in diamond. Sensors **23**(3), 1119 (2023). https://doi.org/10.3390/s23031119
10. Johnston, S., Cox, S.: The raspberry pi: a technology disrupter, and the enabler of dreams. Electronics **6**(3), 51 (2017). https://doi.org/10.3390/electronics6030051
11. Kumaran, V., Carney, M.: Diamonds are for hackers - building the first fully open source and hackable quantum sensor. DEF CON 33 (2025)
12. Kuwahata, A., et al.: Magnetometer with nitrogen-vacancy center in a bulk diamond for detecting magnetic nanoparticles in biomedical applications. Sci. Rep. **10**(1) (2020). https://doi.org/10.1038/s41598-020-59064-6
13. McLellan, C.A., Myers, B.A., Kraemer, S., Ohno, K., Awschalom, D.D., Bleszynski Jayich, A.C.: Patterned formation of highly coherent nitrogen-vacancy centers using a focused electron irradiation technique. Nano Lett. **16**(4), 2450–2454 (2016). https://doi.org/10.1021/acs.nanolett.5b05304
14. Muradoglu, M., et al.: Quantum-assured magnetic navigation achieves positioning accuracy better than a strategic-grade ins in airborne and ground-based field trials (2025). https://doi.org/10.48550/ARXIV.2504.08167
15. Newman, L.H.: A special diamond is the key to a fully open source quantum sensor. WIRED (2025). https://www.wired.com/story/fully-open-source-quantum-sensor-uncut-gem/
16. Nooney, L.: The Apple II Age: How the Computer Became Personal. University of Chicago Press (2023). https://doi.org/10.7208/chicago/9780226816531.001.0001
17. Opaluch, O.R., Oshnik, N., Nelz, R., Neu, E.: Optimized planar microwave antenna for nitrogen vacancy center based sensing applications. Nanomaterials **11**(8), 2108 (2021). https://doi.org/10.3390/nano11082108
18. Rondin, L., Tetienne, J.P., Hingant, T., Roch, J.F., Maletinsky, P., Jacques, V.: Magnetometry with nitrogen-vacancy defects in diamond. Rep. Prog. Phys. **77**(5), 056503 (2014). https://doi.org/10.1088/0034-4885/77/5/056503

19. Shammah, N., et al.: Open hardware solutions in quantum technology. APL Quant. **1**(1) (2024). https://doi.org/10.1063/5.0180987
20. Stegemann, J., et al.: Modular low-cost 3d printed setup for experiments with nv centers in diamond. Eur. J. Phys. **44**(3), 035402 (2023). https://doi.org/10.1088/1361-6404/acbe7c
21. Webb, J.L., et al.: Optimization of a diamond nitrogen vacancy centre magnetometer for sensing of biological signals. Front. Phys. **8** (2020). https://doi.org/10.3389/fphy.2020.522536

Session 2 Quantum Sensors – Quantum sensors in detection and sensing

Molecular Spectroscopy Experiments for Direct Detection of Dark Matter

Florin Lucian Constantin$^{(\boxtimes)}$

Laboratoire PhLAM, UMR 8523, CNRS, 59655 Villeneuve d'Ascq, France
FL.Constantin@univ-lille1.fr

Abstract. Synchronized spectroscopy measurements for acetylene and molecular iodine reference transitions performed with an ultrastable laser signal may be exploited to constrain jointly couplings of dark matter to many fundamental constants. Enhanced sensitivity is expected from precision spectroscopy with quantum light.

Keywords: Ultralight Bosonic Dark Matter · Optical Spectroscopy · Time Synchronization

1 Introduction

Dark matter (DM) makes apparently the majority of the matter in our Universe, but its nature is unknown. Precision measurements in atomic, molecular and optical physics were exploited to probe DM formed of 0-spin, sub-eV-mass particles. This ultralight bosonic dark matter (UBDM) [1], that is expected to behave as a classical field coherently oscillating at the corresponding Compton frequency, couples to the Standard Model (SM) particles and yields oscillations of the fundamental constants (FC) [2]. Atomic and molecular spectroscopy [3–5] and atomic clock comparisons [6, 7] provided stringent limits on the UBDM field couplings.

This contribution proposes an experiment to probe dark matter fields that is based on precision molecular spectroscopy of reference transitions of acetylene at 1.5 μm and of molecular iodine at 0.5 μm. These transitions enabled high signal to noise ratios in spectroscopy and were exploited to build secondary frequency standards [8]. In addition, quantum enhancement was demonstrated in spectroscopy by exploiting squeezed light [9]. Particularly, homodyne detection of squeezed light at 1550 nm enabled noise reduction by 12 dB beyond the standard quantum limit [10] and broad bandwidth [11]. Quantum advantage is already exploited in interferometry for gravitational wave detection [12] and may open the way to a new generation of precision spectroscopy measurements.

F. Barbaresco and G. François (Eds.): QUEST-IS 2025, CCIS 2743, pp. 67–75, 2026.
https://doi.org/10.1007/978-3-032-13852-1_8

2 Probing Variations of Fundamental Constants by Optical Network Spectroscopy

2.1 Principe

The variation of a fundamental constant $\mathcal{F}$ (e.g. fine structure constant α, electron or quark masses $m_i = \{m_e, m_u, m_d, m_s\}$, quantum chromodynamics energy scale Λ) may be quantified from the coupling of a UBDM field φ to the SM with the relevant coupling coefficient $d_{\mathcal{F}}$:

$$\mathcal{F}(\varphi) = \mathcal{F}(1 + d_{\mathcal{F}}\varphi) \tag{1}$$

The UBDM field oscillates as $\varphi(t) = \varphi_0\cos(2\pi f_\varphi t)$ in function of the Compton frequency $f_\varphi = m_\varphi c^2/h$, related to the particle's mass, and an amplitude φ_0 depending on the local density of the DM halo [13].

The variations of the nuclear, atomic or molecular transition frequencies and of the resonance frequencies of the Fabry-Perot cavities may be parametrized in function of the variations of the fundamental constants:

$$\frac{\Delta f_X}{f_X} = Q_\alpha^X \frac{\Delta\alpha}{\alpha} + Q_\mu^X \frac{\Delta\mu}{\mu} + Q_q^X \frac{\Delta(\widehat{m}/\Lambda)}{\widehat{m}/\Lambda} \tag{2}$$

in function of the sensitivity coefficients to the variations of the fine structure constant Q_α^X, proton-to-electron mass ratio ($\mu = m_p/m_e$) Q_μ^X, and mean quark mass ($\widehat{m} = m_u/2 + m_d/2$) Q_q^X. The sensitivity coefficients are calculated using nuclear [14], atomic [15], molecular [16], or solid-state [17] theories.

The fractional variation between the frequency of a Fabry-Perot cavity-stabilized laser f_L and the frequency of a molecular transition f_{mol} is sensitive to the variations of α and μ:

$$\frac{f_L - f_{mol}}{f_L} = \left(Q_\alpha^L - Q_\alpha^{mol}\right)\frac{\Delta\alpha}{\alpha} + \left(Q_\mu^L - Q_\mu^{mol}\right)\frac{\Delta\mu}{\mu} \tag{3}$$

which are modulated with the differences between the relevant sensitivity coefficients $Q_{\alpha,\mu}^{L,mol}$ of the laser and of the molecular frequency. FC variability is parametrized here conveniently in function of the fine structure constant and the proton-electron mass-ratio, while other small contributions from nuclear structure are neglected. A time variation of the fractional frequency that is higher than the experimental uncertainty indicates detection of the dark matter field. Otherwise, the experimental uncertainty may be translated into constraints of the coupling coefficients of the UBDM field to the SM.

The variations of the fundamental constants are correlated. The proposed approach allows performing many measurements exploiting many molecular transitions with different sensitivity coefficients in order to constrain time variation of each fundamental constant. Such result is necessary to constrain New Physics theories beyond the Standard Model.

2.2 Proposed Experimental Approach

The proposed experiment is based on a laser at 1542 nm, stabilized on a Fabry-Perot cavity resonance with 10^{-15} level fractional stability at 1 s and continuously referenced against a hydrogen maser at Observatoire de Paris [18]. The laser signal is disseminated to remote geographical locations through REFIMEVE's stabilized optical fiber links [19]. This research infrastructure is exploited here to set up an optical network of molecular spectrometers (Fig. 1) that will perform synchronized laser absorption spectroscopy measurements. At each spectrometer location, the laser signal is amplified and suitably tuned at near-resonance with the molecular line by frequency shifting to probe acetylene by 1542 nm or by frequency tripling to probe molecular iodine by 514 nm. Additionally, each spectrometer may be quantum-enhanced by squeezing of radiation field that probes the molecular absorption (see next section).

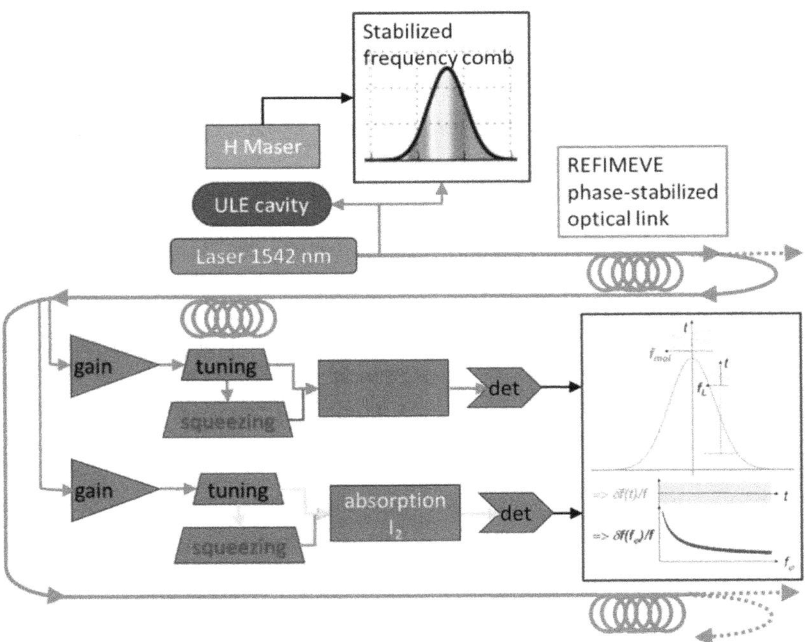

Fig. 1. An optical network of quantum-enhanced spectrometers exploiting the REFIMEVE reference signal at 1542 nm. The Laser, stabilized against the ULE cavity, is measured against the H maser with a Stabilized frequency comb. Laser signal, transmitted through the REFIMEVE phase-stabilized optical link (red line), is exploited directly in acetylene or molecular iodine spectroscopy. The parts indicated ensure different functions for spectroscopy: amplification (gain), frequency tuning (tuning), light quantum state preparation (squeezing), probing molecular transition in a cell (absorption), detection (det). The right box indicates the principle of the measurement, time dependence of the absorption signal and its Fourier spectrum.

The time dependence of each laser absorption signal is recorded and Fourier-transformed. The fractional absorption signal is proportional with the fractional frequency detuning, expressed as:

$$\frac{f_L(t) - f_{mol}(t)}{f_L(t)} = \left(Q_\alpha^L h_L\left(f_\varphi\right) - Q_\alpha^{mol} h_{mol}\left(f_\varphi\right)\right)\frac{\Delta\alpha(t)}{\alpha}$$
$$+ \left(Q_\mu^L h_L\left(f_\varphi\right) - Q_\mu^{mol} h_{mol}\left(f_\varphi\right)\right)\frac{\Delta\mu(t)}{\mu} \tag{4}$$

by introducing the response functions $h_{L,mol}\left(f_\varphi\right)$ of the experimental setup at the Compton frequency f_φ of the oscillating fundamental constants. These response functions display low-pass frequency dependencies imposed by several experimental constraints: finite speed of propagation of the acoustic perturbations in the Fabry-Perot cavity spacer or in the laser's cavity and the linewidth of the molecular lines. The response functions have to be determined experimentally by calibrated measurements of the laser absorption dependence on the laser frequency fluctuation. In addition, both linear absorption spectroscopy and pump-probe Doppler-free saturation spectroscopy measurements are required in order to ensure sensitivity to oscillations of fundamental constants over a broad frequency domain. Some experimental setups were already developed for spectroscopy of acetylene [20] and molecular iodine [4].

3 Improved Sensitivity in Spectroscopy with Squeezed Light

The sensitivity of the proposed experiment is limited by the amplitude noise of the laser beam that is used for spectroscopy. Differential detection schemes, exploited in previous experimental setups [3, 4, 20], enable sensitivity limited by the shot noise in photodetection. Here is proposed a dark mater sensing strategy where the molecular absorption is probed with amplitude-squeezed light to improve the sensitivity beyond the standard quantum limit. The absorption signal is monitored using a balanced homodyne detection scheme, that amplifies the amplitude of the signal by beating with a strong local oscillator, while the impact of the associated shot noise is reduced by subtracting the two photodetection signals [21].

The approach, schematically depicted in Fig. 2, exploits a laser emitting in a mode at angular frequency ω and wavevector $\vec{k}$ with the following electric field operator:

$$\widehat{E}_i\left(\vec{r}, t\right) = \widehat{\alpha}_i u\left(\vec{r}\right) e^{i\left(\vec{k}\,\vec{r} - \omega t\right)} \tag{5}$$

in function of the complex amplitude $\widehat{\alpha}_i = \overline{\alpha_i} + \delta\widehat{\alpha}_i$, expressed in terms of the mean value $\overline{\alpha_i}$ and the fluctuation $\delta\widehat{\alpha}_i$, and the normalized spatial profile $u\left(\vec{r}\right)$. $\overline{\alpha_i}$ is normalized such as $I_i = |\overline{\alpha_i}|^2$ is the intensity of the laser in a transverse section to the laser propagation direction. The laser beam is assumed strong, that is $\left|\delta\widehat{\alpha}_i\right| \ll \overline{\alpha_i}$. The electric field may be also expressed in function of the quadrature operators:

$$\widehat{E}_i = \widehat{X}_i \cos(\omega t) + \widehat{Y}_i \sin(\omega t) \tag{6}$$

that are non-commuting $\left[\widehat{X}_i, \widehat{Y}_i\right] \neq 0$ and their fluctuations obey to the Heisenberg uncertainty relation $\Delta\widehat{X}_i \bullet \Delta\widehat{Y}_i \geq |\overline{\alpha_i}|^2$. The fluctuations are equal to the standard quantum limit (SQL) $\Delta\widehat{X}_i = \Delta\widehat{Y}_i = |\overline{\alpha_i}|$ for a laser field in a coherent state. The squeezed light has one quadrature with fluctuation compressed below the SQL.

In direct detection of the molecular absorption with a coherent field (Fig. 2(a)), gas absorption, assumed small, is modelled as a beamsplitter with amplitude transmission factor $t = 1 - a/2$ and introduction of a vacuum field $\hat{c}$ with amplitude $\sqrt{1 - |t|^2}$:

$$\hat{\alpha}_{out} = t\hat{\alpha}_{in} + \sqrt{1 - |t|^2}\hat{c} \tag{7}$$

For small absorption and ideal detection and optics, the signal is reduced in proportion with the absorption and the additional noise induced by the absorption may be neglected. The signal to noise ratio in optical power units may be expressed as [22]:

$$\frac{S}{N} = \frac{(\mathrm{Re}(a))^2 \overline{I}_{in}}{2(1 + Ex)\Delta v} \tag{8}$$

in function of the mean photodetection current $\overline{I}_{in}$ of the laser beam without absorption, the fraction of excess noise Ex beyond the shot noise in the detection of the laser beam, and the detection bandwidth Δv.

In a differential measurement with a coherent field (Fig. 2(b)) an ideal 50:50 beamsplitter is exploited to separate the laser beam in two parts: one beam pass through the molecular gas and is detected with detector 1, the other beam is detected with detector 2. The signal to noise ratio in power units may be expressed as [22]:

$$\frac{S}{N} = \frac{(\mathrm{Re}(a))^2}{4} \frac{\overline{I}_{in}}{2\Delta v} \tag{9}$$

in function of the mean photodetection current of the laser beam. A loss of a factor of 4 to the shot-noise limit $Ex = 0$ of Eq. (8) is explained by the 50% loss of light intensity by using the beamsplitter.

In this proposal, bright squeezed light is generated for spectroscopy (Fig. 2(c)) by superposing on the 50:50 beamsplitter the coherent laser field $\hat{\alpha}_{in}$ and broadband squeezed vacuum field described with the operator [23]:

$$\hat{a}_s(\omega) = \hat{a}(\omega)\cosh(s) - e^{2i\theta_s}\hat{a}^\dagger(\omega)\sinh(s) \tag{10}$$

where $\hat{a}(\omega)$ is the annihilation operator for the squeezed wave at the same angular frequency that the laser field, $\hat{a}^\dagger(\omega) = [\hat{a}(\omega)]^\dagger$ the creation operator, s the squeezing factor and θ_s the squeezing angle. Squeezed vacuum is generated here using a sub-threshold OPO pumped by the second harmonic of the monochromatic laser, approach that ensures suitable spectral purity and tunability. The emerging bright squeezed beam probes the molecular absorption and the measurement is performed by homodyne detection. A part of the coherent laser beam is phase shifted and used as local oscillator. The difference between the two current intensities is monitored on a spectrum analyzer. To implement saturation spectroscopy measurements, another part of the coherent laser (not shown on Fig. 2(c)) is superposed in the absorption cell on the squeezed beam with an orientation in the opposite direction.

For homodyne detection at the same optical intensities, the signal to noise ratio in optical power units with squeezed light $\left(\frac{S}{N}\right)_{sq}$ is improved to the value $\left(\frac{S}{N}\right)_{cl}$ obtained with classical light. That may be expressed as [24]:

$$\left(\frac{S}{N}\right)_{sq} = \left(\frac{S}{N}\right)_{cl} / (1 + \xi\rho S(\Omega, \theta_s)) \tag{11}$$

in function of the spectrum of squeezing of the output field of the OPO $S(\Omega, \theta_s)$, a term ρ describing transmission of the OPO field, and a term describing field transmission in the homodyne detection ξ.

At efficient propagation and detection $\xi \to 1$, for large squeezing $S(\Omega, \theta) \to -1$, a large enhancement is expected, that scales $\propto 1/(1 - T_0)$ in function of the transmission T_0 of the OPO output coupler.

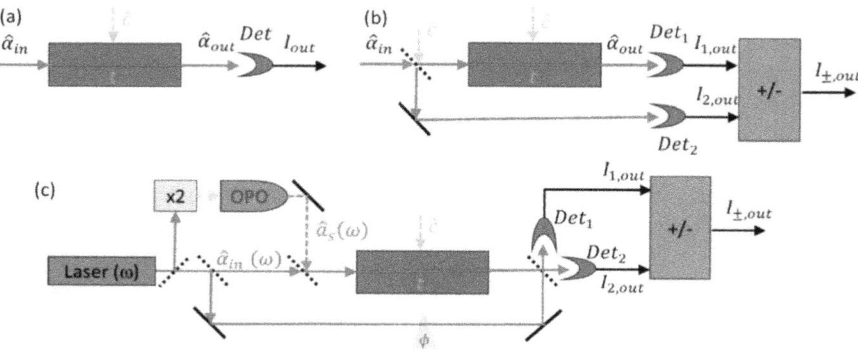

Fig. 2. Detection of a molecular absorption with quantum light. (a) direct detection, (b) differential detection, (c) experimental setup for spectroscopy with squeezed light. Laser emitting at angular frequency ω, x2 second harmonic generation, OPO sub-threshold pumped optical parametric oscillator, ϕ phase-shifter. *Det* detectors providing photodetection currents I_{out}, +/- homodyne detection electronics.

4 Expected Performances

First are evaluated the constraints from measurements on molecular iodine and acetylene transitions by saturation absorption spectroscopy using classical light from the cavity-stabilized laser remotely distributed to both experiments.

The proposal aims to perform time-synchronised measurements of the transitions P(27) of the $v_1 + v_3$ band of $^{12}C_2H_2$ at 1542 nm, and R(35) 44–0 of $^{127}I_2$ at 514 nm by suitably tuning the ultrastable optical signal disseminated in the REFIMEVE optical fiber link network. The metrological performances of the molecular clocks [25, 26] suggest that these measurements may provide fractional frequency uncertainties of $5 \times 10^{-13}/\sqrt{\tau}$ and $3 \times 10^{-14}/\sqrt{\tau}$, respectively, that are expressed in function of the averaging time τ in seconds. The sensitivity of the molecular transitions to FC variations was described previously [27, 28]. The cavity-stabilized laser is mainly sensitive to α-variation [17]. The uncertainties of the spectroscopy measurements are translated subsequently into estimations of the constraints for the coupling coefficients $d_{\alpha,\mu}$ of the UBDM field to α and μ. In doing that, one should take into account high-frequency cutoffs in the experimental response to FC oscillations imposed by the linewidths of the molecular lines and the finite speed of propagation of acoustic perturbations in the laser stabilization cavity. The UBDM field is associated here to the Galactic DM halo on Earth and the estimations for the bounds are shown in Fig. 3.

Squeezing of light enabled measurements with homodyne detection of the reduction of the power spectrum below the shot-noise limit by squeezing factors of 18.2 at 1550 nm [10] and 3.6 at 532 nm [29]. Exploiting squeezed light with these performances in saturated spectroscopy measurements may improve both fractional frequency uncertainties at $3 \times 10^{-14}/\sqrt{\tau}$ for the acetylene clock, and at $8 \times 10^{-15}/\sqrt{\tau}$ for the molecular iodine clock. These estimations are in proportion with the power squeezing factors. The estimations for the bounds for the coupling of the UBDM field to the SM from saturation spectroscopy with squeezed light are also shown in Fig. 3. Quantum advantage translates here into improved bounds by a factor of 40 for the α-coupling and of 21 for the μ-coupling, respectively.

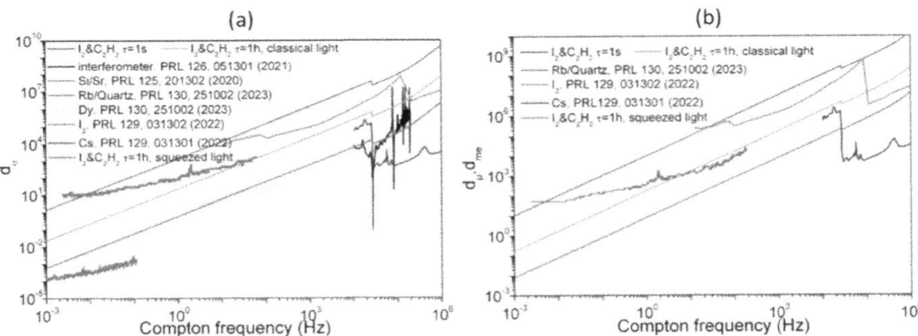

Fig. 3. Joint bounds for UBDM field couplings to (a) fine structure constant α, (b) proton-electron mass ratio $\mu = m_p/m_e$ (averaging time: $\tau = 1$ s red line, classical light, $\tau = 1$ h orange line, classical light, and $\tau = 1$ h pink line, squeezed light). Area above each dependence indicates the excluded range. For comparison, these plots include bounds set on the UBDM couplings to α and m_e from other experiments with classical light.

5   Conclusion

These results indicate that molecular spectroscopy may improve the description of the couplings of an UBDM field to the SM particles by constraining jointly time variations of α and μ. The bounds in the radiofrequency range are competitive with those from previous spectroscopy experiments and enable access to Compton frequencies higher that those were probed in atomic clock comparisons. These bounds may be significantly improved by implementing saturation spectroscopy with squeezed light.

Acknowledgement. This work is supported by the Programme National GRAM of CNRS/INSU and by the French ANR agency under contract No. ANR 10-LABX-0048, Labex First-TF. Laboratoire PhLAM is member of the Equipex+ T-REFIMEVE (French contract PIA 3 Equipements Structurants pour la Recherche of the Ministry of Higher Education, Research and Innovation and the General Secretariat for the Investment Plan).

References

1. Jackson Kimball, D.F., van Bibber, K.: The Search for Ultralight Bosonic Dark Matter. Springer, New York (2023)
2. Arvanitaki, A., Huang, J., van Tilburg, K.: Searching for dilaton dark matter with atomic clocks. Phys. Rev. D **91**(1), 015015 (2015)
3. Tretiak, O., et al.: Improved bounds on ultralight scalar dark matter in the radio-frequency range. Phys. Rev. Lett. **129**(3), 031301 (2022)
4. Oswald, R., et al.: Search for dark-matter-induced oscillations of fundamental constants using molecular spectroscopy. Phys. Rev. Lett. **129**(3), 031302 (2022)
5. Zhang, X., et al.: Search for ultralight dark matter with spectroscopy of radio-frequency atomic transitions. Phys. Rev. Lett. **130**(25), 251002 (2023)
6. Filzinger, M., et al.: Improved limits on the coupling of ultralight bosonic dark matter to photons from optical atomic clock comparisons. Phys. Rev. Lett. **130**(25), 253001 (2023)
7. Sherrill, N., et al.: Analysis of atomic-clock data to constrain variations of fundamental constants. New J. Phys. **25**(9), 093012 (2023)
8. BIPM Recommended values of standard frequencies. https://www.bipm.org/en/publications/mises-en-pratique/standard-frequencies. Accessed 18 June 2025
9. Polzik, E.S., Carri, J., Kimble, H.J.: Spectroscopy with squeezed light. Phys. Rev. Lett. **68**(20), 3020–3023 (1992)
10. Shajilal, B., et al.: 12.6 dB squeezed light at 1550 nm from a bow-tie cavity for long-term high duty cycle operation. Opt. Express **30**(21), 37213–37223 (2022)
11. Ast, S., Mehmet, M., Schnabel, R.: High-bandwidth squeezed light at 1550 nm from a compact monolithic PPKTP cavity. Opt. Express **21**(11), 13572–13579 (2013)
12. Lough, J., et al.: First demonstration of 6 dB quantum noise reduction in a kilometer scale gravitational wave observatory. Phys. Rev. Lett. **126**(4), 041102 (2021)
13. McMillan, P.J.: Mass models of the Milky Way. Mon. Not. R. Astron. Soc. **414**(3), 2446–2457 (2011)
14. Flambaum, V.V., et al.: Limits on variations of the quark masses, QCD scale, and fine structure constant. Phys. Rev. D. **69**(11), 115006 (2004)
15. Dzuba, V.A., Flambaum, V.V.: Relativistic corrections to transition frequencies of Ag I, Dy I, Ho I, Yb II, Yb III, Au I, and Hg II and search for variation of the fine-structure constant. Phys. Rev. A **77**(1), 012515 (2008)
16. Jansen, P., Bethlem, H.L., Ubachs, W.: Perspective: Tipping the scales: Search for drifting constants from molecular spectra. J. Chem. Phys. **140**(1), 010901 (2014)
17. Pašteka, L.F., et al.: Material size dependence on fundamental constants. Phys. Rev. Lett. **122**(16), 160801 (2019)
18. Xie, X., et al.: Phase noise characterization of sub-hertz linewidth lasers via digital cross correlation. Opt. Lett. **42**(7), 1217–1220 (2017)
19. Cantin, E., et al.: An accurate and robust metrological network for coherent optical frequency dissemination. New J. Phys. **23**(5), 053027 (2021)
20. Oswald, R., Vogt, V., Schiller, S.: Search for oscillating fundamental constants using a paired detector and vibrational spectroscopy. Phys. Rev. D **111**(5), 055003 (2025)
21. Yuen, H.P., Chan, V.W.S.: Noise in homodyne and heterodyne detection. Opt. Lett. **8**(3), 177–179 (1983)
22. Schwob, C.: Utilisation des faisceaux corrélés au niveau quantique produits par un Oscillateur Paramétrique Optique en spectroscopie de grande sensibilité. Thèse de doctorat. Sorbonne Université, Paris (1997). https://hal.science/tel-00011780/. Accessed 18 June 2025
23. Loudon, R., Knight, P.L.: Squeezed light. J. Mod. Opt. **34**(6–7), 709–759 (1987)

24. Polzik, E.S., Carri, J., Kimble, H.J.: Atomic spectroscopy with squeezed light for sensitivity beyond the vacuum-state limit. Appl. Phys. B **55**(3), 279–290 (1992)
25. Barbarat, J., et al.: Compact and transportable iodine frequency-stabilized laser. In: Sodnik, Z., Karafolas, N., Cugny, B. (eds.) International Conference on Space Optics; 2018, SPIE, vol. 11180, p. 111800T. SPIE, Bellingham (2019)
26. Hald, J., et al.: Fiber laser optical frequency standard at 1.54 μm. Opt. Express **19**(3), 2052–2063 (2011)
27. Constantin, F.L.: Sensitivity to electron-to-proton mass ratio variation from $^{12}C_2H_2$ rovibrational transitions to v_1+v_3 and $v_1+v_2+v_4+v_5$ interacting levels. Vib. Spectrosc. **85**, 228–234 (2016)
28. Constantin, F.L.: Towards probing a variation of fundamental constants with optical clock transitions of $^{127}I_2$. In: Weinstein, D., et al. (eds.) Joint Conference of the IEEE IFCS and ISAF; 2020, vol. CFP20ISA-POD, pp. 443–445. IEEE, Piscataway (2020)
29. Baune, C., et al.: Strongly squeezed states at 532 nm based on frequency up-conversion. Opt. Express **23**(12), 16035–16041 (2015)

Session 1 Quantum Communications – Test platforms for QKD

The Gilles Brassard Laboratory: An Open Research Platform for Secure Quantum-Era Communications

Antoine Dierick[✉], Lahatra Rakotondrainibe, and Jonathan Pisane

Thales Belgium, Rue des Frères Taymans, 28, 1480 Tubize, Belgium
`antoine.dierick@be.thalesgroup.com`

Abstract. The advent of quantum computing poses a significant threat to the security of today's communication networks, making it urgent to protect both current and future data exchanges. While theoretical countermeasures—such as post-quantum cryptography (PQC) and quantum key distribution (QKD)—exist, these technologies still require substantial development before being integrated into real-world infrastructures. The Gilles Brassard Laboratory (GBL) addresses this challenge by providing a research and development platform dedicated to advancing, testing, and validating key components of quantum-resistant communication systems.

The GBL is structured around two main segments: a Quantum Cryptography Infrastructure (QCI) for the implementation and evaluation of PQC and QKD, and a radio communication segment that emulates both terrestrial (LTE/5G/6G) and space-based links (DVB-S, CCSDS, 5G NTN). This dual framework enables researchers and engineers to investigate secure end-to-end communication solutions within a realistic testing environment.

Initial use cases demonstrate proof-of-concept validations, including the implementation of secure end-to-end communications by combining QKD technologies with standardized radio link protocols. Transmissions are encoded in accordance with DVB-S and CCSDS transport layers (broadcast mode only), paving the way toward the deployment of secure, future-proof communication networks.

Keywords: Post Quantum Cryptography (PQC) · Quantum Key Distribution (QKD) · Quantum Communications infrastructure (QCI)

1 Introduction

Recent technological advances, particularly in quantum computing, pose significant challenges to the security of current information systems. While robust countermeasures—such as post-quantum cryptography (PQC) and quantum key distribution (QKD)—have been theoretically established, their practical development and deployment remain complex. To address this gap, the Gilles Brassard Laboratory (GBL) was created as a dedicated platform for research and experimentation in quantum cryptography. The laboratory is named after Gilles Brassard, who, together with Charles Bennett,

F. Barbaresco and G. François (Eds.): QUEST-IS 2025, CCIS 2743, pp. 79–87, 2026.
https://doi.org/10.1007/978-3-032-13852-1_9

introduced the seminal BB84 protocol [1], recognized as the first QKD scheme and a cornerstone for secure communications based on quantum mechanics.

The purpose of this paper is to present the objectives, architecture, and initial experimental results of the GBL. In particular, we describe how the laboratory enables the integration and validation QKD technologies within realistic communication environments, including satellite-based scenarios. While the platform focuses on QKD; PQC integration is foreseen in future releases.

This paper is structured as follows: Sect. 2 outlines the main threats faced by contemporary information systems. Section 3 describes the objectives and architectural design of the GBL. Section 4 presents initial experimental results focused on secure satellite communication scenarios using DVB-S (Digital Video Broadcasting – Satellite) and CCSDS (Consultative Committee for Space Data Systems) protocols. Finally, Sect. 5 concludes the paper and discusses directions for future work.

2   Context

2.1   The Quantum Threat

The security of contemporary information systems relies on a key assumption: that no adversary possesses sufficient computational resources to decrypt protected communications within a reasonable timeframe. Under this paradigm, the security of a system depends both on the adversary's available resources and on the required duration of data confidentiality.

This approach faces a critical limitation. If technological or algorithmic advances occur within the confidentiality period, data can be compromised. An adversary may simply store encrypted communications and later decrypt them once improved capabilities become available. This threat becomes particularly acute with the emergence of quantum computers, which exploit principles of quantum mechanics to solve certain computational problems more efficiently than classical machines.

Two quantum algorithms illustrate this. Grover's algorithm [2] accelerates brute-force search in unstructured spaces, providing a quadratic speedup. More critically, Shor's algorithm [3] efficiently solves the integer factorization problem in polynomial time, undermining the hardness assumptions on which RSA and elliptic-curve cryptography (ECC) are based. Since these schemes are widely deployed in current infrastructures, their security is directly endangered by the rise of large-scale quantum computing [4].

This looming vulnerability highlights the urgency of developing quantum-resistant alternatives to secure communication systems.

2.2   Possible Solutions: PQC and QKD

In response to the quantum threat, two major strategies are being pursued. The first, Post-Quantum Cryptography (PQC), seeks to secure communications through cryptographic schemes based on mathematical problems believed to remain intractable even for quantum computers. Prominent candidates include lattice-based, code-based, multivariate, and hash-based constructions [5]. While the security of PQC is conditional, depending on the absence of future algorithmic or technological breakthroughs, its main advantage lies in compatibility: PQC can often be integrated into existing communication infrastructures with minimal hardware changes. This makes it a pragmatic solution for rapidly mitigating "store now, decrypt later" attacks, serving as a near-term safeguard while more advanced technologies mature.

The second strategy, Quantum Key Distribution (QKD), leverages the laws of quantum physics to achieve information-theoretic security. QKD enables two parties to establish a shared secret key by exchanging quantum states, which serve as physical carriers of information. A defining feature of QKD is its inherent ability to detect eavesdropping: any measurement of the exchanged states by a third party induces anomalies, due to the fundamental principle that measurement disturbs a quantum system [6].

While QKD offers unconditional security in theory, practical deployment faces challenges. Quantum states are highly sensitive to environmental disturbances, which severely limits long-distance communication without additional technologies. Although quantum repeaters, whose foundational concept was introduced in 1998 [7], are regarded as the key solution to overcome the distance limitations of optical fibers, this technology has not yet reached maturity. Current quantum communication networks therefore rely on alternative approaches, such as trusted relays [8] or satellite-based links [9].

At the GBL, research efforts are directed toward both strategies, aiming to combine the near-term deployability of PQC with the long-term security guarantees of QKD to build quantum-resilient communication systems.

2.3 An Open Quantum Cryptography Laboratory for Researchers

The primary objective of the GBL is to deploy an end-to-end communication network integrating both software and hardware components in order to test, assess, and validate quantum-based cybersecurity and quantum-resilient solutions under realistic conditions. In particular, the laboratory enables:

- **Characterization and assessment of cyber threats**, such as vulnerabilities and risks associated with quantum-enabled attacks, as well as more conventional cybersecurity risks including jamming and spoofing;
- **Deployment and validation of quantum-based security solutions**, evaluated against the performance requirements and characteristics of modern communication networks.

The GBL is conceived as an open and collaborative research facility, accessible to Belgian and European universities, as well as to local and national small and medium-sized enterprises (SMEs) seeking to validate technologies. It also serves as a framework for cooperation with other European laboratories, supporting joint efforts in the context of European Commission initiatives and missions.

The design and management of the GBL have been entrusted to the Thales Group, ensuring professional expertise and operational support. The facility is intended for researchers, engineers, and specialists working on quantum technologies and their integration into secure communication infrastructures.

3   Architectural Design of the GBL Quantum Research Platform

The GBL quantum research platform is functionally divided into three complementary segments:

- **Radio communications segment** (terrestrial and space), responsible for emulating end-to-end communication links;
- **User segment**, which provides tools for control, configuration, definition of measurement scenarios, data collection, and visualization of results;
- **Quantum Cryptography Infrastructure (QCI) segment**, dedicated to quantum key distribution and secure key management.

This modular architecture allows the integration of heterogeneous data streams such as quadrature signals (IQ), network traffic (IP), and quantum samples (QC), into a unified environment accessible through a dynamic graphical user interface.

3.1   Radio Communications Segment

The radio segment emulates end-to-end communication links with both terrestrial (LTE/5G/6G) and space-based (DVB-S, CCSDS, 5G NTN) standards. It combines virtualized environments and software-defined radios (SDRs), supported by high-throughput interconnects.

A channel emulator reproduces realistic propagation effects such as delay, Doppler, attenuation, and interference, enabling the evaluation of secure transmissions under degraded conditions.

Signal processing is performed in real time using GNU Radio, supporting both Frequency Division Duplex (FDD) and Time Division Duplex (TDD) configurations.

3.2   Quantum Cryptography Infrastructure (QCI) Segment

The QCI segment ensures secure key distribution between users through a time-bin implementation of the BB84 protocol. It operates as a key provider, delivering quantum-generated keys to consumers via the standard ETSI GS QKD 014 interface. Keys are subsequently used in symmetric encryption modules (e.g., AES), which may be implemented as dedicated devices or integrated directly into SDRs.

A Key Management System (KMS) is integrated at each QKD node to arbitrate key allocation, ensuring flexibility in adding, removing, or transmitting keys across the network. The encryption modules retrieve keys through the ETSI REST interface and apply them to secure the data streams.

The following Fig. 1 provides a high-level view of the GBL research platform architecture.

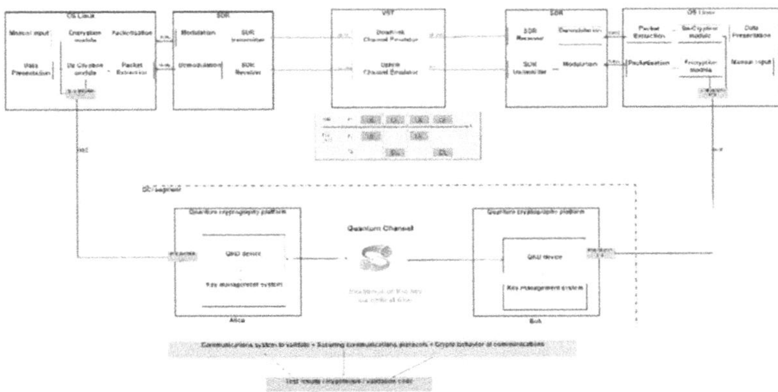

Fig. 1. Block diagram of GBL research platform

4 Experimental Results

The experimental validation of the GBL platform was carried out by demonstrating secure end-to-end communications combining the radio and QCI segments. Two layers of validation were considered:

- Physical-level validation, assessing the performance of the QKD link through key generation metrics, error rates, and hardware stability.
- System-level validation, focusing on the integration of QKD keys into communication protocols such as DVB-S and CCSDS;

4.1 Physical-Level Validation of the QKD Link

The performance of the quantum channel was assessed using data collected from the Alice and Bob devices, as well as from the FPGA processing stage.

Figure 2 shows the FPGA-reported quantum bit error rate (QBER) measured over time. Values remained within the expected range (<5%), validating the security of the generated keys. Figure 3 presents the FPGA-reported sifted key rate, which confirmed the continuous availability of correlated bits for key delivery. As expected, the final secret key rate (after error correction and privacy amplification) is lower, but still sufficient to support encryption for the DVB-S and CCSDS scenarios.

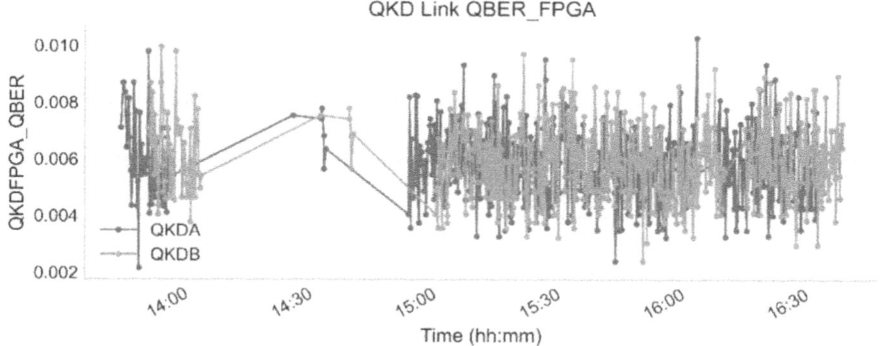

Fig. 2. Quantum Bit Error Rate as logged at the FPGA stage (sifted key error, i.e. after basis sifting but before reconciliation and privacy amplification). The plotted series reflects physical impairments (detector dark counts, timing jitter, visibility fluctuations). For BB84 the theoretical one-way security bound is $\approx 11\%$ [10]; in our implementation a conservative operational threshold of $\sim 5\%$ was adopted to ensure efficient error correction and acceptable finite-key performance.

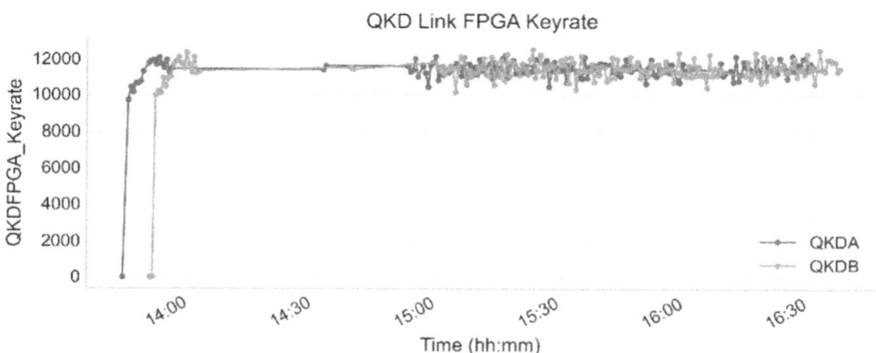

Fig. 3. Sifted key rate reported by the FPGA processing stage. The sustained throughput demonstrates the continuous availability of correlated key material for secure encryption, ensuring integration with higher-layer communication protocols.

Additional hardware monitoring confirmed proper operating conditions. Board and detector temperatures remained stable, and optical visibility provided further evidence of link quality. Variations in laser modulation and detector bias were found to correlate with small fluctuations in QBER, but without impacting overall key delivery.

Together, these results demonstrate that the QKD segment operated within expected physical tolerances and successfully generated cryptographic key material suitable for integration into higher-layer communication protocols. The combined analysis of detections, QBER, key rate, and stability parameters provides a comprehensive validation of the quantum link under realistic operating conditions.

4.2 System-Level Validation

Two representative use cases were implemented to illustrate the platform's capabilities:

4.2.1 DVB-S Scenario

The first use case emulates a satellite video broadcast using the DVB-S (Digital Video Broadcasting – Satellite) standard. A video file was transmitted in real time between two machines, with simultaneous playback at the receiver. Two configurations were tested:

- Unencrypted DVB-S transmission, serving as a baseline scenario;
- Encrypted DVB-S transmission, where the data stream was secured using QKD-generated keys.

Successful validation was demonstrated by receiving uninterrupted video streams in real time (see Fig. 4.a). Furthermore, channel impairments were introduced through the VST (e.g., gain loss, delay, Doppler shift, and attenuation) to simulate realistic conditions. Even under degraded conditions (e.g., 20 dB attenuation), secure transmissions were maintained (see Fig. 4.b).

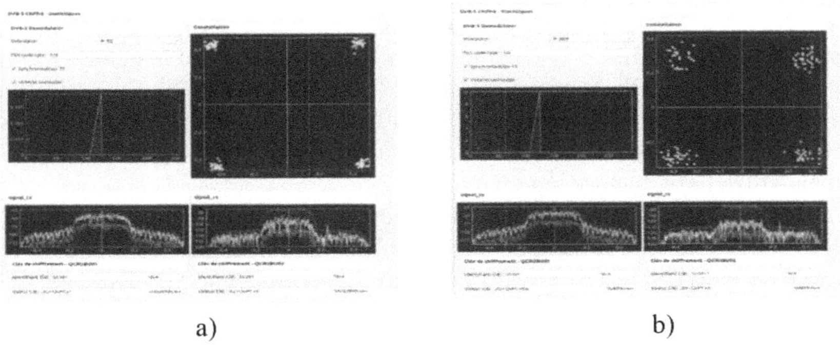

a)   b)

Fig. 4. Demonstration of a secure real-time video broadcast over a DVB-S link. The stream was encrypted with QKD-generated keys and successfully received in real time, even under degraded channel conditions (attenuation and Doppler shift) – a) VST no degradation, b) VST with 20 dB attenuation

4.2.2 CCSDS Scenario

The second use case targeted file transfer in space communication contexts, using the CCSDS (Consultative Committee for Space Data Systems) protocol. The setup involved transmitting binary files from a source PC to a target PC under two configurations:

- Standard CCSDS file transfer, without encryption;
- Encrypted CCSDS transfer, where data confidentiality was ensured by QKD-generated keys.

Validation consisted of confirming reliable file delivery without interruption. As with the DVB-S case, simulated channel degradations (e.g., signal-to-noise ratio loss, frequency shift) were applied to test the resilience of the secured link (see Fig. 5).

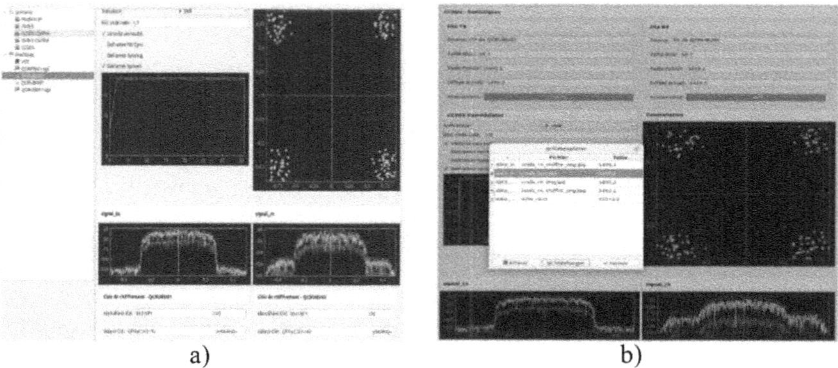

a)   b)

Fig. 5. Validation of a secure file transfer using the CCSDS protocol. QKD keys were integrated into the link encryption process, ensuring reliable and confidential end-to-end delivery despite channel impairments. – a) VST no degradation, b) VST with 20 dB attenuation/1MHz frequency shift.

5   Conclusion and Future Work

This paper has presented the design, implementation, and initial validation of the Gilles Brassard Laboratory (GBL), an open research platform dedicated to quantum cryptography and quantum-resilient communications. The platform combines a radio communications segment, capable of emulating both terrestrial and space-based links, with a Quantum Cryptography Infrastructure (QCI) segment implementing the BB84 protocol and standardized key management interfaces.

The first experimental results demonstrated the feasibility of integrating QKD with classical communication standards. In particular, successful proof-of-concept validations were achieved for secure end-to-end transmissions under DVB-S and CCSDS protocols. In both scenarios, QKD-generated keys were delivered through the ETSI GS QKD 014 interface and used to encrypt real-time video streams or file transfers. Importantly, transmissions remained secure and reliable even when subject to degraded channel conditions such as attenuation, delay, or frequency shifts. These results confirm the capability of the GBL to provide a realistic environment for testing and validating quantum-resilient communication systems.

Looking ahead, several research directions are envisioned:

- Extension to new communication standards: integration of 5G, 6G, and non-terrestrial networks (NTN) within the GBL testbed.
- In future iterations of the platform, PQC implementations will be integrated and validated alongside QKD, enabling comparative studies and hybrid secure communication scenarios.
- Advanced quantum technologies: investigation of quantum repeaters, trusted-node architectures.
- Enhanced resilience testing: characterization of system performance under more complex attack models, including classical cyber threats and combined quantum–classical adversaries.

- Training and education: providing a platform for researchers, engineers, and students to experiment with and learn about next-generation secure communications.

By combining a realistic radio environment with quantum key distribution and post-quantum cryptography research, the GBL paves the way for future secure communication networks that remain resilient in the quantum era.

Acknowledgement. The Gilles Brassard quantum research Laboratory is located at the IDELUX site in Transinne, and is funded by the Walloon Region of Belgium.

References

1. Bennett, C.H., Brassard, G.: Quantum cryptography: public key distribution and coin tossing. Theoret. Comput. Sci. **560**, 7–11 (2014). https://doi.org/10.1016/j.tcs.2014.05.025
2. Grover, L.K.: A fast quantum mechanical algorithm for database search. In: Proceedings of the Twenty-Eighth Annual ACM Symposium on Theory of Computing. Association for Computing Machinery, Philadelphia, Pennsylvania, USA, pp 212–219 (1996)
3. Shor, P.W.: Polynomial-time algorithms for prime factorization and discrete logarithms on a quantum computer. SIAM J. Comput. **26**, 1484–1509 (1997). https://doi.org/10.1137/s0097539795293172
4. Mosca, M.: Cybersecurity in an era with quantum computers: will we be ready? IEEE Secur. Priv. **16**, 38–41 (2018). https://doi.org/10.1109/MSP.2018.3761723
5. Chen, L., et al.: Report on Post-Quantum Cryptography. US Department of Commerce, National Institute of Standards and Technology, vol. 12, Gaithersburg (2016). https://doi.org/10.6028/NIST.IR.8105
6. Gisin, N., Ribordy, G., Tittel, W., Zbinden, H.: Quantum cryptography. Rev. Mod. Phys. **74**, 145–195 (2002). https://doi.org/10.1103/RevModPhys.74.145
7. Briegel, H.-J., Dür, W., Cirac, J.I., Zoller, P.: Quantum repeaters: the Role of imperfect local operations in quantum communication. Phys. Rev. Lett. **81**, 5932–5935 (1998). https://doi.org/10.1103/PhysRevLett.81.5932
8. Sasaki, M., Fujiwara, M., Ishizuka, H., et al.: Field test of quantum key distribution in the Tokyo QKD Network. Opt. Express **19**(11), 10387–10409 (2011). https://doi.org/10.1364/OE.19.010387
9. Liao, S.-K., et al.: Satellite-to-ground quantum key distribution. Nature **549**, 43–47 (2017). https://doi.org/10.1038/nature23655
10. Shor, P.W., Preskill, J.: Simple Proof of security of the BB84 quantum key distribution protocol. Phys. Rev. Lett. **85**, 441–444 (2000). https://doi.org/10.1103/PhysRevLett.85.441

Metrology for Real-World QKD

Alice Meda[1]($\boxtimes$), Salvatore Virzì[1], Marco Gramegna[1], Giorgio Brida[1], Marco Genovese[1,2], and Ivo Pietro Degiovanni[1]

[1] Istituto Nazionale di Ricerca Metrologica (INRIM), Strada delle Cacce 91, 10135 Torino, Italy
a.meda@inrim.it
[2] Istituto Nazionale di Fisica Nucleare (INFN), Sezione Torino, Via Giuria 1, 10125 Torino, Italy

Abstract. Quantum Key Distribution (QKD) is a technology that enables the sharing of secret cryptographic keys between two distant users (Alice and Bob), with intrinsic security guaranteed by the fundamental laws of nature.

QKD has become a mature technology, and in Europe, all 27 member states are collaborating on a European Commission initiative (EuroQCI) to design, develop, and deploy a quantum communication infrastructure. In Italy, the QUID project is responsible for implementing the Italian segment of EuroQCI.

QKD relies on single photons to secure the distribution of the keys and, to become a viable real-world solution, the metrological characterization of optical components and systems is fundamental. To obtain the appropriate security requirements, test and evaluation methods at single-photon level need to be developed; in particular, since the single-photon detectors represent the most vulnerable part of a QKD system, their characterization in terms of operating parameters (quantum efficiency, dead time, jitter, afterpulsing..) is of the utmost importance.

We present the INRIM efforts in the quantum efficiency calibration of single-photon avalanche detectors (SPADs), focusing on QKD application. The detection efficiency is evaluated for a fibre-coupled InGaAs/InP-SPAD and for a free-space Si-SPAD. The calibration is performed using different experimental setups and reference standards with proper traceability chains at the wavelength of 1550 nm and 850 nm respectively. Dependence of detection efficiency on polarization in superconducting nanowire single-photon detectors (SNSPDs) is also reported.

The work is fundamental to align the Italian deployment of QKD, in the framework of QUID, with validation needs, providing test services for the characterization, validation and certification for QKD.

Keywords: Quantum Key Distribution · Single-photon Avalanche Detectors · Quantum Efficiency

1 Introduction

1.1 QKD and Single-Photon Avalanche Diodes

Quantum Key Distribution (QKD) is currently the most advanced technology for sharing cryptographic secret keys with a level of security that is fundamentally independent of computational power [1, 2]. Practical consideration of QKD systems and network

F. Barbaresco and G. François (Eds.): QUEST-IS 2025, CCIS 2743, pp. 88–93, 2026.
https://doi.org/10.1007/978-3-032-13852-1_10

architectures based on specific use cases has already begun [3, 4], and QKD metropolitan networks have been successfully demonstrated across the globe [5, 6].

The European Union (EU) is actively advancing in this domain, having signed a declaration with all 27 member states to jointly pursue the development of the European Quantum Communication Infrastructure (EuroQCI) [7]. The EuroQCI aims to establish secure quantum communication networks across all EU countries. As part of this initiative, the Italian project QUID (Quantum Italy Deployment) [8] has been launched to initiate the deployment of Italy's contribution to the EU-wide network by implementing QKD systems and building Quantum Metropolitan Area Networks (QMANs) in various cities.

Single-photon avalanche diodes (SPADs) represent the most widely adopted commercial solution for single-photon detection technologies. Among their many applications, QKD relies heavily on SPADs for photon detection [9].

As with all cryptographic systems, practical implementation of QKD demands rigorous investigation of the components involved to ensure their correct operation and integration [10–12]. It is therefore crucial to develop theoretical models that accurately describe the behavior of each system element and establish boundaries on the acceptable error rates during key exchange. Deviations from the ideal device behavior can open side channels or backdoors, which may be exploited by a potential eavesdropper to compromise the communication. In particular, numerous detection-based attack strategies have been documented, including backflash-related attacks [13] and the more severe class of detector control attacks [14].

To maintain QKD security and prevent such tailored attacks, it is essential to model and characterize the SPAD response under various operational conditions, taking into account key parameters such as detection efficiency, dead time, and dark count rate.

Here we focus on the calibration of the detection efficiency of a fibre-coupled InGaAs/InP-SPAD [15–17] and of a free-space Si-SPAD [18].

2 Detection Efficiency Characterization

2.1 Fibre-Coupled InGaAs/InP-SPAD

The detection efficiency is the probability that a photon of a specific energy (wavelength) incident at the optical input will be detected within a detection gate. The most exploited measurement approach in metrological institutes is based on the traditional substitution method [15], where the detection efficiency is inferred by comparing the output of the single photon detector and the output of a calibrated power meter, illuminated by the same light but with different levels of attenuations.

We develop a facility for the traceable measurement of detection efficiency for the new generation single-photon detectors based on semiconductor (InGaAs/InP SPAD).

The measurement protocol relies on a detection efficiency measurement of detectors that takes advantage of an experimental set-up whose schematic description is reported in Fig. 1.

A pulsed laser at 1550 nm is externally triggered, with a pulse generator, that triggers also a Time-Correlated Single Photon Counting (TCSPC) device, is connected to a

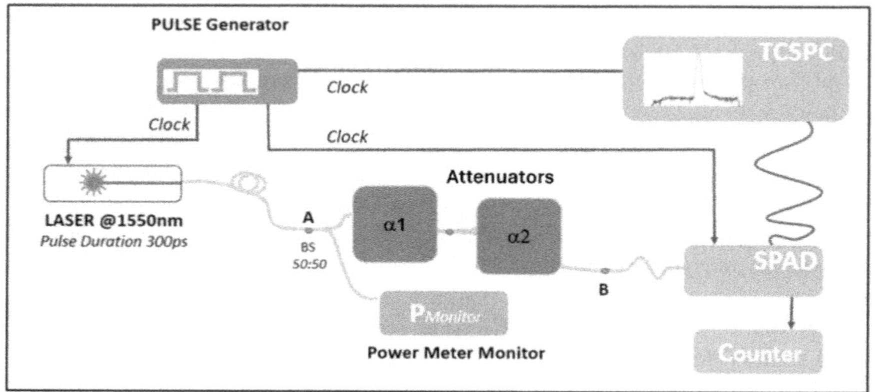

Fig. 1. Experimental setup of INRiM InGaAs/InP SPAD detection efficiency calibration facility

first variable attenuator that provides a nominal transmissivity α_1 and a second variable attenuator that provides a nominal transmissivity α_2, for a total transmissivity $\tau = \alpha_1\alpha_2$. The detection efficiency of the DUT is evaluated according to the formula:

$$\eta = \frac{P_c}{\tau P_0} \tag{1}$$

where P0 is the power in point B and PC is the average power of the effective photons measured by the detector, calculated from the photon rate absorbed by the SPAD corrected for dark counts, dead time and considering the energy of the photon at $\lambda = $ 1550 nm.

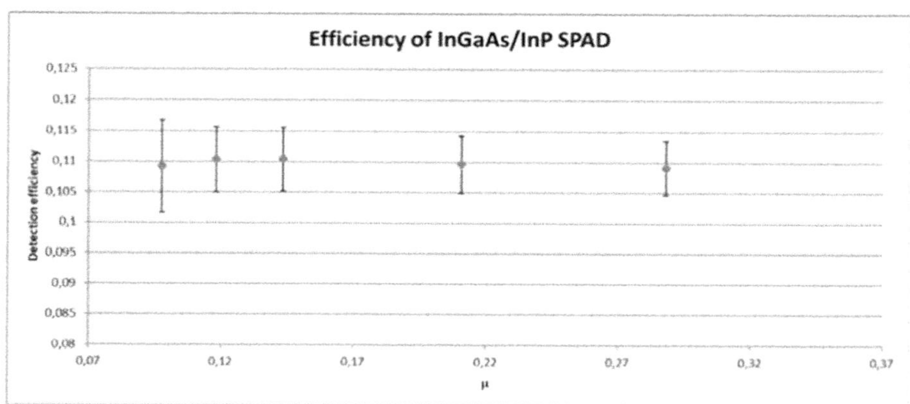

Fig. 2. Detection efficiency of an InGaAs/InP SPAD

Figure 2 reports detection efficiency as a function of the mean photon number per pulse μ. A proper model for the correction for the effects of dead time on the detection efficiency is applied; details can be found in [15–17].

The setup dedicated to the detection efficiency measurement is now also suitable for light-polarization-sensitive detectors as SNSPDs, thanks to the introduction of a dedicated polarization measurement and control system, that allows to reconstruct the estimated efficiency distribution across the Poincaré sphere's surface and with a relative uncertainty $< 5\%$ (k $= 2$) [18].

2.2 Free Space Si-SPAD

A second INRiM measurement facility is designed to determine the detection efficiency at a wavelength of 850 nm, where photons are addressed in free space with a spatial distribution diameter smaller than 40 μm.

The calibration setup is shown in Fig. 3. The photon source is a Ti:sapphire (Ti:Sa) laser with tunable wavelength, set to $\lambda = 850.711 \pm 0.006$ nm. Optical power is modulated by a polarization-based variable attenuator composed of a half-wave plate and a polarizer, allowing precise control of the single-photon rate.

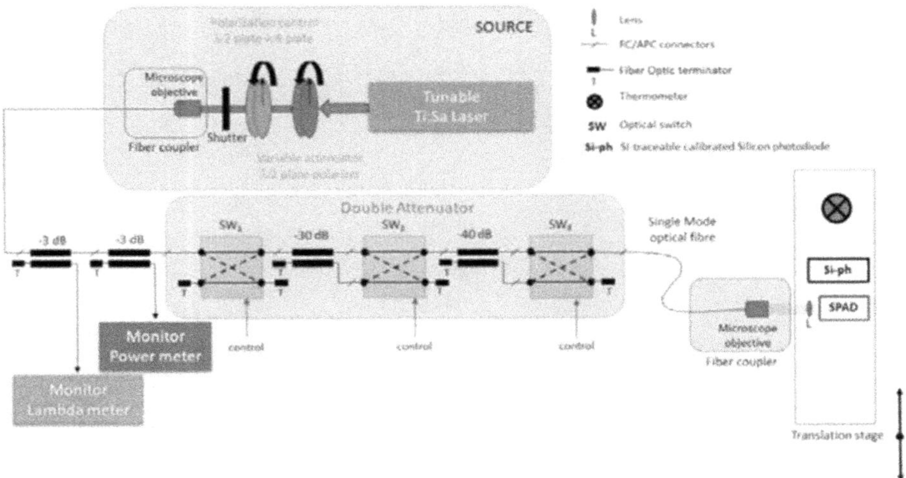

Fig. 3. Experimental setup of INRiM Si-SPAD detection efficiency calibration facility

The photon beam subsequently passes through a mechanical shutter and is coupled into a single-mode fiber via a 20 × microscope objective. Two 50:50 beamsplitters are employed to continuously monitor both the optical power and the source's wavelength stability. The photons then enter a secondary attenuation stage, composed of two strongly unbalanced beam splitters (999:1 and 9999:1), which introduce fixed attenuations of 30 dB and 40 dB respectively. This double-attenuator setup achieves a total attenuation of 70 dB when the appropriate optical path is selected via three PC-controlled optical switches, matching the total transmissivity τ.

After attenuation, photons are out-coupled and collimated in free space with a Gaussian spatial profile and sent into a shielded enclosure to suppress environmental photon noise. A motorized translation stage positions the selected detector based on the optical power level: a SI-traceable silicon photodiode (Hamamatsu S1337-1010BQ) is used for measuring the input optical power P_0, , while the silicon SPAD (DUT) is used to count the detected photons N when the photon flux is reduced to the single-photon regime via the calibrated attenuation τ and to estimate P_c.

Since the typical active area of a free-space silicon SPAD is on the order of hundreds of micrometers, a focusing lens is used to reduce the spatial mode diameter to 40 μm at the detector.

The DUT alignment is optimized using three actuators for translation in the x, y, and z directions. All instruments in the setup are controlled via a LabVIEW interface, and measurement procedures are fully automated.

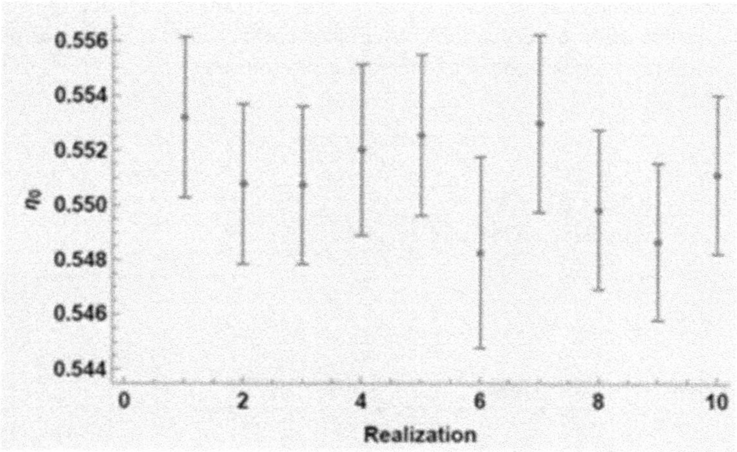

Fig. 4. Quantum efficiency measurement of a Perkin-Elmer Si-SPAD

We tested detection efficiency of a Perkin-Elmer Si-SPAD. The results are illustrated in Fig. 4. The resulting estimations of efficiency in different runs of measurement exhibit strong agreement. The average value obtained for the detection efficiency is: $\langle \eta_0 \rangle = 0.551 \pm 0.003$ [19].

References

1. Bennett, C.H., Brassard, G.: In Proceedings of IEEE International Conference on Computers, Systems and Signal Processing IEEE, Bangalore (1984)
2. Gisin, N., Ribordy, G., Tittel, W., Zbinden, H.: Rev. Mod. Phys. **74**, 145 (2002). https://doi.org/10.1103/RevModPhys.74.145
3. Lewis, A., Travagnin, M.: A secure quantum communications infrastructure for Europe: technical background for a policy vision (2022).https://doi.org/10.2760/180945
4. Meda, A., Mura, A., Virzì, S., et al.: QKD protected fiber-based infrastructure for time dissemination. Sci. Rep. **15**, 13419 (2025). https://doi.org/10.1038/s41598-025-97480-8

5. Peev, M., et al.: The SECOQC quantum key distribution network in Vienna. New J. Phys. **11**(7), 075001 (2009). https://doi.org/10.1088/1367-2630/11/7/075001

6. Stucki, D., et al.: Long-term performance of the swissquantum quantum key distribution network in a field environment. New J. Phys. **13**(12), 123001 (2011). https://doi.org/10.1088/1367-2630/13/12/123001

7. https://digital-strategy.ec.europa.eu/en/policies/european-quantum-communication-infrastructure-euroqci

8. https://quid-euroqci-italy.eu/it/

9. Hadfield, R.H.: Nat. Photonics **3**, 696–705 (2009). https://doi.org/10.1038/nphoton.2009.230

10. Liang, Y., Zeng, H.P.: Sci. China: Phys., Mech. Astron. **57**, 1218 (2014). https://doi.org/10.1007/s11433-014-5450-0

11. Chen, T.-Y., et al.: Opt. Express **18**, 27217 (2010). https://doi.org/10.1364/OE.18.027217

12. Stucki, D., et al.: New J. Phys. **13**, 123001 (2011). https://doi.org/10.1088/1367-2630/13/12/123001

13. Meda, A., Degiovanni, I.P., Tosi, A., Yuan, Z., Brida, G., Genovese, M.: Light: Sci. Appl. **6**, e16261 (2017). https://doi.org/10.1038/lsa.2016.261

14. Lydersen, L., Wiechers, C., Wittmann, C., Elser, D., Skaar, J., Makarov, V.: Nat. Photonics **4**, 686 (2010). https://doi.org/10.1038/nphoton.2010.214

15. López, M., et al.: A study to develop a robust method for measuring the detection efficiency of free-running InGaAs/InP single-photon detectors. EPJ Quantum Technol. **7**, 14 (2020)

16. Georgieva, H., et al.: Detection of ultra-weak laser pulses by free-running single-photon detectors: Modeling dead time and dark count effects. Appl. Phys. Lett. **118**, 174002 (2021)

17. Raupach, S.M.F., et al.: Detection rate dependence of the inherent detection efficiency in single-photon detectors based on avalanche diodes. Phys. Rev. A **105**, 042615 (2022)

18. Manuscript in preparations

19. Virzì, S., et al.: Detection efficiency characterization for free-space single-photon detectors: Measurement facility and wavelength-dependence investigation. Appl. Phys. Lett. **125**(22), 221108 (2024)

Development of National Quantum-Safe Network Testbed in Singapore

Hao Qin[1]([✉]), Jing Yan Haw[1]([✉]), Tosanut Rimprongern[1], Matthew Wee[1], Nelly Ng[2], Christian Kurtsiefer[1], Michael Kasper[3], and Alexander Ling[1]

[1] Centre for Quantum Technologies, National University of Singapore, Singapore, Singapore
{hao.qin,jingyan.haw}@nus.edu.sg
[2] School of Physical and Mathematical Sciences, Nanyang Technological University, Singapore, Singapore
[3] Fraunhofer Singapore Research Centre@NTU, Nanyang Technological University, Singapore, Singapore

Abstract. We present the strategic framework and technical foundations of the National Quantum-Safe Network (NQSN) Testbed—a resilient and fully interoperable quantum-safe network. Built on a star-topology architecture over production-grade fiber infrastructure, the testbed supports multi-protocol quantum key distribution (QKD), post-quantum cryptography (PQC) and accommodates diverse applications from multiple vendors. Interoperability is enabled by a centralized key and network management system, ensuring seamless integration across heterogeneous technologies. We showcase a range of reference applications, including secure data center interconnects, edge computing, hybrid QKD–post-quantum cryptography (PQC) encryption, and multi-layer integration within the OSI stack. These implementations underscore the feasibility and adaptability of deploying quantum-safe technologies in complex, multi-input, multi-output network environments.

Keywords: Quantum key distribution · Quantum key distribution network · Quantum-safe · Quantum-network · Testbed

1 Introduction

As the quantum computing technology landscape undergoes rapid evolution, there is an urgent need to fortify the communication infrastructure that handles high-value assets or requires long-term protection with quantum-safe security enhancements. In this respect, it is desirable to have a robust quantum-safe network [2,7,9,11,13] that is highly interoperable, supporting various types of quantum key distribution (QKD) protocol implementations while ensuring seamless compatibility with post-quantum cryptography (PQC).

H. Qin and J. Y. Haw—Equal contribution.

© The Author(s), under exclusive license to Springer Nature Switzerland AG 2026
F. Barbaresco and G. François (Eds.): QUEST-IS 2025, CCIS 2743, pp. 94–102, 2026.
https://doi.org/10.1007/978-3-032-13852-1_11

In this work, we present the strategic approach and delve into technical considerations integral to the construction of a resilient and fully interoperable quantum-safe network (National Quantum-Safe Network, Singapore). Our quantum-safe network testbed employs a star-type network topology, deployed over existing production-grade fiber infrastructure (Fig. 1(a)), with more than 1250 km of dark fibers in total. The testbed features the interoperability of multiple QKD protocols and full-scale Open Systems Interconnection (OSI) layers of applications integration from various vendors.

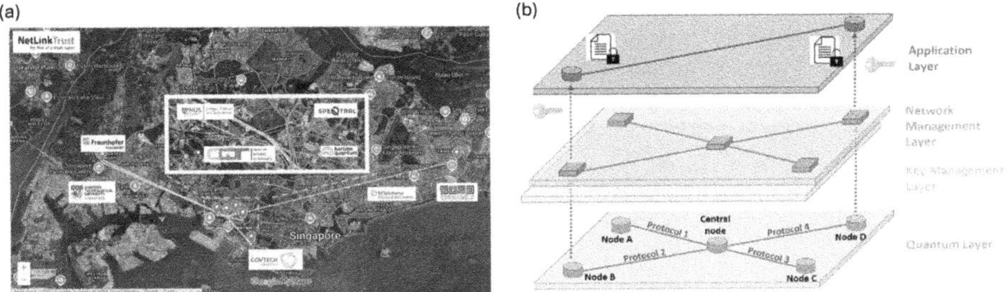

Fig. 1. (a) NQSN Fiber Network, (b) Different layers in NQSN testbed.

The systematic construction of the testbed unfolds across several logical layers [4,5]: the quantum layer, key management (KM) layer, QKD network management layer, and application (App) layer (Fig. 1(b)). This layered approach facilitates the configuration and tailoring of the testbed for scalability and interoperability, both horizontally and vertically. Such a structured framework also enables PQC seamless integration at KM and application layer, and also helps to understand the integration and security at each layer of the network architecture. In the following, we introduce some key features of the NQSN testbed.

2   Quantum Layer: Multi-QKD Protocol

In the Quantum layer, 5 commercially ready QKD devices are deployed to connect the central node with each remote node through the existing production-grade fiber infrastructure. The total loss of all QKD links reaches about 50 dB, with only passive components in between. Commercial QKD devices implementing protocols include BB84, coherent one-way (COW), continuous variable Gaussian modulation coherent state, and entanglement-based BBM92. These QKD devices are evaluated through their functions, performances, and long-term stability over the fiber infrastructure under realistic conditions. Figure 2 gives an example of the secure key rate performance over the loss changes and Quantum Bit Error Rate (QBER) manipulation of one of the BB84 QKD systems. Figure 2 (a) measures the secure key rate and QBER behavior with a variable optical attenuator (VOA) connected to the deployed fiber with loss of 15.45 dB.

Figure 2 (b) measures the secure key rate over the QBER manipulation using a dedicated tool to delay the pulses to introduce errors over QKD signal pulses. At last, the performance of different QKD systems in terms of secure key rate are summarized in the Fig. 3.

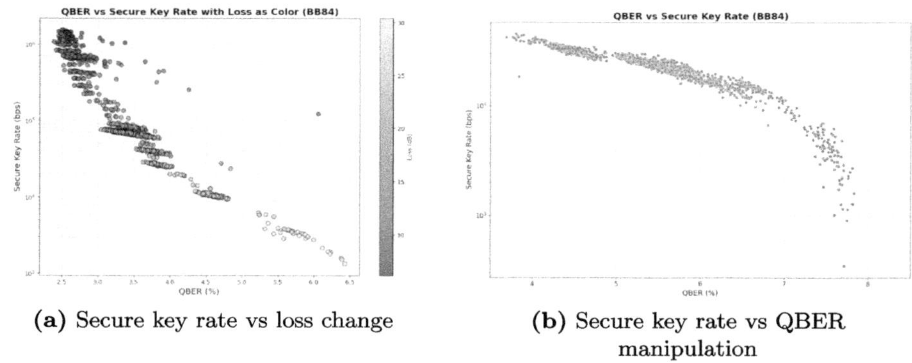

(a) Secure key rate vs loss change

(b) Secure key rate vs QBER manipulation

Fig. 2. Secure key rate under different quantum channel conditions.

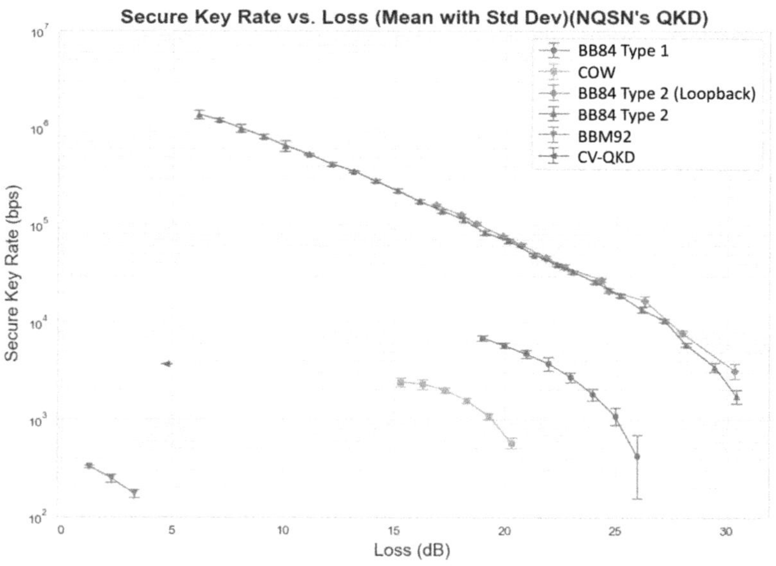

Fig. 3. Secure key rate versus channel loss for multiple QKD systems in NQSN. Error bars indicate measurement variability.

3   KM Layer: Interoperable Key Management Services and Interfaces

At the heart of our network is an interoperable key management service (KMS) solution developed for the NQSN testbed deployed over fiber infrastructure. The KMS ensures network interoperability, scalability, resilience, and crypto-agility, supporting end-to-end symmetric key applications. It is designed to work with diverse network topologies and to integrate with various QKD devices in the quantum layer and applications in the App layer through standardized interfaces [3,6].

Within the NQSN testbed, the KMS operated with five commercial QKD systems using distinct protocols from different vendors and demonstrated compatibility with industry-grade symmetric key applications across OSI layers. Key challenges—such as key synchronization, key storage, relaying, interface compatibility, key refresh, and entity authentication—are identified and analyzed. Figure 4 (Left) demonstrates the key database that stores QKD keys from different QKD devices.

We also evaluated the KMS interfaces, including ETSI QKD GS 014 standards [3] between QKD-KMS and KMS-application layers, using metrics such as response time and latency (Fig. 4 (Right)). Interoperability issues arising from implementation variations were also assessed to enhance compatibility and performance.

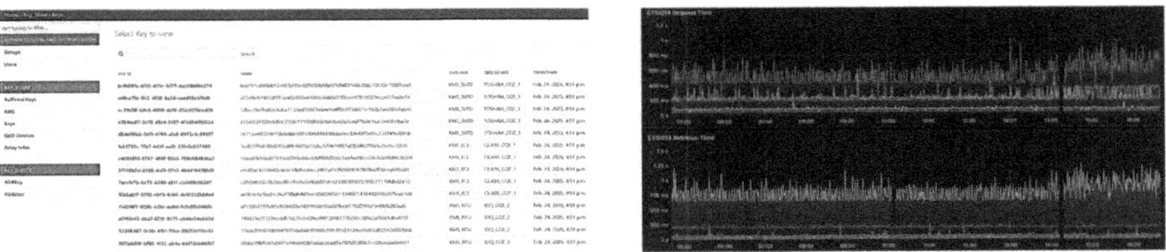

Fig. 4. (Left) NQSN KMS Key Database (Right) ETSI GS QKD 014 Request Time and Retrieval Time.

We further showcased a commercial KMS solution (evolutionQ) with four types of QKD systems (plus a QKD simulator) in a linear topology spanning five QKD nodes, validated by an end-to-end QKD virtual private network (VPN) application (Fig. 5, top). Advanced approaches, including Distributed Symmetric Key Establishment based on secret sharing (Quantum Bridge), were also demonstrated for cross-border connectivity and QKD integration. (Fig. 5, bottom)

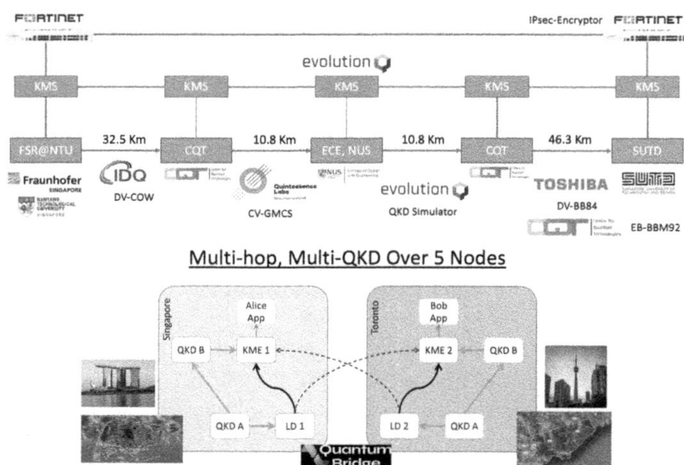

Fig. 5. Key management layer use cases.

4 Application Layer: Versatile Reference Use Cases & Hybrid QKD-PQC Integration

In the application layer, various quantum-secured use cases, spanning data center applications, edge computing, hybrid QKD-PQC encryption, and integration at different OSI layers, are explored from a technical perspective, providing critical insights into the deployment in QKDN with the multi-input/multi-output KM architecture. We focus on practical deployments within the NQSN, showcasing key integration scenarios that highlight the synergy between QKD and PQC. We demonstrated QKD-secured applications across various OSI layers, including physical, data link and network layers, illustrating how secure keys are provisioned and utilized in various network environments (Figs. 6 and 7).

For the evaluation of quantum-safe technologies in critical infrastructure, we performed two representative reference deployments:

1. QKD-secured Data Centre Interconnect [10]
   ID Quantique Cerberis XGR QKD system (COW Protocol) is deployed over NetLink Trust commercial fibre between STTelemedia-Global Data Centres sites, producing >2 Gbit of QKD-derived key material and enabling AES-256 QKD-VPN secure data transfers.
2. QKD-secured Edge Cloud Computing [8]
   In this implementation, field-deployed fiber optics are utilized to connect AWS Edge Compute devices with on-premises Fortinet devices. QKD generates symmetric keys, which are used to establish a quantum-secured VPN that protects edge cloud workloads.

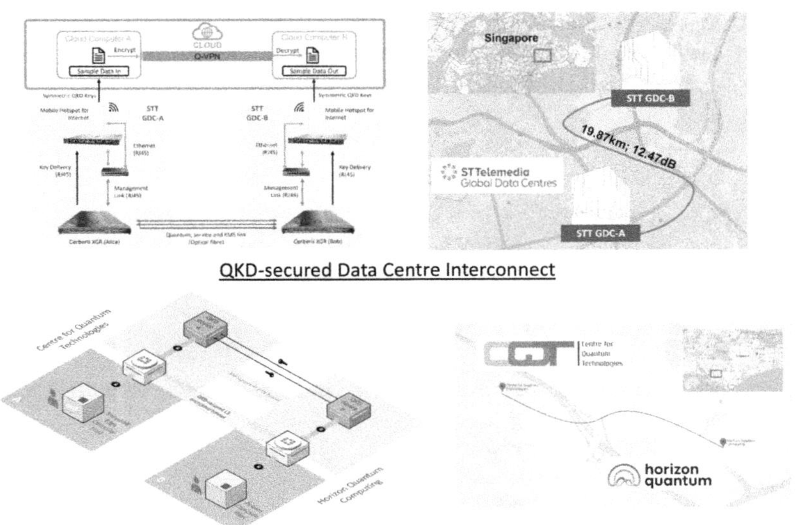

Fig. 6. Quantum-secured infrastructure use cases.

For OSI layer use cases, below are some examples organized within the NQSN Testbed (Fig. 8):

1. QKD-secured video communications
   We deployed Fortinet FortiGate for OSI Layer-3 routing/firewalling and Hitachi Energy MPLS-TP encryption cards for Layer-2.5 transport encryption [1]. Video traffic is protected using symmetric keys from a QKD system over 32.48 km (11.52 dB of loss) of fibre links between Centre for Quantum Technologies (CQT) and Fraunhofer Singapore Research Centre@NTU (FSR@NTU).
2. Three-node QKD-encrypted communications
   We deployed an end-to-end Layer-2 encryption across three sites (CQT, FSR@NTU, GovTech) using ST Engineering OSI Layer-2 encryptors. Symmetric keys are distributed and relayed by the ST Engineering Key Management System (KMS) and are sourced from two QKD different systems (BB84 and COW Protocol).
3. QKD-secured streaming over 5G backbone
   We setup streaming traffic between CQT and Singapore University of Technology and Design (SUTD), which is protected end-to-end using QKD as the primary source of symmetric keys with a PQC fallback, delivered to Layer-2 encryptors over a 5G backbone infrastructure. The deployment uses BB84 QKD System over 46.6 km link (>19 dB loss)and Thales CN6010 encryptor performing AES-256 encryption with keys constantly refreshed.

4. Quantum Encryption at Optical Transport Layer
   We are organsing physical-layer encryption using Ciena L1 hardware encryptors, consuming QKD keys via the NQSN KMS.

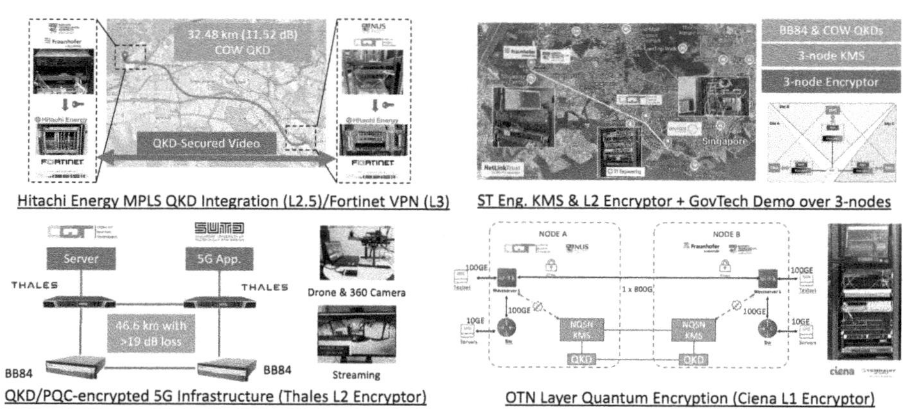

Fig. 7. Quantum-secured OSI layers use cases.

These demonstrations reveal how these quantum-safe technologies can be integrated into modern communication infrastructures.

Additionally, we explored hybrid QKD-PQC implementations, which combine the strengths of both quantum and classical cryptographic techniques to enhance security in various applications. By integrating QKD's ability to securely distribute encryption keys with PQC's resilience against quantum attacks, these implementations provide tailored solutions for a diverse range of security use cases. Several hybrid QKD-PQC use cases evaluated on the NQSN includes

1. QKD-PQC Hybrid VPNs
   We deployed a three-site IP-Sec virtual private networks (VPNs) link using Fortinet appliances integrated with a QKD key source and Check Point appliances using PQC keys. Test plan covers speed tests and file transfer validation across the three nodes.
2. Hybrid QKD-PQC Quantum-Safe application developed an application with layered security model, combining quantum-secure primitives (QKD, OTP) plus Open-SSH (with PQC option from Open Quantum Safe [12]) and conventional symmetric ciphers (e.g. AES). This comprehensive approach ensures information-theoretic confidentiality for short and high-value messages.
3. QKD-PQC Defense in depth deployed Fortinet Fortigate L3 VPN with NQSN-KMS and QuintessenceLabs CV-QKD (qOptica$^T M$ 100). Additionally, we implement the Viavi TerraVM PQC Key Exchange Mechanism (KEM) to secure HTTPS traffic on top of this infrastructure.

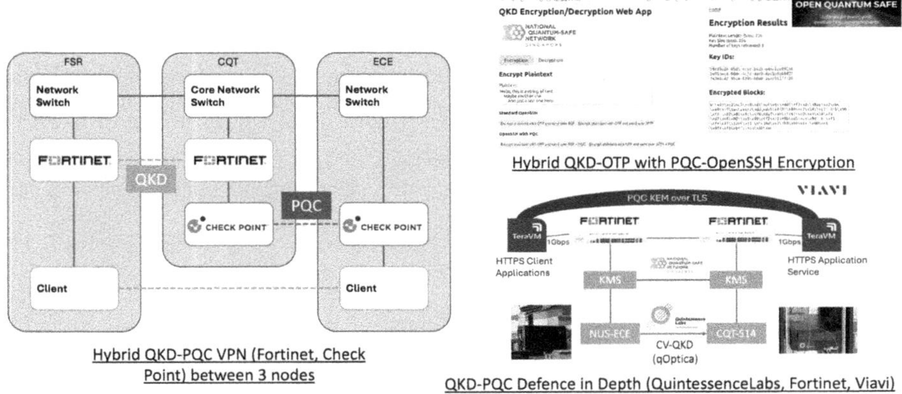

Fig. 8. Hybrid QKD-PQC use cases.

While NQSN hosts QKD technologies sourced from a diverse range of vendors, which form the secure backbone of the network, the inclusion of PQC further complements QKD by offering cryptographic algorithms designed to resist quantum attacks. This dual-layered approach enhances the network's security, serving as a last-mile connectivity solution, a resiliency mechanism, or an additional security layer, depending on the specific application requirements.

5 Conclusion

In this work, our objective is to discuss the challenges encountered during deployment of a practical quantum-safe network, such as interoperability issues, scalability constraints, and integration complexities. Lessons learned from these experiences will provide valuable insights into optimizing hybrid architectures for future applications. Ultimately, these insights underscore the potential of hybrid QKD-PQC architectures in ensuring resilient and scalable quantum-safe networks, paving the way for secure communication in the quantum era.

Acknowledgments. We acknowledge NQSN partners (www.nqsn.sg) for their support. We acknowledge funding support from A*STAR via National Quantum Office (NQO) under its Quantum Engineering Programme 3.0 (National Quantum-Safe Network Testbed, W24Q3D0001), National Research Foundation, Singapore and A*STAR under its Quantum Engineering Programme (National Quantum-Safe Network, NRF2021-QEP2-04-P01), Infocomm Media Development Authority (National Quantum-Safe Network Plus) and start-up grant for Nanyang Assistant Professorship awarded to Ng NHY of Nanyang Technological University, Singapore.

Disclosure of Interests. The authors declare no competing interests.

References

1. Bächli, R., Shoaie, M.A., Floeter, R., Palangadan, V., Foery, A.: Quantum key distribution for mpls-tp traffic encryption. In: CSE N°36—CIGRE. CIGRE—CSE (2025)
2. Dynes, J., Wonfor, A., Tam, W.S., Sharpe, A., Takahashi, R., Lucamarini, M., Plews, A., Yuan, Z., Dixon, A., Cho, J., et al.: Cambridge quantum network. npj Quant. Inf. **5**(1), 101 (2019)
3. Institute, E.T.S.: ETSI GS QKD 014. Quantum key distribution, Protocol and data format of key delivery API to Applications (2019)
4. Infocomm Media Development Authority: IMDA Reference Specification Quantum Key Distribution Network (Singapore) (2023)
5. International Telecommunication Union: ITU-T Recommendation Y.3800, Overview on networks supporting quantum key distribution, plus Corrigendum 1 (2020)
6. International Telecommunication Union: ITU-T Recommendation Y.3803, Quantum key distribution networks – Key management (2020)
7. Martin, V., et al.: Madqci: a heterogeneous and scalable sdn-qkd network deployed in production facilities. npj Quant. Inf. **10**(1), 80 (2024)
8. Moreno, J., Proctor, C.: Implementing a Quantum-Secured Network in a Metropolitan Area (2023) https://aws.amazon.com/blogs/quantum-computing/implementing-a-quantum-secured-network-in-a-metropolitan-area/, aWS Quantum Technologies Blog
9. Peev, M., Pacher, C., Alléaume, R., Barreiro, C., Bouda, J., Boxleitner, W., Debuisschert, T., Diamanti, E., Dianati, M., Dynes, J.F., et al.: The secoqc quantum key distribution network in vienna. New J. Phys. **11**(7), 075001 (2009)
10. Qiu, K., Haw, J.Y., Qin, H., Ng, N.H.Y., Kasper, M., Ling, A.: Quantum-secured data centre interconnect in a field environment. J. Surveill. Sec. Saf. **5**(3) (2024)
11. Sasaki, M., Fujiwara, M., Ishizuka, H., Klaus, W., Wakui, K., Takeoka, M., Miki, S., Yamashita, T., Wang, Z., Tanaka, A., et al.: Field test of quantum key distribution in the tokyo qkd network. Opt. Express **19**(11), 10387–10409 (2011)
12. Stebila, D., Mosca, M.: Post-quantum Key Exchange for the Internet and the Open Quantum Safe Project. In: Avanzi, R., Heys, H. (eds.) SAC 2016. LNCS, vol. 10532, pp. 14–37. Springer, Cham (2017). https://doi.org/10.1007/978-3-319-69453-5_2
13. Wang, S., Chen, W., Yin, Z.Q., Li, H.W., He, D.Y., Li, Y.H., Zhou, Z., Song, X.T., Li, F.Y., Wang, D., et al.: Field and long-term demonstration of a wide area quantum key distribution network. Opt. Express **22**(18), 21739–21756 (2014)

QuaNTUM: A Modular Quantum Communication Testbed for Scalable Fiber and Satellite Integration

Julien Chénedé[1,2(✉)] [ID], Tjorben Matthes[1,2] [ID], Josefine Krause[1,2,3],
Asli Cakan[1,2] [ID], and Tobias Vogl[1,2] [ID]

[1] Technical University of Munich, Arcisstraße 21, 80333 Munich, Germany
`julien.chenede@tum.de`
[2] Munich Center for Quantum Science and Technology, Schellingstrße. 4,
80799 Munich, Germany
[3] Friedrich-Schiller-Universität Jena, Albert-Einstein-Straße 5, 07745 Jena, Germany
`https://www.tum.de/, https://www.mcqst.de/`

Abstract. Secure communication is a cornerstone of modern society, underpinning everything from financial transactions to critical infrastructure. As classical encryption faces growing threats from advancing computational power, quantum communication offers a fundamentally secure alternative, leveraging the laws of physics to protect data.

We introduce QuaNTUM (Quantum Network at the Technical University of Munich), a modular and extensible quantum communication testbed designed to enable scalable, flexible, and secure quantum communication across fiber-based campus networks and satellite-ground links. QuaNTUM integrates early deployments of solid-state quantum emitters in small satellites, bridging terrestrial and free-space channels. As an open-access platform, it supports experimental quantum communication protocols, quantum device benchmarking, and hybrid network integration.

The terrestrial network connects universities and research institutes across the high-tech campus in Garching near Munich, using single-mode fibers in a star-shaped topology. Each node features polarization-maintaining components, multiplexers, and time-synchronized analysis modules, with a central switching hub for dynamic routing. Active polarization control and real-time feedback ensure low error rates and stable qubit transmission, enabling high-fidelity quantum key distribution (QKD) and entanglement distribution.

A central focus of QuaNTUM is its use of deterministic single-photon sources based on optically active defects, such as in hexagonal boron nitride (hBN) or excited erbium atoms. These sources are currently being deployed in orbit, marking one of the first demonstrations of a solid-state quantum emitter in space. By uniting fiber and free-space links with scalable hardware and open protocols, QuaNTUM offers a basis that could adapt to future hybrid quantum networks, supporting both current testbed research and the long-term vision of a global quantum internet.

F. Barbaresco and G. François (Eds.): QUEST-IS 2025, CCIS 2743, pp. 103–112, 2026.
https://doi.org/10.1007/978-3-032-13852-1_12

Keywords: quantum communication · hBN single-photon emitter · CubeSat · fiber-based testbed · quantum network architecture

1 Introduction

The arrival of the quantum computer threatens today's encrypted networks by efficiently solving complex mathematical problems (like factoring large integers), on which nowadays most of our secured communication system are relying on [1]. One replacement candidate is quantum communication which guarantees a safe communication method against any eavesdropper or even the operator of the network. Several projects were done during the past years proving the feasibility of such links using long distance communication via satellites [2] and shorter ones via fibers [3]. Before quantum repeaters become technically feasible and enable a full "quantum internet", hybrid networks can be created using both fiber and satellite links, with satellites serving as trusted nodes. Ultimately, QuaNTUM aims to enable long-distance quantum networking, linking remote quantum processors and supporting reliable quantum information exchange.

Quantum key distribution (QKD) is a remarkable technology: it uses the fundamental laws of physics to secure communication. By encoding keys onto single photons, it makes eavesdropping impossible, thanks to the no-cloning theorem [1] and the fact that any attempt to measure the photons disturbs them. Yet, a limitation remains: when we rely only on fiber optics, QKD hits a wall after a few hundred kilometers, or even less for some implementations. The signal weakens exponentially, and unlike classical signals, amplification alone is impossible.

Realizing a true "quantum internet" relies on quantum repeaters, which are still nowadays at an early stage [3]. In the meantime, there is a workaround: a hybrid approach that combines the best of both worlds. Satellites can bridge vast distances through free space, where signal loss is almost negligible above 10 km, while fiber networks handle the last kilometers, connecting buildings, campuses, or even cities.

Despite these advancements, a critical gap remains: the absence of a large-scale, real-world open testbed that integrates these technologies into a cohesive framework. The QuaNTUM project addresses this integration challenge by establishing a star topology fiber network across the Garching campus near Munich. This infrastructure interconnects various institutes and laboratories, enabling shared access to quantum resources. Unlike trusted-node QKD networks that have already been demonstrated in Boston, Vienna, Tokyo, and China, QuaNTUM is designed as an open research testbed that directly connects laboratories for end-to-end quantum experiments without relying on intermediate trusted relays. Researchers can thus conduct experiments involving single-photon emission, for instance, from hBN [4] color centers or erbium-doped materials [5], optical quantum memories [6,7], and diverse QKD protocols [3], as well as investigate quantum entanglement distribution [8].

In a subsequent phase, the network will be extended to Munich's city center and to a ground station to support optical satellite links. Preparatory work is already underway with the low-Earth-orbit CubeSat QUICK[3] [9], which is performing foundational quantum optics experiments in microgravity. Upon completion of the commissioning phase, the QuaNTUM network will be made accessible to the Munich Center for Quantum Science and Technology (MCQST), the Munich Quantum Valley (MQV), and external partners. This will facilitate collaborative research in areas such as secure quantum communication, quantum sensing, and the interfacing of quantum memories.

2 QuaNTUM Testbed Architecture

2.1 Fiber-Optic Backbone

The QuaNTUM fiber-optic backbone is implemented as a star topology single-mode network centered at the TUM-MI node (Departments of Mathematics and Computer Science), roughly located in the center of the campus. Some standard Telecom-Band SMF-28 fibers form the primary campus backbone and interconnect the principal sites described in Fig. 1 with typical link lengths of 1 km to 2 km. At the central node, a quantum reconfigurable add-drop multiplexer (q-ROADM) is specified to provide wavelength de/multiplexing, low-loss optical switching and a consolidated point for polarization control hardware and timing distribution. The network design preserves the option of direct fiber-to-fiber links to minimize insertion loss for critical quantum channels.

Cable composition and connectorization are chosen to balance compatibility with telecom infrastructure and the specific wavelength requirements of quantum systems. Each main cable contains multiple SMF-28 fibers (optimized for telecom O-/C-bands at 1330 nm and 1550 nm) together with dedicated short-wavelength fibers (780-HP and 1060-XP) to support key quantum technologies [10]:

- Solid-state qubits: Color centers in diamond (e.g. SiV centers at 738 nm [11]) and silicon carbide (e.g. silicon vacancy centers emitting at longer near-infrared wavelengths [12]) require near-visible to near-infrared wavelengths, but fiber attenuation below 780 nm is prohibitive for kilometer-scale links– hence the 780 nm lower bound in our design.
- Quantum memories: Rare-earth-doped crystals (e.g. Tm^{3+}:Y_2SiO_5 at 793 nm [13]) and alkali vapors (e.g. Rb D2 transitions at 780 nm [10]) often operate in the 700–800 nm range, motivating the inclusion of 780-HP fibers.
- Telecom compatibility: Long-distance quantum communication leverages the near-infrared (NIR) free-space window and telecom O-/C-bands (1330/1550 nm).

E2000 APC simplex terminations are specified for quantum ports to minimize back-reflection and allow individual fiber access. Loss budgeting and noise mitigation are central design constraints. The end-to-end optical loss model includes fiber attenuation, splice loss, connector mated loss, and insertion loss of passive/active elements (Wavelength Division Multiplexing [14], switches, polarizers). To protect single-photon channels from classical noise, high-power classical

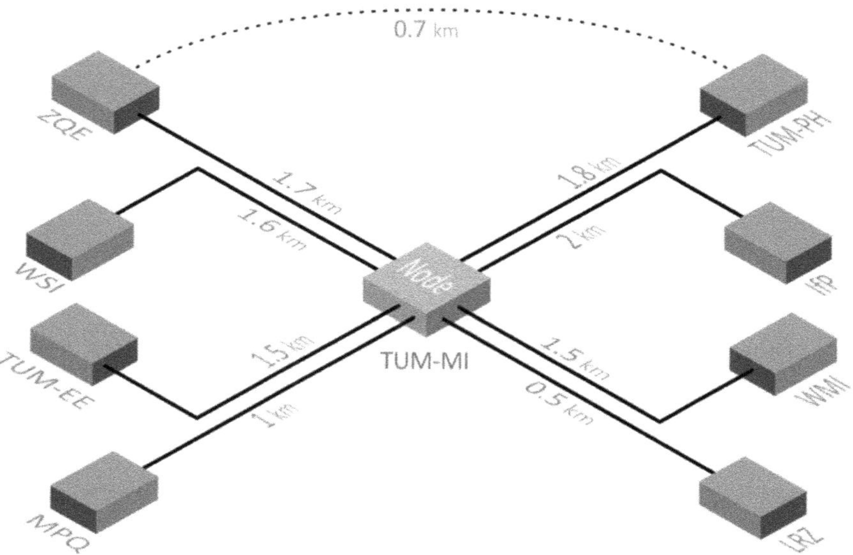

Fig. 1. Topology of the QuaNTUM testbed on the Garching campus near Munich: star-shaped fiber connections using SMF-28 ultra fibers, 780-HP and 1060-XP between the department of physics (TUM-PH), Electrical Engineering (TUM-EE), the Max Planck Institute of Quantum Optics (MPQ), the Leibniz Supercomputing Centre (LRZ), the Walter Schottky Institut (WSI), the Walther-Meißner-Institut (WMI) and with the central node in School of Computation, Information and Technology (TUM-MI). An additional special cable with only SMF-28 ultra fibers is connecting directly ZQE and TUM-PH (dotted-line). – For geographic reference, see the map: https://v.bayern.de/N74hf – ©Bayerische Vermessungsverwaltung 2025

signals are separated spectrally or routed on separate fibers; APC connectors are used to back reflections. For polarization-encoded protocols the backbone incorporates motorized fiber polarization controllers (FPCs) and software feedback to maintain state fidelity under environmental perturbations. Market-ready devices can provide polarization compensation such as silicon-based decoder [15], Pockels cell [16] and fiber squeezing devices [17].

2.2   Timing and Control

Synchronization and deterministic control are fundamental to all QuaNTUM experiments: they enable reliable coincidence detection, accurate protocol timing, and automated feedback for polarization and routing. The timing architecture therefore combines a dedicated classical reference channel, local high-resolution time-tagging at each measurement node, and a coordinated control plane that drives active hardware and manages experiment scheduling.

A low-jitter classical reference channel is provisioned alongside the quantum wavelengths to distribute a 10 MHz clock and a 1 pulse per second (PPS) marker. This provides a stable, common time base and avoids many limitations of satellite-based references for short-range campus networks. Each polarization analysis module (PAM) is paired with a synchronized time-tagger capable of sub-100-ps resolution for arrival-time stamping; these local time stamps are essential for coincidence analysis, quantum bit error ratio (QBER) estimation and post-processing.

For deterministic, low-latency distribution and scalable network timing, the testbed will evaluate precision time protocols such as White Rabbit [18] because they provide sub-nanosecond synchronization and known latency bounds over fiber links.

The control plane is implemented as a hybrid of FPGA/embedded control at the hardware layer and a centralized scheduler at the management layer. FPGA units or microcontrollers co-located with the q-ROADM and at user nodes perform deterministic control of optical switches, electro-optic modulators and motorized FPCs with microsecond actuation times. Timing-critical loops, for example, live polarization compensation, run locally at each node (fast perturbation, local evaluation, accept/reject), while higher-level policies (when to perform a full reference recalibration, scheduling of satellite passes) are managed centrally.

Practical implementation requires explicit calibration and monitoring: per-link fixed delays must be measured and recorded (optical time-domain reflectometer and two-way time transfer), time-tagger offsets and cable dispersive effects characterized, and ongoing telemetry (detector counts, QBER, temperature) integrated into the control dashboard. These elements together ensure that QuaNTUM achieves the temporal precision and operational automation necessary for reliable QKD and entanglement distribution experiments across a campus-scale hybrid fiber network.

2.3 Entanglement and QKD Layers

From a security standpoint, protocols divide into device-independent (theoretically secure against all attacks, with no hardware assumptions) and device-dependent (relying on trusted measurement tools but compatible with off-the-shelf components). In practical terms, implementations split between discrete-variable (e.g. polarized photons) and continuous-variable (e.g. coherent states) approaches. QKD protocols fall into three primary types (Table 1) [19]:

QuaNTUM is implementing two complementary quantum channel layers on the same physical backbone: a wavelength-multiplexed entanglement distribution layer for multi-user and a PM-QKD layer that supports standard schemes.

For entanglement distribution, a broadband spontaneous-parametric-down-conversion (SPDC) source in a Sagnac configuration [20] is specified as the primary photon-pair generator. The SPDC output is spectrally partitioned by a demultiplexer; the q-ROADM in the central node assigns channel pairs to endpoint links, producing a reconfigurable, wavelength-multiplexed entanglement fabric. At each user endpoint, polarization analysis modules (PAMs) implement

Table 1. QKD protocol classification by structure

Type	Key Mechanism
Prepare-and-Measure (PM)	Alice prepares quantum states; Bob measures them to generate a shared key.
Entanglement-Based	Alice and Bob exploit entangled particle correlations for key distribution.
Measurement-Device-Independent	Security remains intact even if measurement devices are untrusted.

random-basis selection and perform polarization projection and detection. Time-stamping of events with sub-100-ps resolution enables coincidence identification and correlation analysis for entanglement verification. The architecture is designed to support multi-hop protocols [18] by routing idler photons to intermediate Bell-state measurement stations when required, enabling tests of networked entanglement distribution.

The PM layer will support different standard QKD implementations (for example B92, BB84, SARG04 [19]). Deterministic single-photon emitters based on localized defects in hBN (at 575 nm) are primarily intended for short-reach or on-chip experiments, as their visible-band emission is incompatible with long-distance transmission over telecom-band fibers. For integration with the testbed's backbone, a frequency conversion strategy is employed: Visible photons are converted to telecom bands via nonlinear optics [21] before entering the network.

For long-distance operation, the testbed provisions alternative sources (e.g. telecom-compatible Spontaneous Parametric Down-Conversion pairs [22] or InGaAs-based single-photon emitters [23]). Real-time operational control closes the loop for secure operation: measured QBER and coincidence statistics drive local feedback (motorized fiber polarization controllers and software routines) for live polarization compensation and calibration.

3   Single-Photon Emitters in hBN

3.1   Deterministic Fabrication

In QuaNTUM we implement deterministic fabrication by localized electron-beam irradiation of multilayer hBN using a standard scanning electron microscope (SEM) [24]. Exposures are targeted to sub-micron coordinates to create emitter arrays with high lateral precision; process parameters (beam energy, current, dose and spot dwell time) are optimized to maximize yield and spectral reproducibility. The produced emitters show a reproducible zero-phonon line (ZPL) near 575 nm and are consistent with carbon-related defect complexes identified by density-functional calculations [28]. This approach is compatible with wafer-scale patterning workflows and with subsequent on-chip or micro-optical integration because the emitter position is known *a priori*. Scalability hinges on automation of exposure patterning, batch handling of hBN flakes or films, and yield optimization to reduce per-device screening. For production of fiber-coupled modules or on-chip arrays, deterministic sites are pre-aligned to lithographically defined photonic structures (waveguides, cavities, grating couplers) so that subsequent assembly and coupling are mainly mechanical rather than optical alignment tasks.

3.2   Optical Performance

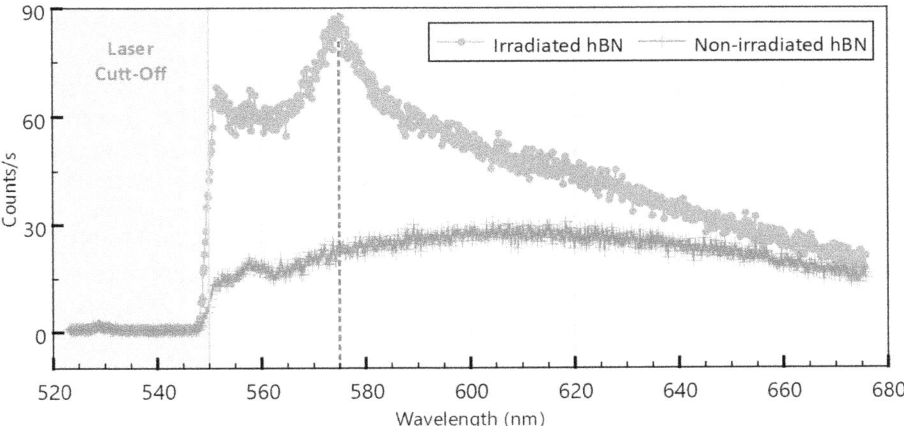

Fig. 2. Room-temperature photoluminescence spectrum of a single hBN emitter fabricated by localized electron irradiation, featuring peak emission near (orange line). Comparison with a non-irradiated hBN (blue line). Laser excitation at 530 nm and long-pass filter at 550 nm. (Color figure online)

hBN hosts a diverse family of quantum emitters with zero-phonon lines (ZPLs) spanning UV to NIR wavelengths [25], enabling compatibility with a wide range of quantum systems. For this work, we focus on defects produced by localized irradiation that, under 530 nm femtosecond laser excitation, exhibit a peak emission 575 nm with a typical linewidth (FWHM) on the order of tens of nanometers at room temperature (Fig. 2). The second-order correlation function $g^{(2)}(0) < 0.1$ confirms single-photon emission [24], while the reported excited-state lifetime of 3.83(1) ns supports high repetition-rate operation [24]. Long-term photostability has been observed in many sites even 3 years post-irradiation, though device-level qualification–including performance under variable temperature and optical pump conditions–remains necessary for network deployment or space applications. Visible-band ZPLs and room-temperature linewidths are advantageous for bright, high-rate operation but pose compatibility challenges with standard SMF-28 telecom infrastructure and with narrowband quantum memories. To address these issues, two engineering strategies are relevant: (1) spectral narrowing and tuning via photonic engineering (microcavities, Purcell enhancement, strain or electric-field tuning) to reduce the effective linewidth and stabilize the ZPL; and (2) quantum frequency conversion (QFC) to translate emitted photons to telecom wavelengths for low-loss fiber transmission or to memory-matched atomic transitions. Microcavity coupling can also increase the collection efficiency and reduce the required source brightness for a target key rate or coincidence rate, while QFC preserves quantum statistics if conversion noise is sufficiently suppressed.

3.3   Integrated Photonics

Integration of hBN emitters with photonic platforms (e.g. fiber integrated platforms [26]) addresses three objectives: improve collection/coupling efficiency into single-mode fibers, control emission dynamics (Purcell effect) and enable on-chip routing and switching for compact, stable modules [4]. In practice, deterministic emitter sites are aligned to on-chip waveguides, microcavities or grating couplers using lithographic markers and high-precision assembly. Finite-difference time-domain method (FDTD) simulations guide the resonator and waveguide design to maximize coupling efficiency and to position cavity resonances relative to the emitter ZPL [27]. Thanks to photonic integration, stricter constraints are typically designed for satellite deployment. The QUICK^3 payload strategy uses a miniaturized version of a different emitter platform based on $MoSe_2$ [9]: the chip is mounted within a compact, thermally managed optical bench that includes the pump laser, stabilization optics, fiber-coupling stage and pointing interface. Packaging choices emphasize mechanical resilience (vibration and shock tolerance), thermal control over the expected orbital temperature range, radiation tolerance of active components, and minimal moving parts.

4   Conclusion

In conclusion, the QuaNTUM project addresses a critical gap in the field of quantum communication by providing a large-scale, real-world open testbed that integrates fiber and satellite links into a cohesive framework. This platform, akin to a campus-wide hybrid network, offers researchers and engineers a controlled yet dynamic environment to experimentally validate, refine, and benchmark system performance under realistic conditions. By leveraging advanced technologies such as solid-state quantum emitters and polarization control, QuaNTUM aims to overcome the limitations of current quantum communication systems. The successful integration of these technologies could pave the way for a global quantum internet, enabling secure communication and advanced quantum applications. Future work will focus on expanding the network, improving the performance of quantum emitters, and exploring new applications in quantum communication and sensing.

Acknowledgments. This research is part of the Munich Quantum Valley, which is supported by the Bavarian state government with funds from the Hightech Agenda

Bayern Plus. This work was funded by the Deutsche Forschungsgemeinschaft (DFG, German Research Foundation) under Germany's Excellence Strategy- EXC-2111-390814868 (MCQST). The authors acknowledge support from the Federal Ministry of Research, Technology and Space (BMFTR) under grant number 13N16292 (ATOM-IQS).

Disclosure of Interests. The authors declare that they have no competing interests relevant to the content of this article.

References

1. Gisin, N., Ribordy, G., Tittel, W., Zbinden, H.: Quantum cryptography. Rev. Mod. Phys. **74**, 145–195 (2002)
2. Lu, C.-Y., Cao, Y., Peng, C.-Z., Pan, J.-W.: Micius quantum experiments in space. Rev. Mod. Phys. **94**, 035001 (2022)
3. Granados, G., Velasquez, W., Cajo, R., Antonieta-Alvarez, M.: Quantum key distribution in multiple fiber networks and its application in urban communications: a comprehensive review. IEEE Access **13**, 100446–100461 (2025)
4. Vogl, T., Lecamwasam, R., Buchler, B.C., Lu, Y., Lam, P.K.: Compact cavity-enhanced single-photon generation with hexagonal boron nitride. ACS Photonics **6**(8), 1955–1962 (2019)
5. Gritsch, A., Weiss, L., Früh, J., Rinner, S., Reiserer, A.: Narrow optical transitions in erbium-implanted silicon waveguides. Phys. Rev. X **12**, 041009 (2022)
6. Lvovsky, A.I., Sanders, B.C., Tittel, W.: Optical quantum memory. Nat. Photonics **3**, 706–714 (2009)
7. Çakan, A., et al.: Quantum optics applications of hexagonal boron nitride defects. Adv. Opt. Mater. **13**(7), 2402508 (2025)
8. Neumann, S.P., Buchner, A., Bulla, L., Bohmann, M., Ursin, R.: Continuous entanglement distribution over a transnational 248 km fiber link. Nat. Commun. **13**, 6134 (2022)
9. Ahmadi, N., et al.: QUICK^3 - design of a satellite-based quantum light source for quantum communication and extended physical theory tests in space. Adv. Quantum Technol. **7**(4), 2300343 (2024)
10. Cholsuk, C., Suwanna, S., Vogl, T.: Tailoring the emission wavelength of color centers in hexagonal boron nitride for quantum applications. Nanomaterials, **12**(14) (2022)
11. Bradac, C., Gao, W., Forneris, J., Trusheim, M.E., Aharonovich, I.: Quantum nanophotonics with group iv defects in diamond. Nat. Commun. **10**, 5625 (2019)
12. Castelletto, S.: Silicon carbide single-photon sources: challenges and prospects. Mater. Quantum Technol. **1**, 023001 (2021)
13. Simon, C., et al.: Quantum memories. Eur. Phys. J. D **58**, 1–22 (2010)
14. Bathaee, M., Rezai, M., Salehi, J.A.: Quantum wavelength-division-multiplexing and multiple-access communication systems and networks: global and unified approach. Phys. Rev. A **107**, 012613 (2023)
15. Du, Y., et al.: Silicon-based decoder for polarization-encoding quantum key distribution. Chip **2**(1), 100039 (2023)
16. Caputo, C., Simoni, M., Cirillo, G.A., Turvani, G., Zamboni, M.: A simulator of optical coherent-state evolution in quantum key distribution systems. Opt. Quantum Electron. **54**, 689 (2022)

17. Kaiser, F., Fedrici, B., Zavatta, A., D'Auria, V., Tanzilli, S.: A fully guided-wave squeezing experiment for fiber quantum networks. Optica **3**, 362–365 (2016)
18. Alshowkan, M., Lukens, J.M., Lu, H.-H., Peters, N.A.: Resilient entanglement distribution in a multihop quantum network. J. Lightwave Technol. **43**, 1–8 (2025)
19. Kumar, M., Mondal, B.: A brief review on quantum key distribution protocols. Multimedia Tools Appl. **84**, 33267–33306 (2025)
20. Strömberg, T., Schiansky, P., Peterson, R.W., Quintino, M.T., Walther, P.: Demonstration of a quantum switch in a Sagnac configuration. Phys. Rev. Lett. **131**, 060803 (2023)
21. Fisher, P., Cernansky, R., Haylock, B., Lobino, M.: Single photon frequency conversion for frequency multiplexed quantum networks in the telecom band. Phys. Rev. Lett. **127**, 023602 (2021)
22. Ikuta, R., et al.: Wide-band quantum interface for visible-to-telecommunication wavelength conversion. Nat. Commun. **2**, 537 (2011)
23. Zhang, J., Itzler, M.A., Zbinden, H., Pan, J.-W.: Advances in ingaas/inp single-photon detector systems for quantum communication. Light: Sci. Appl. 4, e286–e286 (2015)
24. Kumar, A., et al.: Localized creation of yellow single photon emitting carbon complexes in hexagonal boron nitride. APL Mater. **11**, 071108 (2023)
25. Cholsuk, C., Zand, A., Çakan, A., Vogl, T.: The HBN defects database: a theoretical compilation of color centers in hexagonal boron nitride. J. Phys. Chem. C **128**(30), 12716–12725 (2024)
26. Vogl, T., Lu, Y., Koy Lam, P.: Room temperature single photon source using fiber-integrated hexagonal boron nitride. J. Phys. D: Appl. Phys. **50**, 295101 (2017)
27. Ji, P., Qian, C., Finley, J.J., Yang, S.: Thickness insensitive nanocavities for 2d heterostructures using photonic molecules. Nanophotonics **12**(17), 3501–3510 (2023)
28. Cholsuk, C., Suwanna, S., Vogl, T.: Comprehensive scheme for identifying defects in solid-state quantum systems. J. Phys. Chem. Lett. **14**(29), 6564–6571 (2023). https://doi.org/10.1021/acs.jpclett.3c0147

Session 2 Quantum Communications – Quantum Key Distribution

Network-Wide Quantum Key Distribution with Onion Routing Relay

Pedro Otero-García, David Pérez-Castro, Manuel Fernández-Veiga,
and Ana Fernández-Vilas[(✉)]

atlanTTic - Research Center, Universidade de Vigo, Vigo, Spain
`{pedro.otero,dperezcastro,mveiga,avilas}@det.uvigo.es`

Abstract. The advancement of quantum computing threatens classical cryptographic methods, necessitating the development of secure quantum key distribution (QKD) solutions for QKD Networks (QKDN). In this paper, a novel key distribution protocol, Onion Routing Relay (ORR), that integrates onion routing (OR) with post-quantum cryptography (PQC) in a key-relay (KR) model is evaluated for QKDNs. This approach increases the security by enhancing confidentiality, integrity, authenticity (CIA principles), and anonymity in quantum-secure communications. By employing PQC-based encapsulation, ORR aims to avoid the security risks posed by intermediate malicious nodes and ensures end-to-end security. Our results show a competitive performance of the basic ORR model, against current KR and trusted-node (TN) approaches, demonstrating its feasibility and applicability in high-security environments maintaining a consistent Quality of Service (QoS). The results also show that while basic ORR incurs higher encryption overhead, it provides substantial security improvements without significantly impacting the overall key distribution time. Nevertheless, the introduction of an end-to-end authentication extension (ORR-Ext) has a significant impact on the Quality of Service (QoS), thereby limiting its suitability to applications with stringent security requirements.

1 Introduction

The development of quantum computing is rapidly evolving [1], with recent advancements in quantum computation that brings algorithms [2] like Shor's and Grover's closer to reality. These advances pose a potential risk to existing cryptographic systems [3], as a quantum adversary could eventually break the security of classical cryptography, especially asymmetric algorithms.

The scientific community has focused on two main approaches to address quantum threats: quantum cryptography (QC), which uses quantum systems for security, and post-quantum cryptography (PQC), which seeks algorithms resilient to quantum attacks. QC leverages quantum mechanics to ensure security, with Quantum Key Distribution (QKD) being a prominent solution [4], relying on quantum superposition or entanglement [5]. However, QKD faces implementation challenges [6]: The lack of commercial quantum repeaters or scalability constrains, both of which limit its range of applicability.

F. Barbaresco and G. François (Eds.): QUEST-IS 2025, CCIS 2743, pp. 115–124, 2026.
https://doi.org/10.1007/978-3-032-13852-1_13

In contrast, PQC does not require specialized hardware and offers a more practical, cost-effective solution, although it is based on complex mathematical problems that could eventually be broken by sufficiently powerful quantum computers [7]. NIST is leading the effort to standardize PQC algorithms, including CRYSTALS-Kyber, CRYSTALS-Dilithium, HQC or FALCON.

Despite the hardware limitations of QC, QKD remains the only solution proven to be unconditionally secure [8], proving to be safe even against quantum attacks. However, scalability remains a challenge, with models like key-relay (KR) [9] and trusted-node (TN) [10] not ensuring end-to-end confidentiality across intermediate nodes. Thus, a hybrid QC and PQC approach is required for comprehensive security in quantum key distribution networks (QKDNs).

This paper evaluates the feasibility of our previous work [11] —a secure key distribution approach based on the key-relay (KR) model, enhanced with Onion Routing (OR) [12] and PQC techniques to ensure CIA principles and anonymity— by comparing it with the main existing alternatives: key-relay and trusted-node models. The remainder of the document is structured as follows: Sect. 2 reviews existing QKDN security models, Sect. 3 outlines the technical background, Sect. 4 introduces the ORR model, Sect. 5 presents the experimental setup, and Sect. 6 provides a comparative performance analysis. Finally, Sect. 7 summarizes the findings, highlighting ORR(-Ext)'s advantages and drawbacks in security and QoS ending with the future lines of work to follow.

2 Related Work

In QKDNs without commercial quantum repeaters, the primary approaches to overcome distance limitations are KR and TN models. Both require fully trusted nodes, introducing security vulnerabilities [13]. Rass *et al.* [14] emphasize that network security is only as strong as the least trusted intermediate node. De Santis *et al.* [15] propose satellite-based configurations to reduce dependence on multiple intermediaries, assuming only one or two trusted nodes. Calsi *et al.* [16] introduce a KR enhancement where nodes establish QKD links with both nearest and next-nearest neighbors, improving resilience but remaining susceptible to multiple-node attacks. In TN models, Vyas *et al.* [17] suggest trust-level segmentation to relax security assumptions at intermediate nodes, though this requires additional infrastructure for trust verification.

To address malicious node threats, post-quantum cryptography (PQC) has been integrated into classical protocols [18]. Rios *et al.* [19] show that combining Kyber and Dilithium enhances classical crypto performance at high-security levels, albeit with increased traffic overhead. Hybrid QKDâĂŞPQC solutions are increasingly favored for critical infrastructure in the NISQ[1] era [20], as they

---

[1] Noisy Intermediate-Scale Quantum (NISQ) refers to nowadays quantum devices with tens to hundreds of qubits. They are powerful enough to surpass some classical simulations but remain limited by noise and lack of full error correction, making them a transitional stage before fault-tolerant quantum computers.

offer both long and short-term protection. One approach uses QKD to encrypt PQC-based key exchanges [21].

Building on this trend, we adopt a hybrid method combining KR-based QKD with OR and PQC, enabling secure key distribution between distant nodes, as introduced in our earlier work [11].

3   Technological Background

QKD and PQC derive security from fundamentally different principles. QKD offers unconditional security based on quantum mechanics, while PQC achieves computational security against quantum attacks. While QKD ensures point-to-point confidentiality, PQC excels in speed and scalability. By combining both, one can build resilient systems for both short- and long-term security. Our method incorporates OR and PQC-encrypted keys into a KR-based QKDN, achieving layered protection and preserving end-to-end confidentiality.

3.1   Key-Relay Quantum Key Distribution Networks

In KR networks, a sender and receiver share a secret key S via a path of intermediate QKD-enabled nodes. The sender generates S using a QRNG, encrypts it with the shared quantum key, and transmits it through the network. Each node decrypts and re-encrypts S for the next hop. Although simple and effective, this model exposes S to every intermediate node, compromising confidentiality. KR is also vulnerable to spoofing on classical channels. Although intercepting a quantum key is unlikely without compromising a QKD node, such attacks can still disrupt network operations.

3.2   Onion Routing

OR is a layered encryption technique designed to anonymize communication paths. A message is encrypted in successive layers, starting with the public key of the final node and moving backward to the first one. Each node decrypts one layer, learns only the next hop, and forwards the message, preserving sender-receiver anonymity. OR enhances privacy but introduces higher latency and exposes the exit node to potential attacks. Some extensions [22] improve security by verifying message integrity end-to-end, defending against message tampering and overload-based denial-of-service (DoS) attacks on onion routers, however, increasing the computation and the size of the message.

4   Model Principles

The proposed model, Onion Routing Relay (ORR), guarantees unconditionally secure key distribution within QKDN. It builds on the KR model for QKDN and integrates an OR extension that ensures end-to-end authentication. This approach upholds CIA principles and anonymity of the destination node. Figure 1 illustrates the basic operation of ORR.

Since classical OR protocols remain vulnerable to quantum attacks, the model replaces insecure classical cryptography —particularly asymmetric crypto- graphy— with PQC. Rather than using public-key cryptography to encrypt the onion layers, ORR uses PQC Key Encapsulation Mechanism (PQC-KEM) algorithms to distribute keys among nodes. It then applies symmetric encryption algorithms, such as AES,[2] to create the encrypted layers (onions). The ORR model is detailed described in our previous work [11].

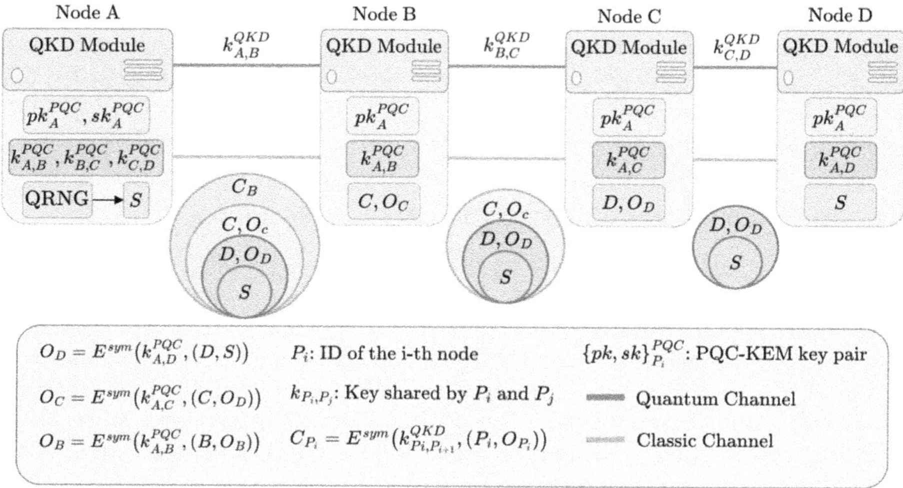

$$O_D = E^{sym}(k_{A,D}^{PQC}, (D, S)) \qquad P_i: \text{ID of the i-th node} \qquad \{pk, sk\}_{P_i}^{PQC}: \text{PQC-KEM key pair}$$

$$O_C = E^{sym}(k_{A,C}^{PQC}, (C, O_D)) \qquad k_{P_i,P_j}: \text{Key shared by } P_i \text{ and } P_j \qquad \text{Quantum Channel}$$

$$O_B = E^{sym}(k_{A,B}^{PQC}, (B, O_B)) \qquad C_{P_i} = E^{sym}(k_{P_i,P_{i+1}}^{QKD}, (P_i, O_{P_i})) \qquad \text{Classic Channel}$$

Fig. 1. Simplified example of the Onion Routing Relay key distribution model. In this scheme, it can be seen a circuit formed by three nodes (B, C and D) in addition to the initiator node (A). Node A possesses a KEM key-pair (pk_A^{PQC}, sk_A^{PQC}), which is used to compute shared keys with the nodes in the circuit ($k_{A,B}^{PQC}$, $k_{A,C}^{PQC}$, $k_{A,D}^{PQC}$). Using these keys, along with the QKD key shared with the first intermediate node B ($k_{A,B}^{QKD}$) A is able to create the onion ciphertext that then sends to B, thereby the ORR key distribution.

5   Test Scenario and Implementation

To compare the different models, a series of scripts were developed in C to simulate their behavior. A GitHub repository[3] is available to readers, providing

---

[2] Due to Grover's algorithm halving the security of symmetric cryptography, AES-256 should be used in a post-quantum world to maintain the same security level as AES-128 today.

[3] https://github.com/pedrotega/ORRvsTNvsKR.

the source code and execution instructions. It is important to note that due to the performance differences between the ORR model and its extended version with end-to-end authentication (ORR-Ext), it has been decided to evaluate each approach separately.

For the experiment, only a pair of QKD nodes was available –specifically the Cerberis QKD XGR [23] developed by ID Quantique– which provided the quantum functionalities for the simulations. Each simulated node in the scripts requests QKD keys from one of the real nodes, which works as a key-managment-system, and retrieves the key by its identifier from the other real node. All the request follows a RESTful API format specified in ETSI-014 [24].

In the code, each node is executed in a different thread that allows the execution of code concurrently or sequentially, as appropriate, at any given time. Since a QRNG was not available for this project, a PRNG (Pseudo-Random Number Generator) was used as a replacement. Table 1 lists the algorithms and functions employed in the implementation, along with the respective software suppliers.

Table 1. Cryptographic Algorithms and Their Implementations

Category	Implementation	Category	Implementation
Symmetric Encryption	AES-256-CBC (OpenSSL)	PQC-KEM	Kyber-768 (LibOQS)
PRNG	RAND_bytes() (OpenSSL)	XOR	Directly in C

The parameters measured in the tests include the encryption time and the key distribution time, which is calculated from the moment the secret is generated at the initiator node until it reaches the destination node. For the encryption time, the measurement varies depending on the model: In ORR the longest encryption time in this is spent encrypting the initial onion at the initiator node. In the version with authentication end-to-end (ORR-Ext) the time to create the authentication extension is included. For the TN model, the encryption time is measured during the calculation of the ciphertext to be sent to the destination node at the trusted node. Finally, in the KR model it is only necessary to measure the time of XOR operation when encrypting the secret in the initiator node.

For the implementation of ORR-Ext, the HMAC-256 algorithm was employed as signature scheme. This choice is motivated by the fact that it is still considered secure even in the presence of Grover's algorithm, offering post-quantum resistance without the overhead associated with PQC alternatives. In contrast to HMAC, PQC signature schemes are asymmetric –which necessitates the inclusion of the public key in each message– significantly increasing the overall message size. Table 2 presents a comparison of the message lengths in both ORR and ORR-Ext models under different signature algorithms, highlighting the trade-offs in communication efficiency introduced by PQC signature methods.

Table 2. Ciphertext length (Bytes) for a circuit of 5 nodes

Variant	Onion	Public Key	Signature	Ciphertext
ORR	416	0	0	432
ORR-Ext-HMAC-256	416	0	32	1472
ORR-Ext-Falcon-1024	416	1793	1280	16612
ORR-Ext-Dilithium3	416	1952	3293	27537

6 Results and Discussion

The results presented in this section have been obtained by averaging 100 iterations with each simulated model for circuits[4] with 3, 5, 7, 9, and 11 nodes. Figure 2 illustrates the average time required for the encryption procedure in each model. The KR model exhibits a highly consistent and minimal encryption time, ranging from 1.5 to 1.93 μs, remaining largely unaffected by the number of nodes due to its simple XOR-based encryption using QKD keys. The TN model shows a moderate increase, starting at 4.76 μs for a 3-node circuit and reaching 29.7 μs at 11 nodes, as encryption requires handling more ciphertexts with each additional node. The ORR model, which implements layered encryption, maintains moderate growth in encryption time, starting at 33.4 μs and increasing up to 47.25 μs with 11 nodes. Meanwhile, the ORR-Ext model introduces a significant overhead due to the additional operations required to ensure end-to-end authentication. It starts at 238.78 μs for 3 nodes and escalates sharply to 5327.31 μs for 11 nodes.

Regarding the key distribution time, represented in Fig. 3, the growth trend across models reflects the increasing communication and coordination overhead as the number of nodes rises. The KR model again stands out as the fastest, beginning at 167.7 μs for 3 nodes and reaching 272.31 μs for the largest circuit. The TN model, which starts slightly above KR at 224.24 μs, experiences more dramatic growth, requiring 862.55 μs at 11 nodes. The ORR model, while initially slower than KR and TN (227.62 μs), maintains a steadier increase, finishing at 389.02 μs. In contrast, the ORR-Ext model incurs the highest key distribution times due to the authentication processes involved and the major length of their ciphertexts, with the average time rising from 465.8 μs for 3 nodes to 6571.89 μs for the 11-node configuration.

From the results obtained in the previous section, it can be concluded that the inclusion of layered encryption and authentication mechanisms results in significantly different performance profiles among the four models. As expected, the ORR-Ext model introduces the most substantial encryption overhead, requiring several orders of magnitude more time compared to the TN and KR models. Specifically, the encryption time in ORR-Ext can be more than 1800 times higher than in KR and over 150 times higher than in TN for the 11-node circuit. This

---

[4] A circuit is formed by the intermediates and destination nodes.

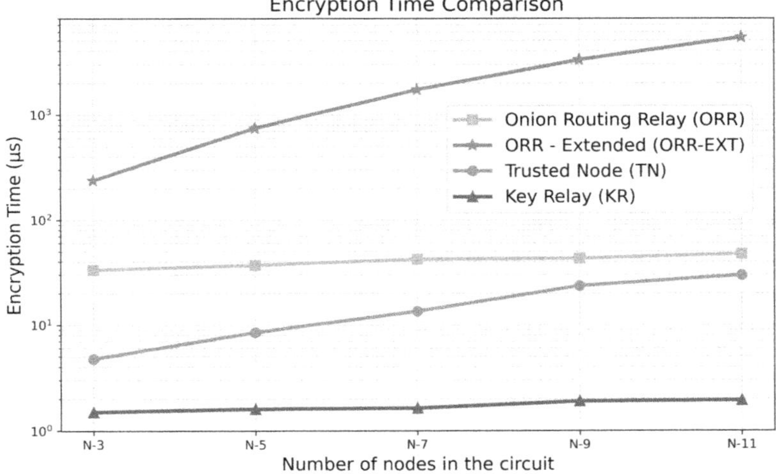

Fig. 2. Encryption time time comparison of the different models.

is due to the added authentication extension operations needed to guarantee end-to-end security, which accumulate across layers.

The TN model shows a moderate but clear growth in encryption time as the number of nodes increases, stemming from the need to aggregate and process ciphertexts at the trusted node. This processing overhead is absent in the KR model, which maintains consistently minimal encryption times, as its XOR operation does not depend on the circuit size. The basic ORR model without the authentication extension, while implementing layered encryption, shows a more controlled increase, demonstrating its balance between added security and computational efficiency compared to ORR-Ext.

Nevertheless, the most impactful distinction across models lies in the key distribution time, which directly affects the system's QoS. The ORR-Ext model, in particular, suffers from very high distribution times due to both the authentication-related message exchanges and the context switching between threads simulating node-to-node communication. These results suggest that in practical deployments, where network delays and message propagation times are even more significant, the additional burden introduced by ORR-Ext could represent a limiting factor in terms of responsiveness. TN also exhibits a steep rise in key distribution time, particularly for larger circuits, primarily due to the additional communication step required to forward the key to the final destination.

In contrast, the KR and basic ORR models display a more favorable trend, especially at higher node counts. Notably, while basic ORR's encryption time is higher than that of KR, its key distribution time remains relatively competitive. For example, in the 11-node case, the difference between ORR and KR in key distribution is approximately 117 μs, which is consistent with the expected com-

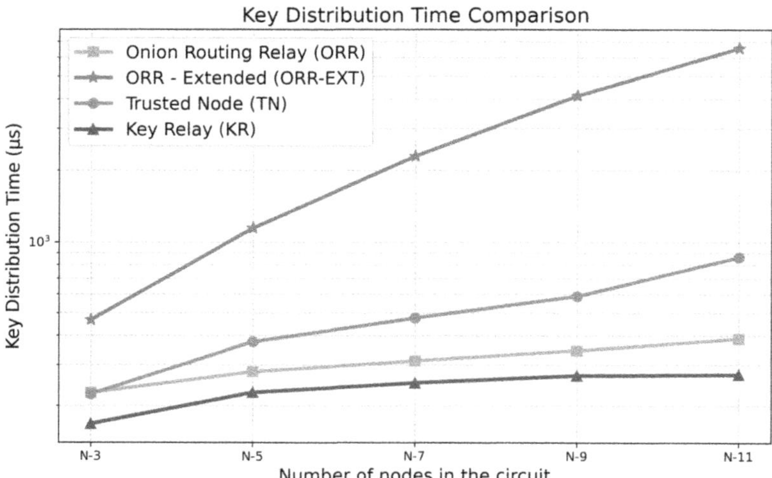

Fig. 3. Key distribution time comparison of the different models.

munication overhead introduced by layered routing, yet still acceptable within practical bounds.

Therefore, it can be concluded that although ORR-Ext significantly enhances security guarantees through authentication, it does so at the cost of scalability and performance. In realistic environments where encryption times become negligible compared to network latency, models like ORR—despite their layered encryption—can achieve a competitive QoS while offering enhanced protection against malicious intermediaries, presenting a viable compromise between efficiency and security.

7 Conclusions and Future Work

In this paper, we have proposed and tested a key relaying solution based on the principles of onion routing, to ensure the anonymity of end-to-end keys and the sequence of intermediate nodes. Our evaluation shows that the cost of nested PQC encryption with an ORR scheme is high but affordable if the extra level of security in relaying must be achieved. Therefore, ORR and ORR-Ext both strengthen the security of key relaying in QKD networks at the cost of some extra bandwidth only. Future work will focus on implementing these models in a real QKDN to evaluate whether the proposed solutions can meet the requirements of practical deployment scenarios. In particular, the objective is to determine whether ORR-Ext can sustain a minimum key distribution rate that would make it suitable for highly security-sensitive applications.

Acknowledgments. This work was supported under the grants TED–2021–130369B–C31 funded by MICIU/AEI/ 10.13039/501100011033 by the "European Union NextGenerationEU/PRTR"; PID2023-148716OB-C31 funded by MICIU/AEI/10.13039/501100011033; grant ED431B 2024/41 (GPC) by the Galician Regional Government; and the regional agreement for laboratories and demonstration centers in cybersecurity within the RETECH program.

References

1. Aasen, D., et al.: Roadmap to fault tolerant quantum computation using topological qubit arrays (2025). https://arxiv.org/abs/2502.12252
2. Jordan, S.P., Liu, Y.-K.: Quantum cryptanalysis: Shor, Grover, and beyond. IEEE Secur. Priv. **16**(5), 14–21 (2018)
3. Syed, S.A.: The quantum threat: preparing for the impending impact on cyber security. Int. J. Eng. Technol. Res. Manage. (IJETRM) **7**(03) (2023)
4. Bennett, C.H., Brassard, G.: An update on quantum cryptography. Springer, Berlin Heidelberg (1984)
5. Ekert, A.K.: Quantum cryptography based on bell's theorem. Phys. Rev. Lett. **67**(6), 661 (1991)
6. Stanley, et al.: Recent progress in quantum key distribution network deployments and standards. J. Phys.: Conf. Ser. (2022)
7. Alagic, et al.: Status report on the fourth round of the NIST post-quantum cryptography standardization process. National Institute of Standards and Technology, Tech. Rep. NIST IR 8545 (2025)
8. Scarani, V., Bechmann-Pasquinucci, H., Cerf, N.J., Dušek, M., Lütkenhaus, N., Peev, M.: The security of practical quantum key distribution. Rev. Mod. Phys. **81**(3), 1301–1350 (2009)
9. Elliott, C.: Building the quantum network. New J. Phys. **4**(1), 46 (2002)
10. International Telecommunication Union, Quantum key distribution networks – Key management, ITU-T, Tech. Rep. Y.3803 (2020)
11. Otero-García, P., et al.: Onion routing key distribution for QKDN. arXiv preprint: arXiv:2502.06657 (2025)
12. Goldschlag, D., Reed, M., Syverson, P.: Onion routing. Commun. ACM **42**(2), 39–41 (1999)
13. Huttner, et al.: Long-range QKD without trusted nodes is not possible with current technology. NPJ Quantum Inf. (2022)
14. Rass, S., Mehic, M., Voznak, M., König, S.: Hacking the least trusted node: indirect eavesdropping in quantum networks. IEEE Access (2024)
15. De Santis, et al.: Parallel trusted node approach for satellite quantum key distribution. arXiv preprint: arXiv:2406.08562 (2024)
16. Calsi, D.L.: End-to-end QKD network with non-localized trust (2025). https://arxiv.org/abs/2411.17547
17. Vyas, N., Mendes, P.: Relaxing trust assumptions on quantum key distribution networks. arXiv preprint: arXiv:2402.13136 (2024)
18. Shim, H., Kang, B., Im, H., Jeon, D., Kim, S.: qTrustNet virtual private network (VPN): enhancing security in the quantum era. IEEE Access (2025)
19. Rios, R., Montenegro, J.A., Muñoz, A., Ferraris, D.: Towards the quantum-safe web: benchmarking post-quantum TLS. IEEE Netw. (2025)
20. Zeng, et al.: Towards efficient and secure quantum-classical communication networks. In: 2024 IEEE 6th International Conference on Trust, Privacy and Security in Intelligent Systems, and Applications (TPS-ISA), pp. 520–523. IEEE (2024)

21. Djordjevic, I.B.: Joint QKD-post-quantum cryptosystems. IEEE Access **8**, 154708–154712 (2020)
22. Kuhn, C. Beck, M., Strufe, T.: Breaking and (partially) fixing provably secure onion routing (2019). https://arxiv.org/abs/1910.13772
23. ID Quantique, Cerberis XGR QKD system (2021). https://www.idquantique.com/quantum-safe-security/products/cerberis-xgr-qkd-system/
24. ETSI, "Quantum Key Distribution (QKD); Protocol and data format of REST-based key delivery API," European Telecommunications Standards Institute, Tech. Rep. (2019)

Complete Centralized Architecture for QKD Networks

Anne de Chastenet[1]($\boxtimes$), Eric Augé[2], Gonzague Reydet[2], Maxime Coursimault[2], Philippe Legros[1], and Stephanie Molin[3]

[1] Thales, Thales Systemes d'information et Communication Securises, 4 Avenue des Louvresses, 92230 Gennevilliers, France
`anne.dechastenet@thalesgroup.com`
[2] Thales, Thales Systemes d'information et Communication Securises, 110 Avenue Leclerc, 49309 Cholet, France
[3] Thales, Thales cortAIx Labs, 1 Avenue Augustin Fresnel, 91767 Palaiseau, France

Abstract. We propose a complete architecture for a QKD (Quantum Key Distribution) network, from the physical layer up to the network controller. Our design is based on a centralized approach for Quantum Keys Management to mitigate security risks on the trusted relays.

This paper explores the essential components, design principles, and architectural models of QKD networks, which are critical in enabling secure information ex-change amid rising cybersecurity concerns. Security within QKD networks is paramount, encompassing measures to address various threats. This includes the establishment of quantum-safe classical networks, rigorous node authentication, and dynamic eavesdropping countermeasures. Moreover, our architecture accommodates scalability and interoperability to integrate new nodes and technologies seamlessly. All these features are addressed in this paper, with a focus on the centralized Key Management System (KMS) we designed to drive QKD.

Keywords: QKD · Trusted Node · centralized · XOR · security · ITU-T Y.3803

1 Introduction

Quantum Key Distribution (QKD) elementary links on optical fiber are currently limited to a distance of few tens of km due to fiber propagation losses. Therefore, QKD Networks (QKDN) use Trusted Nodes (TN) to digitally relay the shared secret key. QKD keys are established on each elementary link and stored in the TN; a final key (KSA key) is then elaborated end-to-end using the QKD keys generated along the path. Key Management (KM) is therefore at the heart of QKDN. Several acknowledged schemes regarding the key transfer technique are provided by Rec. ITU-T Y.3803 [1].

The first two usual schemes, "classical key transfer" or "key forwarding", consist of a "decentralized approach". The key is generated at one end-node (User Node, UN) or chosen to be the QKD key shared with its Access Node to the QKDN (AN) and forwarded hop-by-hop using the One-Time-Pad (OTP) encryption technique (Information-Theoretic-Secure process) to the other end node. In the decentralized approach, the key is

F. Barbaresco and G. François (Eds.): QUEST-IS 2025, CCIS 2743, pp. 125–135, 2026.
https://doi.org/10.1007/978-3-032-13852-1_14

thus accessible inside each node along the path, raising the need for duly and expensively protected trusted relay nodes.

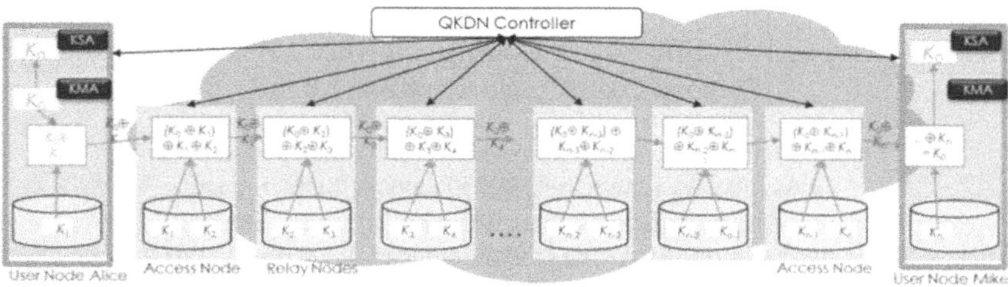

Fig. 1. One classical approach of ITU for key transfer (as per Fig. 6 of ITU-T Y.3803). Each Node includes a local Key Manager, to store KMA-keys and transfer/deliver KSA-keys.

In alternative schemes (Figs. 7 and 8 of Rec. ITU-T Y.3803), the key to share is not forwarded to nodes along the path. Instead, QKD keys generated along the elementary links are used to provide an end-to-end key. The second of those 2 schemes is based on a centralized architecture, with an additional Key Management Agent, part of the KMS (Fig. 1).

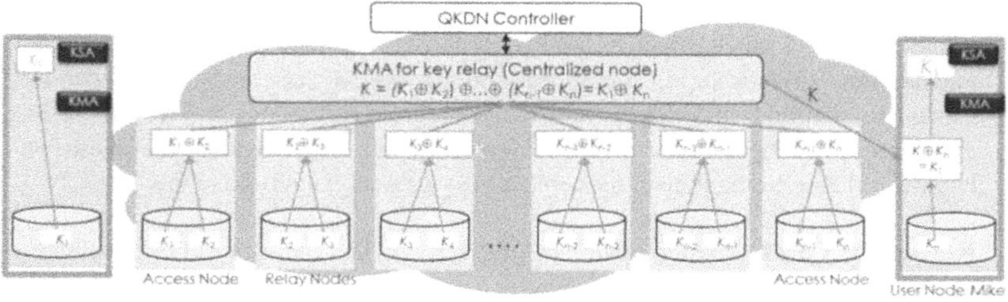

Fig. 2. Centralized approach of ITU for key transfer (as per Fig. 8 of ITU-T Y.3803).

In this configuration, only the "access node" of user A knows directly the key, drastically reducing security weaknesses of the whole network (and the AN of user B could deduce the key, not directly but if it can intercept the contribution sent by user B).

No full and mature standard is available for the moment for QKDN, especially for the centralized approach of key transfer (Fig. 2) which is auspicious to enhancements. Characteristics of this new approach need to be grasped in new perspectives. In the following, we sketch design concepts for an enhanced centralized QKD Network and call it "centralized XOR node" approach.

2   Main Components of Key Management in Centralized Approach

Key Management System (KMS) is the KM layer, a specificity of QKDN (Fig. 2). In the "centralized XOR node" approach, it includes local and centralized instances:

Local KMS (LKMS):

Each Trusted Node hosts a LKMS, including a Key Management Agent (KMA) and a Key Supply Agent (KSA). The KMA formats and stores as KMA-keys the keys generated by QKD on the quantum links with its adjacent nodes, whereas the KSA interfaces with the key consumer and delivers final keys, KSA-keys.

A Centralized KMA for key relay:

The centralized entity in Fig. 2 is not explicitly described by the ITU. We chose to implement it in two distinct instances: the Central Key Management Core (CKMC) and the "XOR node". The CKMC is a key stone within the centralized XOR architecture, behaving like a central hub between all modules (QKDN Manager, QKDN Controller, LKMS). Its primary functions are to trigger the transfer of XORed keys from LKMSs to the centralized "XOR node" and elaborate the metadata of the key shared by user node (Alice) with another (Mike). The CKMC actuates also the centralized "XOR node" in charge of the global XOR computation and its delivery to Mike.

Management and Control functions are similar to the ones of classical networks.

The QKDN Manager is in charge of the configuration and constant monitoring of all network components. It ensures the network's health and stability.

As well, it initiates and automates the configuration of the KMS, i.e. the CKMC, the central XOR node and the LKMS. The QKDN Manager provides the static network topology to the QKDN Controller for the key routing path computation and allowances according to user's SLA and is in charge of the billing function.

The QKDN Controller computes end-to-end route across the QKDN in response to a route request, issued by the CKMC, which should be the optimal path to trigger.

For this computation, the QKDN Controller requires the complete view of the QKDN topology (static and dynamic, like link failure for instance) as well as key resource availabilities per node (key volume in real time, per link to be more specific).

3 Enhancements of the "Centralized XOR Node" Approach

In our approach, the CKMC and "XOR node" could be distinct entities, installed in the Network Operation Centre (NOC). Collocating the NOC with a QKD node is beneficial for authentication and encryption of its exchanges.

The CKMC does not host keys but is in charge of gathering key metadata from KMA storages in the various nodes of the network (metadata, incl. Key IDs, key hash and key length). It is also in charge of checking consistency of storages of all nodes and of providing synthetical information to the QKDN Controller and Manager.

The "XOR node" only receives individual XOR computation from each LKMS and computes a resulting XOR bit string from which Mike will deduce Alice's key.

In this design, the numerous inter-node KMA links (in meshed networks) are replaced by only one KMA link per node connected to the central one. In addition, KSA links between adjacent nodes are useless. Our approach drastically reduces the complexity, optimizing the number of links, especially for architecture going at scale.

Unlike forwarding schemes, this centralized scheme is prone to enhancements:

Simultaneous actuation of all nodes; Once the path has been provided to the CKMC by the QKDN Controller, the CKMC can immediately and simultaneously ask each node on the path to send the XOR of chosen KMA keys with its 2 adjacent nodes.

Such parallel process minimizes the transfer delay, which is not increasing with the number of nodes of the path, but also the probability of flop, as failure of a node before it could transfer the key is much more unlikely than in classical hop-by-hop transfers, and as, all KMA keys of the path being reserved simultaneously, no conflict between various requests for keys in the network will occur.

Symmetrisation of process, for a doubled capacity of transfer;

The XOR node delivers a resulting string to Mike, which is the XOR of Alice's key and Mike's key. Our XOR node sends this same string to Alice, so that she can get Mike's key: they share 2 keys instead of 1, sparing transport keys.

Centralization of storage and elaboration of metadata, for a better system check and reduction of metadata transfer and divulgation;

Full metadata of the transferred KSA key are sensitive: they give attackers too much information on the transfer process (between which addressees it is shared, which nodes and KMA keys were used...). In classical key transfer, they usually are elaborated and transferred progressively with the key, which is a vulnerability.

The CKMC will easily elaborate and archive full metadata of transferred keys and provide addressees with only non-sensitive synthetical metadata (key Id, key length, computed key age, computed end-to-end security parameter).

Limitation of information on a need-to-know basis, for a mitigation of risks due to attacks from a specific entity in the QKD network that would turn out to be malicious;

For instance, the QKDN Controller is asked by the CKMC to find a path between access nodes and does not know about the concerned users' nodes.

Relay nodes do not know about the concerned users' nodes: they are only requested by the CKMC to send the XOR of KMA key with Id xx and KMA key with Id yy, with tag MsgId to the XOR Node. The XOR node knows which nodes send him contributions but does not know about their role in the path: access or user or pure relay nodes (Fig. 3).

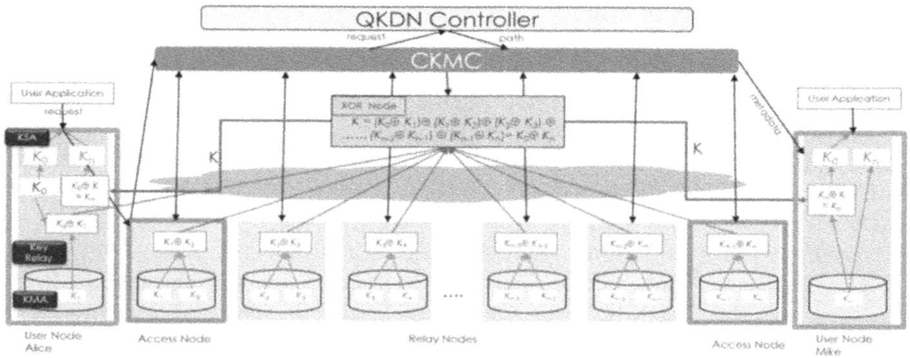

Fig. 3. Simplified process and exchanges of a key request

Mitigating the Risk on Relay Node Compromission;

Security of Users' and Access Nodes is paramount. They correspond to the definition of fully Trusted Nodes. Notably, their exchanges with CKMC and XOR node will be encrypted and authenticated.

Accessing a relay node does not give enough information, the attacker must also identify the other nodes involved in the key transfer, intercept their communications with the XOR node, be able to correlate them with the actual transfer, deduce each KMA key involved to retrieve the transferred key. Such an attack would be all the more complicated since XORs sent by the access (and users) nodes are encrypted.

Possible additional mechanisms, as a 3rd centralized entity for parallel and independent security features, for further in-depth protection (Patented by Thales);

Ease of hiding the transfer process to external attackers;

The CKMC affects different various Message Identifiers for the responses of the various nodes, to make traffic observers unable to correlate these responses to a same key transfer. Only the XOR node has the adequate list of these MsgIds.

Use of a QRNG (Quantum Random Number Generator) key generated at Alice's node;

It can be used as KSA key instead of a KMA key shared with her access node so that the access node is, in this case, unaware of the final KSA key (the interest of this choice has to be evaluated in some implementations where users' nodes are chosen to be unknown to the XOR node) or combined with quantum keys to resist to attacks on integrity (patented by Thales).

4   Risk Mitigation in a Centralized QKD Network

Note that we consider Quantum Communication Infrastructures which deal with one unique sensitivity level, as mixing various sensitivity levels would induce potential security risks.

One of the security loopholes in the non-centralized architecture is the key transportation. Indeed, the confidentiality of the QKD key exchanged between two neighbours is preserved. It is no longer the case when a third-party node is located in the middle of the path. The key to share goes through this node in plain text or at best encrypted with a local key. So, the third party has access to the key. It becomes a greater problem when the third party is located in a foreign country. The centralized architecture prevents this need-to-know issue. All parties involved in the key transportation know only the key exchanged with its neighbourhood (which is intrinsically the case in a QKDN).

4.1   Authentication

The security brought by QKD technology is based on two principles:

- The key material exchange performed by the quantum layer;
- The authentication of the classical link for key negotiation.

The key material exchange is recognized to have Information-Theoretic Security (ITS). In order to keep this property, the authentication should get it too. ITS can be obtained by pre-shared keys (for the very first time) renewed by reserving exchanged QKD keys for it or by loading new shared key by a trusted-courier.

European National Cyber-Security Agency wrote a common paper [2] saying that to preserve ITS property, the Service Application should use ITS algorithm when it uses the QKD keys, One-Time-Pad for encryption etc. This will be possible only for very sensitive communication with low volume of transmitted data.

Therefore, another axis for authentication may be the use of Post-Quantum Cryptography (PQC). Indeed, authentication is not concerned by the threat Store Now and Decrypt Later (SNDL) and PQC covers Quantum Computer threats. PQC implementation can provide additional security by encrypting the exchanges too.

If some elements of the QKDN cannot be connected with a quantum layer; QKDN Manager, QKDN Controller, CKMC etc.; PQC will have to be used for the protection of their exchanges (authentication and/or encryption).

4.2 Complexification of Attacks

The security of the nominal centralized architecture can be easily improved.

Limiting information on a need-to-know basis: e.g.

- Nodes do not know for whom they are transferring a key, the CKMC only knows the service applications involved and asks for a route between the nodes where the applications are connected without telling the subscriber Id. This improves the confidentiality in case of remote spying on the network.
- msgIds are all different, when the CKMC sends the XOR request, it uses a unique msgId for each XOR request/answer and gathers them on the same internal ID. Indeed, in case of one node compromission, the XOR sent from the compromised node to the CKMC can be decrypted with the local keys and the traffic interception of other involved nodes can help to get the final transferred key (getting one key after the other from the intercepted messages). Without common Id, the correlation is harder to be done, the XOR messages can be considered as a part of a whole exchange from external view.

  Adding layers: e.g.

- Addition of a "Secret Layer" on top of the described design can also be easily implemented in a centralized way, as an additional security service (Thales patent) and a way to provide protection against errors inserted by malicious or failing QKDN nodes or XOR node (Thales patent).
- Splitting centralized KMA into CKMC and XOR node and limiting their exchanges on a need-to-know basis will prevent a malicious XOR node to collaborate with a corrupted access node for attacks

Dummy messages: in order to disturb traffic interception, dummy requests can be sent from CKMC to LKMS. The fake XOR sent by LKMS is not used by CKMC but can corrupt the result of the spy.

5 Results and Discussions

5.1 Security Assessment

A ground-breaking aspect of QKD Security is that it is mastered via a security parameter, **ε-security**, as an upper bound for the probability of the QKD key to be wicked, (either too much information available to Eve or errors in key deduced by Bob). This notion has to be extended when several QKD links intervene for the transfer. Let's note:

- **P** the probability of a transferred key to be wicked (errors or compromised)
- $\mathbf{n_j}$ the probability for node j to be hacked (all data and operations performed by this node are known by the attacker or can be corrupted)
- $\mathbf{p_j}$ the probability for a KMA-key of link j to be wicked (including ε-security parameter of QKD but also probability of compromission in storage).
- $\mathbf{q_j}$ the probability that transmissions from node j can be understood by a hacker (including the ability to correlate this message with a given key transfer)
- $\mathbf{a_j}$ the probability of transmissions from node j being altered (by hackers or because of transmission errors)
- Nodes 0 and n are the Users' Nodes (UN), Nodes 1 and (n-1) are the Access Nodes (AN), Nodes 2 to (n-2) are the relay nodes on the path
- K_0 is the KSA key to be transferred, and generated at the user's node

In the case of classical transfer, the KSA key is not wicked if and only if none of the steps in the transfer is wicked (notably, one hacked node is enough for that). Thus,

$$P = 1 - (1 - n_n).\prod_{j=0}^{n-1} (1 - n_j).(1 - a_j).(1 - p_j.q_j)$$

P increases with the number of nodes.

In the case of the centralized design, the KSA key is wicked when one transfer KMA key is wicked and the attacker also has knowledge of the nodes in the path and accesses all k consecutive XORs sent to the XOR node, deduces all other keys to get K_n or K_1 and is able to get either the XOR sent by Alice or the one received by Mike to get K_0. Thus, the probability is bounded by:

$$P = 1 - (1 - n_n).(1 - a_n).\prod_{j=0}^{n-1} (1 - n_j).(1 - a_j).(1 - p_j.r_j)$$

$$r_k = \prod_{j=0}^{k} q_j + \prod_{j=k+1}^{n} q_j - \prod_{j=0}^{n} q_j \qquad \text{for } 0 \leq k \leq n$$

The r_k coefficients moderate the increase of P with the number of nodes.
Note: this probability P is an upper limit. With the centralized approach,

- a wicked relay node would not be enough to get the KSA key: the nj values for relay nodes should be decreased to consider only corruption of data.
- the knowledge of the nodes in the path is all the more difficult to get for an attacker as no full metadata transit and spying the exchanges of CKMC or XOR node can be fruitless if those exchanges are subtle (e.g. with different msgIds). A probability for such a knowledge should also be set (instead of 1 in the above formula).

As it will be seen, this upper bound is already adequate for the comparison.

The probability of a transferred KSA key to be wicked highly depends on assumptions on the elementary probabilities, as can be shown on the following figures.

For the first case, we assumed 10^{-6} for n_j, 10^{-4} for p_j, 10^{-4} for a_j, and 10^{-1} for q_j. And, when vulnerable relay nodes, 10^{-5} for n_j, 10^{-3} for p_j, and 1 for q_j (Figs. 4, 5).

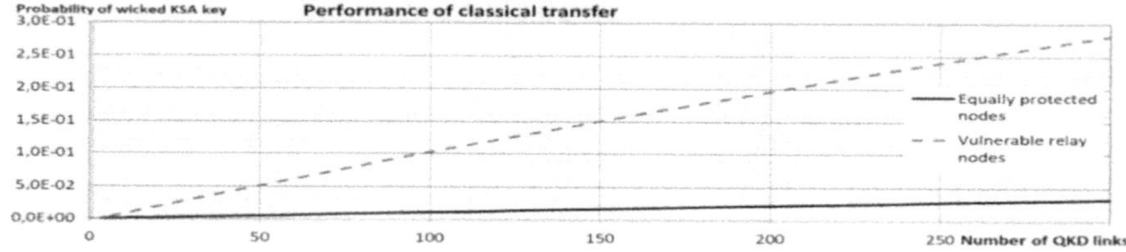

Fig. 4. Performances of classical scheme

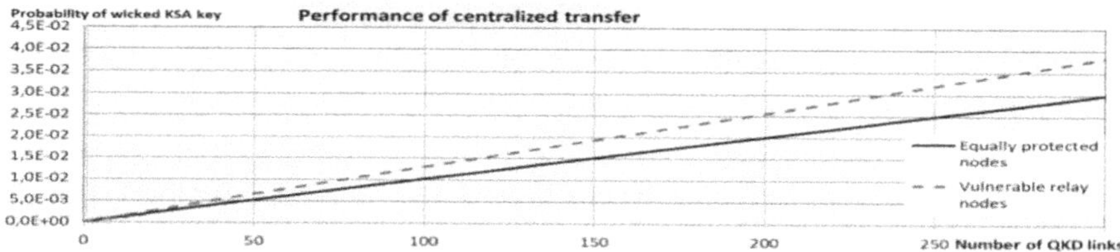

Fig. 5. Performances of the Centralized XOR node approach

For the second case, we assumed 10^{-5} for n_j, 10^{-3} for p_j, 10^{-4} for a_j, and 5.10^{-1} for q_j. And, when vulnerable relay nodes, 10^{-4} for n_j, 3.10^{-2} for p_j, and 1 for q_j (Figs. 6 and 7).

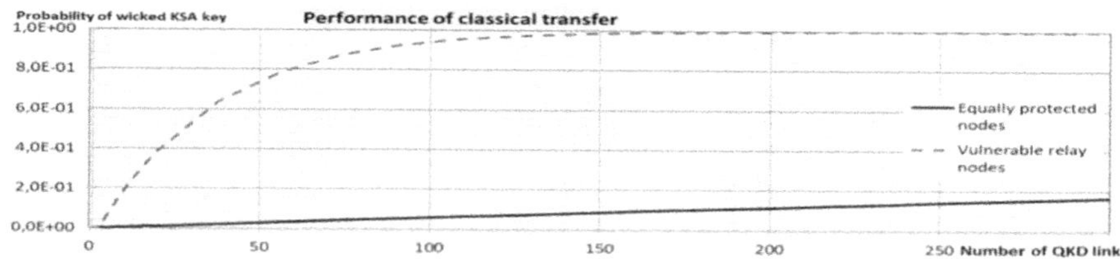

Fig. 6. Performances of classical scheme with less protection

In a situation of all nodes having same protection characteristics, the probability of a wicked KSA key is always decreased with the centralized approach (factor respectively 1.2 and 5 for the two cases here above) with respect to classical key forwarding.

In cases where relay nodes are less protected than other nodes (user and access nodes), degradation in performances of classical transfer as a function of total number of nodes is drastic whereas security of centralized transfer is far less affected.

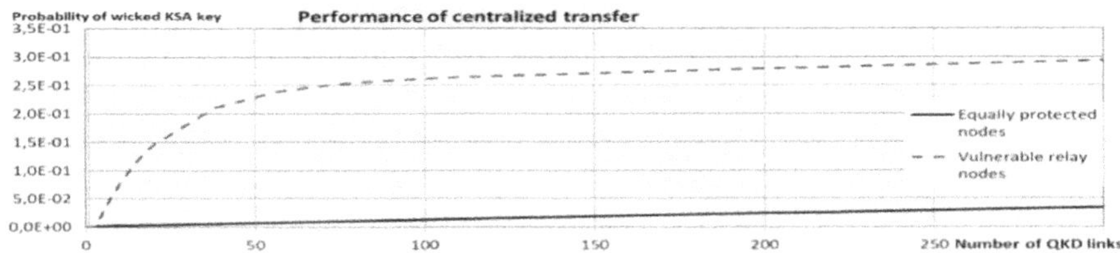

Fig. 7. Performances of centralized scheme with less protection

The comparison of performances of classical and centralized schemes shows drastic advantage of the basic centralized scheme. We did not include yet the impact of the enhancements proposed with our approach, that would increase the advantage.

5.2 Discussion

About the Security Evaluation;

The theoretical evaluations here above show how security is impacted by the hypothesis of vulnerability of the various nodes or of their communication means. This view demonstrates the advantage of the "Centralized XOR node" approach.

The values of the various p_j, n_j, q_j and a_j have to be assumed. They correspond to the probability that vulnerabilities in the system can be exploited, which may be hard to quantify: for instance, p_j and n_j also depend on the confidence we have in the system provider, the system certification or in the system administrators and q_j and a_j depend on the confidence about the links protection.

Assumptions have to be made, depending on the protection level that is considered, for physical access, certification or sophistication of encryption and authentication processes. Then, theoretical computations can be refined to determine impact of enhancements (not quantified yet) and trade-offs between security elements design such as:

– Generation of key with a user QRNG
– Hiding identity of Alice and Mike to the XOR node.

About Foreseen Drawbacks and Advantages of the Centralized Solution;

Load of the CKMC or the XOR node in case of high traffic requests may be a question and should be checked with a simulation of the complete QKD Network. However, this load should be lower than the load of the QKDN Controller that performs path computations. It is to be noted that the principles we described here are also applicable whenever these functions are distributed to a cluster of CKMC or XOR nodes (for instance, considering a geographic or role repartition), that would allow for load balancing and redundancy, provided the inclusion of adequate protocols to run this distribution.

Authentication of CKMC and XOR node exchanges is mandatory (QKDN Controller exchanges should also be authenticated, in any design) to avoid attacks and data alteration. That is what the XOR node and the CKMC entities are preferably users of the QKD network, so that their authentication to each user node can be easily updated with KSA keys.

Duration of key transfer, which is a very important system performance, is minimal and does not increase with the number of nodes, with the centralized approach.

Management of metadata is also eased and more secure, as no transfer of full metadata is necessary. Users will be sent only synthetized metadata whereas full metadata will be centralized managed and archived for potential revocation actions.

About perspectives;

A cluster of "centralized" entities instead of one, with back-up functions, is also a way of dealing with availability and of avoiding single points of failure and shall be considered in future operational QKD Networks.

Splitting responsibilities and knowledge of information can also be helpful for security.

Extending the concept of "centralized XOR node" for the interconnection of various QKD networks, each one with its own design, can lead to attractive solutions (Fig. 8).

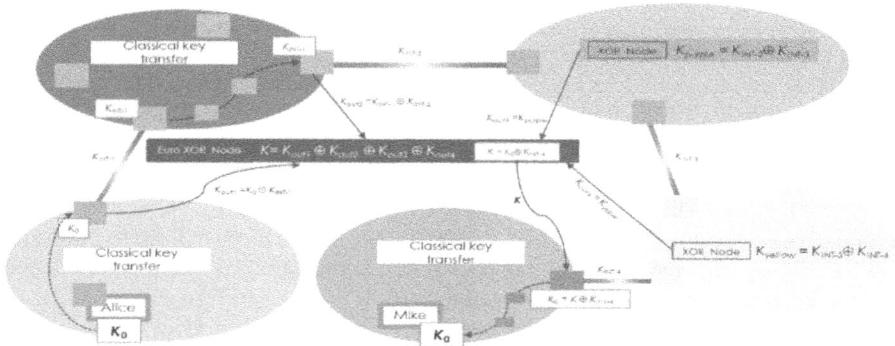

Fig. 8. Example of interconnection of heterogeneous QKD domains with centralized approach

6 Conclusion

We have designed an enhanced centralized XOR node approach of a global QKD network and analysed its performances.

We showed how the security of the network is impacted by the hypothesis of vulnerability of the various nodes or of their transmission means.

Our results show the advantage of the "Centralized XOR node" approach compared to the classical transfer approach in terms of security.

We consider this approach as a good solution also for the interconnection of various QKD networks, that may not use the same scheme and will thus avoid having to trust all QKD networks on the path.

Acknowledgements. The authors thank the European Commission for supporting part of this work within EuroQCI projects (e.g. PETRUS and FranceQCI).

References

1. Rec. ITU-T Y.3803. SERIES Y: GLOBAL INFORMATION INFRASTRUCTURE, INTERNET PROTOCOL ASPECTS, NEXT-GENERATION NETWORKS, INTERNET OF THINGS AND SMART CITIES, "Quantum key distribution networks – Key management" (2020)
2. French Cybersecurity Agency (ANSSI), Federal Office for Information Security (BSI), Netherlands National Communications Security Agency (NLNCSA), Swedish National Communications Security Authority, Swedish Armed Forces, "Position Paper on Quantum Key Distribution" (2024)

QKD-Encryption of Air Traffic Control Data with Negligible Extra Latency

Abraham Sotomayor[1]([✉]) [iD], Paulette Gavignet[1] [iD], Patrice Robert[1],
Sandrine Gravant[1], Luca Calderaro[2] [iD], Alessandro Emanuele[2], Costantino Agnesi[2] [iD],
Gonzague Reydet[3], Olivier Ferrouillat[3], Stephanie Molin[4], Nicolas Courtel[5],
Sylvain Fournier-Bazoune[5], Alexandre Tremon[5], and Thierry Andre[5]

[1] Orange, 2 Av. Pierre Marzin, 22300 Lannion, France
abraham.sotomayor@orange.com
[2] ThinkQuantum S.R.L., Via della Tecnica, 85, IT-36030 Sarcedo, VI, Italy
[3] Thales, 110 avenue du Maréchal Leclerc, 49300 Cholet, France
[4] Thales, 1 avenue Augustin Fresnel, 91767 Palaiseau, France
[5] French State, Ministère charge des Transports, DGAC/DSNA/DTI,
1 Avenue du Dr-Maurice-Grynfogel, 31035 Toulouse Cedex 1, France

Abstract. This study explores the implementation of a Quantum Key Distribution (QKD) system using a Trusted Node (TN) and an optical switch. The TN is used for transmitting the end-to-end encryption key from one QKD link to the next, while the optical switch allows to reduce the number of required Alice/Bob pairs, thereby lowering costs. A testbed was developed within the FranceQCI project to evaluate the performance of the optical switch-based QKD system in the context of Air Traffic Control (ATC) data transmission. The setup involved two polarization-based QKD transmitters, the optical switch, a single receiver, and two encryptors. Results show that the optical switch effectively allows the transmission of end-to-end encryption keys. Furthermore, the availability of encryption keys remains sufficient even during physical link interruptions, ensuring data availability and integrity. User-centric evaluations demonstrated that the QKD-based encryption process did not significantly impact data latency, aligning with operational requirements.

Keywords: QKD · Trusted Node · Optical Switch · VPN Encryption

1 Introduction

The deployment of Quantum Key Distribution (QKD) systems using commercial optical fibers is currently limited by distance due to fiber attenuation, which reduces the number of useful keys after the post-processing step [1]. One approach to extend the reach of a QKD-based key generation is to utilize Trusted Nodes (TN), where quantum keys are established section by section, and a global encryption key is transmitted end-to-end using One-Time-Pad (OTP) encryption with the various quantum keys [2, 3]. However, a significant drawback is the requirement for additional pairs of Alice/Bob, one for each

© The Author(s), under exclusive license to Springer Nature Switzerland AG 2026
F. Barbaresco and G. François (Eds.): QUEST-IS 2025, CCIS 2743, pp. 136–145, 2026.
https://doi.org/10.1007/978-3-032-13852-1_15

additional TN, which increases the overall cost. A proposed strategy, as discussed in [3–7], involves using an optical switch to alternate between different QKD links. The selection of active QKD links is orchestrated by a Key Manager System (KMS) according to a defined configuration.

The effectiveness of the optical switch-based QKD must be evaluated from a user perspective to determine if the complete cryptographic system, comprising QKD, the optical switch, and encryptors, meets user requirements at the application layer, such as throughput, latency, and the availability of encryption keys in the event of a failure or attack on the QKD system.

In this context, within the framework of the FranceQCI [8] European collaborative project, an institutional end-user, the "Direction des services de la navigation aérienne" (DSNA), hosted a test campaign to assess the performance of QKD interconnected to a conventional terrestrial network. The objective was to compare, from a user perspective, both direct unencrypted and encrypted transmissions using QKD of simulated realistic Air Traffic Control (ATC) data.

2 Setup Configuration

2.1 User Platform

The end-user is the Direction générale de l'aviation civile/Direction des services de la navigation aérienne (DGAC/DSNA), the French Air Navigation Service Provider. Air Traffic Control (ATC) data are characterized by low flow rates, currently peaking at a maximum of 100 Mbps in worst-case scenarios. This data primarily consists of real-time information transmitted between remote sites, including sensors, local and central sites, and dedicated networks. The main threat to ATC data is an attack on data integrity, while data availability is a critical factor essential for ensuring public traffic security and confidentiality is less critical. Table 1 provides a general description of ATC data characteristics.

Table 1. Air Traffic Control (ATC) data characteristics.

Traffic data on a site (en-route ATC center, i.e. Area Control Center)	Radar sensor data	~ 20 kbps × 6 radars per zone = > 120 to 160 kbps
	Radar Tracks	TMA4 (Terminal Manoeuvring Area) flow (broadcast to Airport Controls): ~ 50 to 100/150 kbps
		TMA8 flows: ~ 20 kbps only
	Flight Plans[1]	~ 11 kbps

(continued)

---

[1] In certain emergency situations, all flight plans may need to be sent simultaneously, with a maximum of approximately 900 flight plans, resulting in a data rate of about 10 Mbps.

Table 1. (continued)

	Digital Voice for Air Controllers	~ 120 kbps (vs. 64 kbps classically) × simultaneous communications (permanent radio channels); for the largest site: 8400 kbps (telephone) + 70 000 kbps (radio)
Maximal latency	Data for Surveillance	~ 150 ms
	Voice	~ 50 ms (jitter < 5 ms)
Availability	> 99,999%	
Data or integrity loss	$< 10^{-5}$	
Various priorities to be handled	Radio, voice and radar	
	Flight plans and ATFCM (Air Traffic Flow and Capacity Management)	
	Other data only in best effort	

2.2 Testbed Implementation

The test campaign was designed to evaluate the encryption of simulated realistic ATC data. To achieve this, the platform was divided into two Virtual Local Area Network (VLAN) architectures for comparison: one with encryption and one without. This setup facilitated the same data exchanges, as illustrated in Fig. 1.

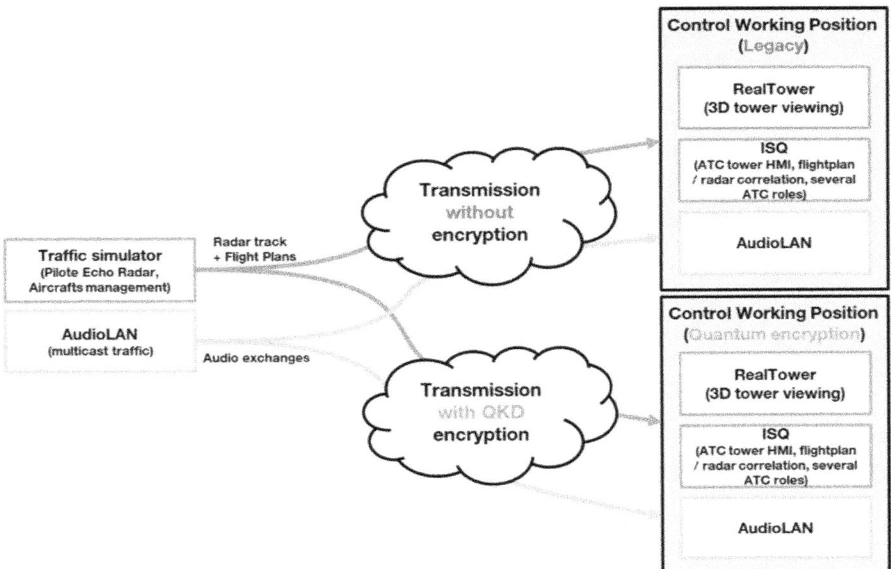

Fig. 1. Testbed implemented at the Direction des Services de la Navigation Aérienne (DSNA).

The transmission system with QKD encryption is illustrated in Fig. 2. The QKD system consists of two polarization-based QKD transmitters, named Alice1 and Alice2, and a single QKD receiver, named Bob, forming two links or channels using G.652 optical fibers. Each transmitter connects to the output ports of the optical switch via the quantum channel port, while its corresponding input connects to Bob's quantum channel port. The optical switch selects one active channel at a time. The QKD transmitters, Alice1 and Alice2, provide secure keys to the encryptors, which then encrypt the data from the traffic simulator.

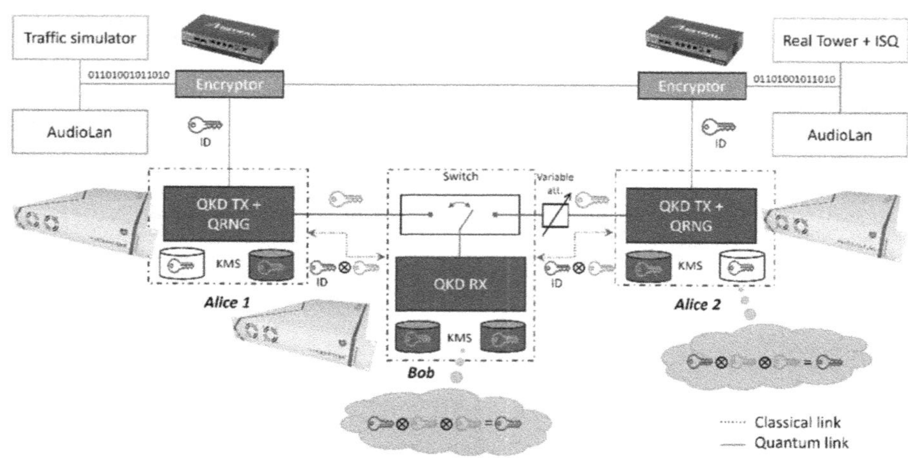

Fig. 2. Encryption system comprising QKD devices, an optical switch, and two encryptors.

The QKD devices used were the commercial Quky developed by ThinkQuantum S.r.l., while the optical switch was sourced from Agiltron. The encryptors employed on the quantum network were the Mistral IP9001 from Thales. In this work, we utilized the standard configuration of the Quky devices, which includes one fiber port for the quantum channel in the C-band and an RJ45 port for the classical channel.

If no switching is employed, the number of QKD transmitter/receiver (Tx/Rx) pairs must match the number of physical links. In a network with n nodes, this leads to a maximum of $\frac{n*(n-1)}{2}$ Tx/Rx pairs. By contrast, the use of optical switching can significantly reduce the number of required QKD devices, since nodes may share common transmitters or receivers through time-multiplexing across multiple physical connections. Furthermore, by employing virtual connections, i.e., trusted-node hops, a full mesh topology can be realized with only n QKD devices. The drawback of this approach is a reduced secret key generation rate, as resources are shared, and link-activation overheads limit the effective time available for key distribution. Such overheads can, however, be mitigated by minimizing the number of switching operations. Ultimately, this introduces a trade-off between system performance and deployment cost that must be assessed with respect to the network requirements. In the present use case, the analysis favored the multiplexed approach.

2.3 QKD System

The QuKy devices implement the efficient three-state BB84 protocol using the polarization degree of freedom of the photon to encode the quantum states. Furthermore, the system implements the one decoy intensity strategy to mitigate Photon-Number-Splitting (PNS) attacks. This simplified protocol guarantees the same level of security of the original BB84 protocol, whereas the one-decoy scheme can provide higher rates in the finite-key scenario compared to the two-decoy scheme.

The system can be seamlessly integrated in a data center thanks to the 19" rackmount enclosure and interface with standard network equipment. The quantum transmission only requires one simplex fiber, since the temporal synchronization is performed via the Qubits4Sync algorithm which uses the qubits themselves as a frequency and absolute time reference [9]. This implies that this QKD system does not requires any synchronization subsystem, which is usually implemented with a second fiber or channel wavelength multiplexing to share the time reference.

The wavelength used for the transmission of the quantum signals is centered at 1550 nm, which is the most convenient when using a dedicated fiber, as this ensures the minimum losses per km of fiber. The transmitter is equipped with a polarization encoder, iPOGNAC, that guarantees stable, low-error, and calibration-free polarization encoding, generating the horizontal and vertical states for the key generation basis and the diagonal polarization for the control basis [10]. Furthermore, these QKD devices perform phase randomization via laser gain switching, and randomness is provided by a source-device-independent Quantum Random Number Generator (QRNG) integrated within the QKD transmitter. Both features are crucial for the secure implementation of the QKD protocol.

At the receiver, the quantum states are projected into two mutually unbiased bases, which are randomly and passively selected by a beam splitter. The bases are aligned before each QKD link is activated; this base agreement procedure takes some minutes to complete. After the projection, the single-photons are detected using InGaAs/InP single-photon avalanche detectors (SPAD). The polarization drift induced by the fiber is compensated by an all-fiber polarization controller. The correction signal applied to this polarization controller is given by a set of public quantum states which are framed between the qubits used for the QKD protocol. These states are uncorrelated with the qubits used for the QKD protocol and therefore do not represent a security threat. The QKD systems perform a real-time post-processing to generate the secure key using an auxiliary classical communication channel.

The auxiliary channel is authenticated using the Message Authentication Code (MAC) Wegman-Carter algorithm. Before the first secure key is generated via QKD, the Pre-shared Key (PSK) is used for authentication. Once secure key is available, the Wegman Carter algorithm is updated with a portion of the QKD key.

2.4 Generation of End-To-End Encryption Key

In this section, we explain the process for delivering end-to-end encryption keys between Alice1 and Alice2 using Bob as a TN (as depicted in Fig. 2). Using the Quky devices, the KMS, which is software embedded in each Quky device, determines, through a global configuration file, the rules to generate end-to-end encryption keys.

- Alice1 generates a secret key for end-to-end encryption (the purple key in Fig. 2) through the QRNG embedded in the QKD transmitter. This key must be securely transmitted to Alice2.
- First, Alice1 and Bob, with the optical switch connecting Bob to Alice1, establish a QKD key (the orange key in Fig. 2).
- Alice1 then encrypts the purple key using the orange key with OTP encryption and sends it to Bob via a classical link. This classical link does not need to be secure (but authenticated) as long as the purple key is fully encrypted, and Bob acts as a trusted node.

- Bob retrieves the purple key using the orange key.
- To securely send the purple key from Bob to Alice2, the switching mechanism should be activated. According to the KMS configuration set in this testbed, the switching mechanism is based on the available key collected in the buffers of each QKD link[2]. When a QKD link has produced a minimum of two key blocks, the system checks the ratio between the amount of key collected in the two physical links and compares it with the ratio of the relative weights defined in the KMS configuration file. Depending on this comparison, the system decides to keep the current active QKD link or to switch and activate the other one. When a QKD link's available keys reach 98% of its buffer size, priority is given to the other link regardless of the weight's comparison. This is to avoid waste of secret keys. If both links have their buffers more than 98% full, the link with the least stored key is activated when the decision is made.
- Therefore, once one of these conditions is met, and Bob and Alice2 are physically connected, they can establish a QKD key (the green key in Fig. 2).
- Then, Bob uses this key to encrypt the purple key with OTP and sends it to Alice2, who retrieves the purple key using the previously established green key.
- Finally, the purple keys are delivered to the encryptors.

2.5 Cryptographic Application

The Mistral solution from Thales enables the establishment of secure IPsec Virtual Private Networks (VPN) between network enclaves to be protected. The Mistral IP9001 encryption device allows for throughput of up to 2×1 Gbps Full Duplex and low latency of less than 1 ms.

For the needs of QKD networks and EuroQCI projects, a specific firmware has been developed for the Mistral IP9001 device to add ETSI 004 and ETSI 014 interfaces, allowing the retrieval of quantum keys from any external Key Management Entity (KME) system compliant with one of these standards. Quantum keys are utilized for the Advanced Encryption Standard (AES)-256 ciphering of VPN data traffic. In this setup, quantum keys replace classical keys obtained through the Diffie-Hellmann algorithm during the Internet Key Exchange (IKE) negotiation protocol. For the DSNA testbed experimentation, the two Mistral IP9001 encryptors were configured to request and use new quantum keys after processing 700 MB of ciphered data.

3 Results and Discussions

We conducted several tests to measure the stability of the QKD system over extended periods, focusing on the Secret Key Rate (SKR) and Quantum Bit Error Ratio (QBER). Additionally, we assessed the impact on user data traffic by specifically measuring the latency for encrypted data and comparing it with the latency for non-encrypted data.

---

[2] All QKD devices have implemented a key buffer storing secret keys (end-to-end encryption keys) and QKD keys. While these buffers are used in the switching mechanism, they also serve to avoid interruptions in the encryption process due to a lack of secret keys.

3.1 Measurements of SKR and QBER

In our experiments, we considered a scenario where the Alice1-Bob and Alice2-Bob links each have a fiber length of 5 km, with a variable optical attenuator installed on the Alice2-Bob link to simulate different span losses. The Alice1-Bob link exhibited a total loss of approximately 4 dB (including the optical switch), while the attenuator for the Alice2-Bob link was set to 5 dB, plus an insertion loss of 4 dB, resulting in a total loss of approximately 13 dB. Figure 3 and 4 show the recovered SKR and QBER for each link, respectively.

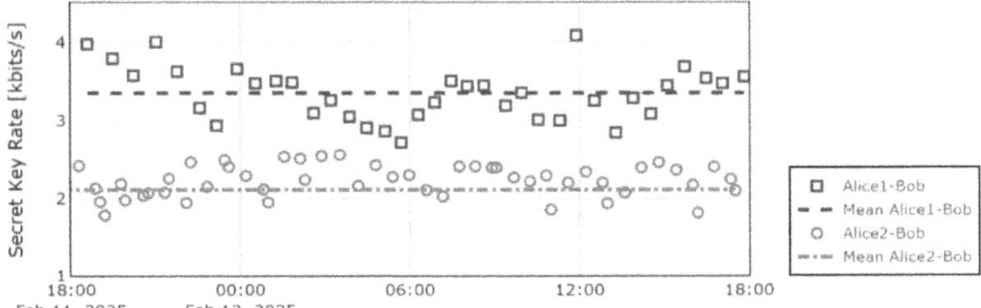

Fig. 3. Secret Key Rate (SKR) measured in Alice1-Bob with total loss ~ 4 dB (blue squares) and Alice2-Bob with total loss ~ 13 dB (red circles). The dashed lines are the mean SKR over 24 h of measurement for Alice1-Bob: ~ 3.35 kbits/s and for Alice2-Bob: ~ 2.11 kbits/s.

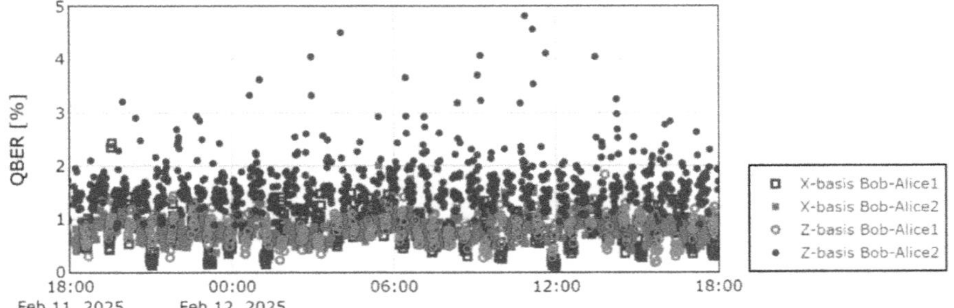

Fig. 4. Quantum Bit Error Ratio (QBER) measured in Alice1-Bob with total loss ~ 4 dB and Alice2-Bob with total loss ~ 13 dB for X and Z basis.

3.2 Interruption of Quantum Links to Check Key Consumption

To evaluate the use of global keys by the encryptors, we observed the number of available keys in Alice1 and Alice2. During normal ATC operation, the available keys are maintained at their maximum of ~ 10 MB, as the QKD system continuously generates encryption keys. To assess the impact on available keys during link failures or events such as Denial of Service (DoS) attacks, we cut the two physical links (Alice1-Bob and Alice2-Bob) and monitored the decrease in available keys in the buffers. Given the low data rate of ATC, we observed a minimal decrease in the available keys, as shown in the left-side and right-side insets of Fig. 5. To challenge the encryption system, we transmitted 50 GB of data within a 10-min timeslot, resulting in a data rate of ~ 85 MBps. The

Mistral encryptors in this setup have a key renewal threshold of 700 Mb, triggering an AES-256 key renewal every second. After 40 h, ~ 170,000 bytes of keys were utilized out of the 10 million available (not shown in Fig. 5). This indicates that data can still be encrypted for an extended period, even during link failures or DoS attacks.

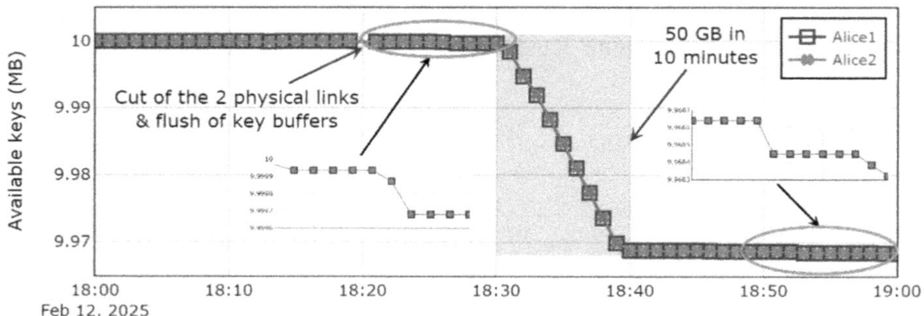

Fig. 5. Evolution of the use of available end-to-end encryption keys.

3.3 User Results

From the user's perspective, maintaining consistent latency was crucial. Therefore, the test focused on measuring the latency of the encrypted data flow and comparing it to that of non-encrypted data. The data used in this test consisted of simulated surveillance and audiolan data. Figure 6 presents these measurements.

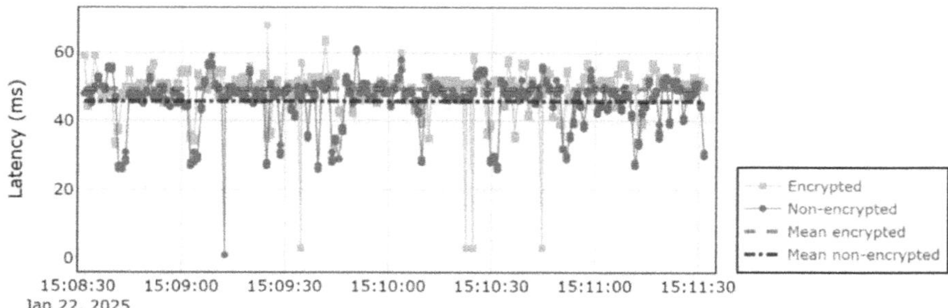

Fig. 6. Comparison of latency between encrypted and non-encrypted data flows.

As observed, there is no visual or auditory difference in latency between the encrypted and non-encrypted data, with no impact on the user application.

4 Conclusions

The suitability of QKD-based encryption was tested from the user's perspective to ensure it meets requirements such as consistent data latency and the availability of end-to-end encryption keys. An optical switch was introduced to allow for a single receiver. After confirming acceptable values for the SKR and a low QBER, we established that the available keys are sufficient to continue encrypting the data flow, even during QKD link failures or DoS attacks. Latency measurements in the user application confirmed that the time required to encrypt messages with QKD and Mistral encryptors is nearly negligible, making this solution suitable for user needs. It is worth mentioning that optical switching reduces the SKR in a single QKD link, but in this case, this was not an issue due to the ample quantity of accumulated encryption keys and the low traffic rate to be encrypted. Data availability is a critical aspect for this use case. In this regard, implementing a full (or partial) redundancy scheme, accounting for other technologies such as post-quantum cryptography (PQC) or classical Diffie-Hellman key exchange, could be necessary. Additionally, a secure key storage and usage policy could be considered, given the significant amount of generated secure keys.

Acknowledgements. This research was co-funded by the European Union under the Digital Europe Program Grant Agreement No. 101091675 (FranceQCI).

References

1. Dušek, M., Lütkenhaus, N., Hendrych, M., Wolf, E. (eds.): Chapter 5 - Quantum cryptography, vol. 49, pp. 381–454. Elsevier (2006)
2. Pincemin, E., Gavignet, P., et al.: 400-Gbps coherent transmission of 100-Gbps QKD-secured data stream over 184-km of standard single mode fiber through three QKD links and two trusted nodes. J. Lightwave Technol. **42**(12), 4302–4309 (2024)
3. Salvail, L., Peev, M., Diamanti, E., Alléaume, R., Lütkenhaus, N., Länger, T.: Security of trusted repeater quantum key distribution networks. J. Comput. Secur. **18**(1), 61–87 (2010)
4. Picciariello, F., Karakosta-Amarantidou, I., Rossi, E., et al.: Intermodal quantum key distribution field trial with active switching between fiber and free-space channels. EPJ Quant. Technol. **12**, 6 (2025). https://doi.org/10.1140/epjqt/s40507-025-00306-9
5. Tayduganov, A., et al.: Optimizing the deployment of quantum key distribution switch-based networks. Opt. Express **29**, 24884–24898 (2021)
6. Karavias, V., Lord, A., Payne, M.C.: Switching in quantum networks: an optimization investigation. J. Opt. Commun. Network. **16**(3), 404–418 (2024). https://doi.org/10.1364/JOCN.513637
7. Brunner, H.H., Fung, C.H.F., Peev, M., et al.: Demonstration of a switched CV-QKD network. EPJ Quant. Technol. **10**, 38 (2023). https://doi.org/10.1140/epjqt/s40507-023-00194-x
8. Petrus Homepage. https://petrus-euroqci.eu/. Accessed 18 Apr 2025
9. Calderaro, L., et al.: Fast and simple qubit-based synchronization for quantum key distribution. Phys. Rev. Appl. **13**, 054041 (2020). https://doi.org/10.1103/PhysRevApplied.13.054041
10. Avesani, M., Agnesi, C., Stanco, A., Vallone, G., Villoresi, P.: Stable, low-error, and calibration-free polarization encoder for free-space quantum communication. Opt. Lett. **45**, 4706–4709 (2020)

Multi-level Parameter Estimation for Receiver Noise in CV-QKD

Guillaume Ricard$^{(\boxtimes)}$, Yves Jaouën, and Romain Alléaume

Telecom Paris - Institut Polytechnique de Paris, 19 Pl. Marguerite Perey,
91120 Palaiseau, France
Ricard@telecom-paris.fr

Abstract. We present an experimental method for continuous-variable quantum key distribution (CV-QKD) that removes the need for conventional receiver noise calibration. By exploiting Multi-Level Parameter Estimation (MLPE) with controlled attenuation at the receiver, our approach directly estimates the total receiver noise, including contributions from the local oscillator, while simultaneously extracting channel-induced excess noise. This calibration-free technique allows continuous operation without halting quantum transmissions, increasing experimental throughput and robustness. We implement the method in a dual-polarization, local-local oscillator CV-QKD system multiplexed with a classical communication channel, demonstrating precise receiver noise estimation. Our results highlight MLPE as a practical tool for scalable, real-time CV-QKD deployment.

Keywords: Continuous-variable quantum key distribution
(CV-QKD) · parameter estimation · multi-level parameter estimation
(MLPE) · Receiver noise

1 Introduction

Continuous Variable Quantum Key Distribution (CV-QKD) has seen significant progress in recent years, with successful field trials [14] and demonstrations over impressive distances [6]. Among the different families of QKD protocols, the continuous variable approach stands out as a particularly promising option for metropolitan-scale deployment, thanks to its cost-effectiveness, compatibility with standard telecommunication hardware, and its ability to coexist with classical optical communication systems [9,17].

In the early development of CV-QKD, system performance was heavily constrained by excess noise, much of which originated from the receiver's electronic components. To mitigate this, researchers treated this electronic noise as trusted [13], an assumption justified by the controlled access to the receiver's laboratory and precise calibration of this noise source. This strategy enabled CV-QKD implementations to extend to much longer distances.

However, electronic noise is not the sole contributor to receiver-induced noise. Technical impairments of the Local Oscillator (LO) also impact the overall

F. Barbaresco and G. François (Eds.): QUEST-IS 2025, CCIS 2743, pp. 146–156, 2026.
https://doi.org/10.1007/978-3-032-13852-1_16

excess noise. These LO-related contributions are typically overlooked or implicitly absorbed either into untrusted noise terms or even in the shot noise estimator in conventional security models.

In this work, we propose a practical method that alleviates the need for conventional receiver calibration by leveraging Multi-Level Parameter Estimation (MLPE). By modulating the amplitude at the receiver using controlled attenuators, MLPE provides independent measurements of both total receiver noise and channel-induced excess noise. This approach allows continuous operation without interrupting key distribution, improving experimental throughput, and enabling real-time noise monitoring.

We implement this method in a dual-polarization CV-QKD system operating with a local-local oscillator (LLO) configuration and multiplexed classical communication channels. Our results demonstrate that MLPE accurately estimates the total receiver noise and separates it from channel contributions, paving the way for full receiver noise calibration-free CV-QKD experiments that maintain high performance over long distances and in realistic experimental conditions.

2 Multi-level Parameter Estimation (MLPE)

We propose a novel approach for parameter estimation in CV-QKD that leverages an amplitude modulator placed at Bob's side. While amplitude modulation at Bob has previously been explored as a countermeasure against side-channel attacks [11,15], our scheme uses it to refine the estimation of key channel parameters, thereby enhancing security and performance.

2.1 Standard Parameter Estimation

In conventional parameter estimation, Alice reveals a subset of her transmitted symbols, allowing Bob to compute statistics over his received signal. With discrete modulation, such as 64-QAM probabilistic constellation shaping (PCS), Bob can compute not only the overall variance of the received signal (V_B), but also the conditional variance for each symbol sent by Alice ($V_{B|A}$). This conditional variance effectively removes the contribution of the modulation, isolating the channel noise. Incorporating this information provides an additional equation that can be exploited for more accurate parameter estimation.

In standard CV-QKD, the receiver noise is typically modeled as the sum of shot noise and electronic noise:

$$\nu_{\mathrm{rec}} = 1 + \nu_{\mathrm{el}}. \tag{1}$$

For heterodyne detection, the total and conditional variances at Bob are:

$$V_{B|A} = \frac{\eta T}{2}\xi_A + \nu_{\mathrm{rec}} \tag{2}$$

$$V_B = \frac{\eta T}{2}V_A + \frac{\eta T}{2}\xi_A + \nu_{\mathrm{rec}} \tag{3}$$

Excess noise and transmittance can then be estimated via:

$$\xi_A = \frac{2}{\eta T}(V_{B|A} - \nu_{\rm rec}) \tag{4}$$

$$T = \frac{2(V_B - V_{B|A})}{\eta V_A} \tag{5}$$

2.2 MLPE with Controlled Attenuation

In our MLPE scheme, Bob introduces controlled optical attenuations η_1 and η_2 using his amplitude modulator. These known attenuations effectively adjust the full channel transmittance, providing two additional equations that allow a more precise determination of channel parameters.

For example, in practice, $\nu_{\rm rec}$ also contains optical contributions from the local oscillator, which are not fully captured by electronic noise and shot noise alone:

$$\nu_{\rm rec} = 1 + \nu_{\rm el} + \nu_{\rm LO} \tag{6}$$

Accurate estimation can therefore be achieved with controlled attenuations at Bob. With two known attenuations η_1 and η_2, the variances become:

$$V_{B,\eta_i} = \frac{\eta_i T}{2}V_A + \frac{\eta_i T}{2}\xi_A + \nu_{\rm rec} \tag{7}$$

$$V_{B|A,\eta_i} = \frac{\eta_i T}{2}\xi_A + \nu_{\rm rec} \tag{8}$$

This leads to direct MLPE estimators:

$$\nu_{\rm rec} = \frac{V_{B,\eta_1} - R_{\rm att}V_{B,\eta_2}}{1 - R_{\rm att}} \tag{9}$$

$$\xi_{A,\eta_i} = \frac{2}{\eta_i T}(V_{B|A,\eta_i} - \nu_{\rm rec}) \tag{10}$$

where $R_{\rm att} = \eta_1/\eta_2$

And finally, if we know the electronic noise variance $\nu_{\rm el}$, we can estimate the local oscillator contribution to receiver noise that is currently wrongly attributed to shot noise:

$$\nu_{\rm LO} = \frac{V_{B,\eta_1} - R_{\rm att}V_{B,\eta_2}}{1 - R_{\rm att}} - 1 - \nu_{\rm el} \tag{11}$$

3 Shot-Noise Unit (SNU) and Calibration Cycle

In practical CV-QKD, the shot-noise unit must be calibrated. The standard procedure typically involves two steps:

1. Measure the electronic noise with both the LO and signal off:

$$\widehat{\sigma}^2_{\mathrm{calib},1} = \sigma^2_{\mathrm{elec}} \tag{12}$$

2. Measure the full receiver noise with the LO on and signal off:

$$\widehat{\sigma}^2_{\mathrm{calib},2} = \sigma^2_{\mathrm{rec}} \tag{13}$$

From these measurements, the shot-noise variance is estimated as

$$\widehat{N_0} = \widehat{\sigma}^2_{\mathrm{calib},2} - \widehat{\sigma}^2_{\mathrm{calib},1} \tag{14}$$

allowing all subsequent data to be expressed in SNU.

This conventional estimator neglects the contribution from the local oscillator, ν_{LO}, which means MLPE cannot separate it accurately unless the shot noise is already known. Explicitly accounting for the LO yields

$$\widehat{N_0} = N_0 + \sigma^2_{\mathrm{LO}} \tag{15}$$

In this framework, the LO contribution is incorrectly attributed to shot noise and cannot be distinguished using only the standard calibration equations: electronic noise calibration (no LO contribution), full receiver noise calibration, and the measured variances V_B and $V_{B|A}$, where both shot noise and LO contributions appear identically.

Nevertheless, MLPE remains valuable. Electronic noise is typically small and stable, requiring infrequent calibration. By contrast, total receiver noise often fluctuates significantly, forcing frequent calibrations in conventional setups and imposing a duty cycle on quantum transmission: time must be spent alternating between calibration, collecting sufficient statistics, and actual quantum communication. MLPE mitigates this limitation by directly estimating ν_{rec} from controlled attenuated signals, eliminating the need to halt transmission for frequent calibration. This enables near-continuous monitoring of receiver noise and improves overall experimental throughput.

4   Experimental System Design

We have designed a coherent optical communication system that can jointly perform classical and shot-noise-limited quantum communications. Both coherent communications channels are multiplexed over the same fiber and operated in the so-called local local oscillator (LLO) [12] configuration. This allows in particular to perform CV-QKD while closing the security loophole related to calibration attacks on transmitted local oscillator [7]. Our approach notably focuses on recovering the carrier phase using a "symbiotic" classical communication channel [2,16], rather than relying on pilot tones [10]. While more complex to perform than sole CV-QKD, this approach has the potential to increase quantum communications (and CV-QKD) pervasiveness by minimizing the resources necessary to deploy them but also by pushing further the conditions under which QKD and classical communications can coexist over the same fiber.

At Alice's side (the emitter), a 100 Hz linewidth laser at telecom wavelength is modulated using a dual-polarization IQ modulator. The classical channel is digitally shifted from the carrier by 4 GHz, and similarly, the quantum channel is shifted by 1 GHz, as illustrated in the spectrum of Fig. 1. It's crucial that the classical signal is accurately pulse-shaped to prevent its frequencies from interfering with the quantum signal. Such filtering is not needed on the quantum signal however, since the DA-LMS algorithm (at reception, as discussed in the next section) naturally converges to an RC-like filter, while also accounting for non-ideal response of electrical devices. Additionally to being spectrally separated, classical and quantum channels are also modulated on orthogonal polarization states and the quantum signal is selectively attenuated until it reaches the desired modulation variance V_A.

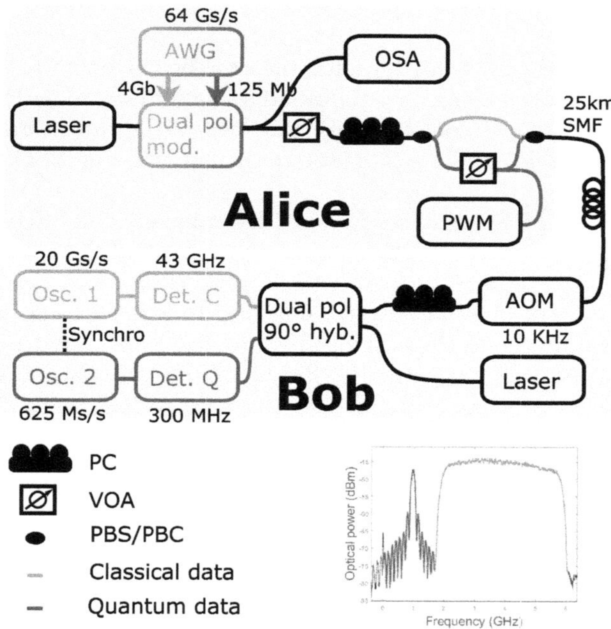

Fig. 1. Block diagram of the experimental setup and optical spectrum read on the OSA.

After being recombined, both classical and quantum signals travel through a 25 km single-mode fiber (SMF) link before reaching Bob's lab (the receiver). Just after the channel, the light encounters an acousto-optic modulator (AOM) driven by a 10 kHz square analog electrical signal. This AOM allows to perform CV-QKD acquisitions at different total transmission levels, hence allowing a more refined parameter estimation procedure.

Finally, light enters our dual polarization 90° hybrid along with the LO, whose optical frequency is tuned to detect the quantum signal in baseband. Coherent detectors measure both quadratures of the resulting optical signals, on both polarizations. The signals are digitally sampled using two distinct oscilloscopes that share the same clock reference and trigger for accurate time synchronisation.

The balanced receivers used for quantum signal detection exhibit a 13 dB clearance over a 350 MHz bandwidth. The losses induced within the detection apparatus consist in the detectors efficiency (0.5 dB), losses in the hybrid (2dB), and losses in the AOM (3 dB). Those induced by our 25 km SMF channel are 5 dB. Shot noise calibration is made by cutting incoming light with LO still on [3]. In our case, this is done before each block acquisition of $1.8 \cdot 10^5$ symbols.

5   Digital Signal Processing (DSP)

We developed a joint digital DSP routine leveraging the high signal-to-noise ratio (SNR) estimates from classical signals to precisely correct in frequency and phase the quantum signals.

The classical signal is generated as a Quadrature Phase Shift Keying (QPSK) signal at 4 Gb, pulse-shaped using a Root Raised Cosine (RRC) filter with a roll-off of 0.1, facilitating sharp frequency cutoff and minimizing leakage across quantum signal frequencies. Meanwhile, the quantum signal is modulated as a 64 states Quadrature Amplitude Modulation (64-QAM) at 125 Mb with Probabilistic Constellation Shaping (PCS) [1,4].

Upon reception, the classical signal undergoes a series of fairly standard processing steps [8], with a customized two-convergence radius (to account for Bob's amplitude modulation) version of the Constant Modulus Algorithm (CMA). We typically observe error-free classical transmission. For the quantum signal, corrections in both frequency and phase are applied using estimations derived from the classical signal. Subsequently, a data-aided least mean squares algorithm (DA-LMS) [5] is employed to optimize a Finite Impulse Response (FIR) filter that will account for imperfections and optimally downsample the QKD signal to a 1 sample per symbol regime, at the cost of sacrificing 4000 quantum symbols (out of $1.8 \cdot 10^5$) for convergence. We carefully tune the learning rate of the LMS algorithm to maximize performance. Any remaining frequency offset is then corrected during parameter estimation by iteratively minimizing excess noise.

6   Results

6.1   Attenuation Accuracy

Calibration of the attenuation ratio R_{att} is central to the success of MLPE. This calibration can be performed once optically, after which the ratio and its stability can be verified from the data itself. In practice, we use either a classical signal that experiences the same attenuation as the quantum channel, or high-intensity quantum data. As shown in Table 1, the ratio of variances between the signals at attenuation levels η_1 and η_2 provides a precise and stable estimate of R_{att}, with tight 95% confidence intervals over several independent data blocks. The values obtained from classical and quantum signals are consistent, confirming both the calibration and the long-term stability of the attenuation settings.

The raw measurement statistics further illustrate this effect. Figure 2(a) shows time-domain quadrature traces under three different conditions: full extinction (receiver noise only), intermediate attenuation η_2, and higher transmission η_1. The corresponding probability density functions are displayed in Fig. 2(b). While a Gaussian model accurately fits the receiver noise (extinction case), it no longer provides a good description of the modulated quadrature distributions at finite transmission, where clear deviations from Gaussianity appear. These observations underline the reliability of the attenuation calibration as well as the statistical distinctness of the different attenuation regimes.

Table 1. Ratio of variances between signals in η_1 and η_2 attenuation configurations with 95% confidence intervals, used to verify attenuation stability over four data blocks, for both classical and (high-intensity) quantum data.

Signal	Ratio of transmittances η_1/η_2	CI 95%	Samples per block
Classical	3.2636	0.0032	5 000 000
Classical	3.2513	0.0032	5 000 000
Classical	3.2455	0.0034	5 000 000
Classical	3.2455	0.0033	5 000 000
Quantum	3.3136	0.0255	156 250
Classical	3.2874	0.0242	156 250
Classical	3.3042	0.0253	156 250
Classical	3.2454	0.0247	156 250

6.2 Noise Estimation

The receiver noise can be calibrated independently, and MLPE provides a direct way to cross-check this calibration. Figure 3(a) shows the estimated excess noise as a function of the modulation variance V_A for both attenuation configurations η_1 and η_2. A linear fit with intercept constrained to zero yields $\xi_{\eta_1} = 0.0043V_A$ and $\xi_{\eta_2} = 0.0013V_A$, fully consistent with the measured attenuation ratios and confirming that the excess noise scales as expected with the input modulation.

In addition, Fig. 3(b) presents the estimate of the receiver noise variance ν_{rec} obtained from MLPE. The result is in excellent agreement with the independently calibrated value, demonstrating consistency between experimental mea-

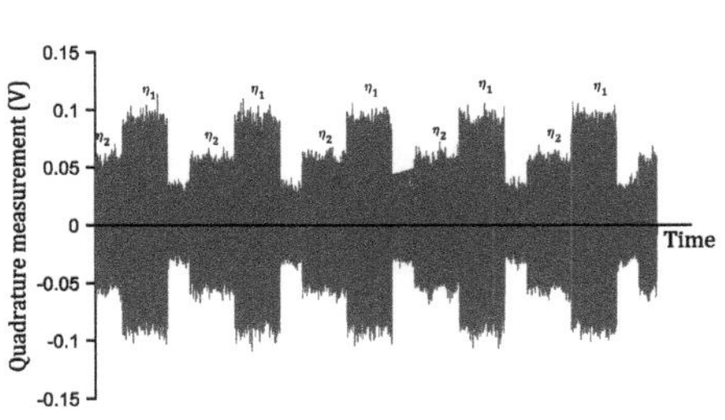

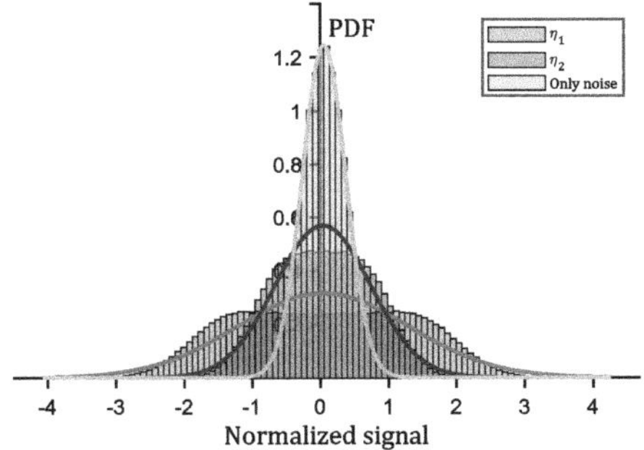

(a) Time-domain quadrature measurements in volts vs. time, showing three distinct attenuation levels: complete extinction (receiver noise only), attenuation η_2, and attenuation η_1.

(b) Histograms of quadrature distributions for η_1, η_2, and noise only. Solid lines: Gaussian fits (good for noise only, poor for modulated signals).

Fig. 2. Quadrature measurements illustrating the effect of attenuation. (a) Time-domain traces show three distinct attenuation regimes. (b) Corresponding PDFs with Gaussian fits highlight that while noise is well described by a Gaussian model, modulated signals deviate significantly.

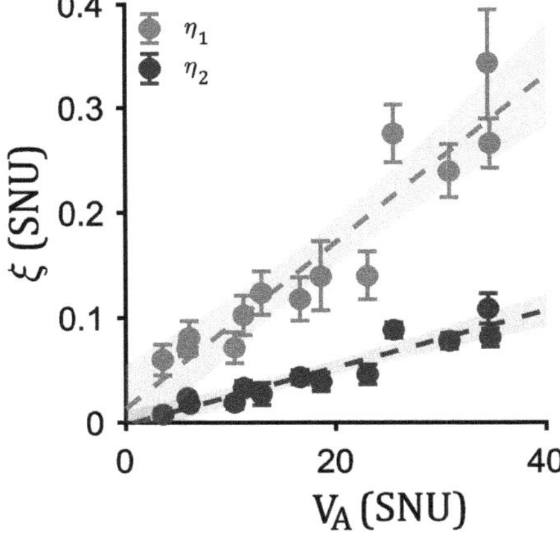

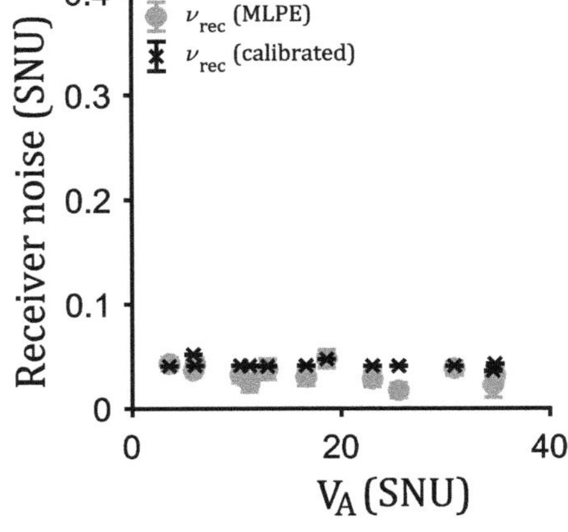

(a) Excess noise ξ vs. modulation variance V_A for the two attenuation settings, with linear fits and uncertainty.

(b) Receiver noise estimate ν_{rec} from MLPE compared with independent calibration, showing excellent agreement.

Fig. 3. Noise estimation using MLPE. (a) Excess noise scales linearly with V_A and matches the expected attenuation ratios. (b) Receiver noise estimate agrees with the independently calibrated value, validating the approach.

surements and theoretical expectations. This agreement validates both the accuracy of the attenuation calibration and the robustness of the MLPE-based noise estimation procedure.

6.3 Performance

The fitted noise model allows us to predict the asymptotic secret key rate in the trusted-receiver scenario. Figure 4 shows the expected performance, with the modulation variance V_A optimized at each distance. The computation relies on the trusted excess noise model obtained by fitting the dependence of ξ on V_A, and the key rate is evaluated using the asymptotic proof of [4]. Our experimental point at 25 km is compatible with a secret key rate of 3.73 Mbps.

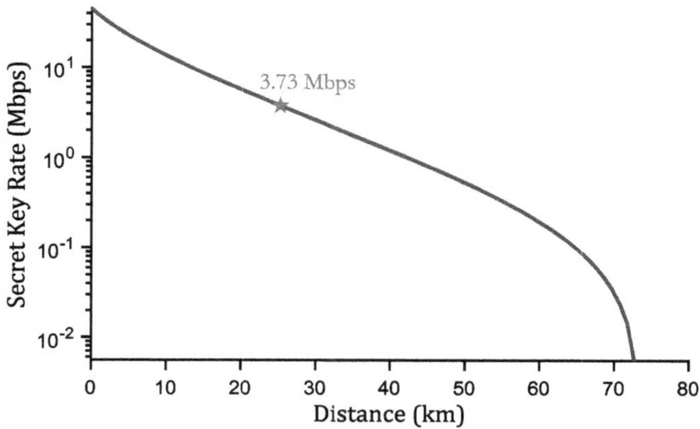

Fig. 4. Asymptotic secret key rate versus distance in the trusted-receiver scenario, with modulation variance V_A optimized at each point.

7 Conclusion and Discussion

We have introduced and experimentally validated the MLPE calibration method, which enables direct estimation of receiver noise from quadrature measurements without requiring additional calibration steps. The method was shown to be robust, providing stable attenuation ratios and noise estimates consistent with conventional calibration procedures. We further demonstrated that the excess noise inferred from MLPE scales linearly with the modulation variance. This model allowed us to predict the asymptotic secret key rate with optimized modulation variance, and our experimental data at 25 km confirmed compatibility with a secret key rate of 3.73 Mbps.

These results establish MLPE as both a practical calibration tool and a means to improve throughput in prepare-and-measure CV-QKD systems, including standard implementations relying on pilot tones for phase recovery. Future work will focus on extending MLPE to the finite-size regime, where calibration accuracy and stability are critical for practical deployment.

Acknowledgements. This work has been supported by the European Union's Horizon Europe research and innovation program, under the project QSNP (grant agreement 101114043), by Digital Europe under the project FranceQCI (grant agreement 101091675) and by French National Research Agency under France 2030 program, project QCommTestbed (grant agreement ANR-22-PETQ-011).

Guillaume Ricard gratefully acknowledges the support of the Paris Region PhD program.

References

1. Aymeric, R., et al.: Quantum key distribution and classical communication coherent deployment with shared hardware and joint digital signal processing. Proc. SPIE Int. Soc. Opt. Eng. **12133**, 22–32 (2022)
2. Aymeric, R., et al.: Symbiotic joint operation of quantum and classical coherent communications. In: Optical Fiber Communication Conference (OFC) 2022. Optica Publishing Group, San Diego (2022). W2A.37. isbn: 978-1-55752-466-9
3. Brunner, H.H., et al.: Precise noise calibration for CV-QKD. In: 2020 22nd International Conference on Transparent Optical Networks (ICTON), pp. 1–4 (2020). ISSN: 2161-2064
4. Denys, A., Brown, P., Leverrier, A.: Explicit asymptotic secret key rate of continuous-variable quantum key distribution with an arbitrary modulation. Quantum **5**, 540 (2021). arXiv:2103.13945 [quant-ph]. issn: 2521-327X
5. Faruk, Md.S., Savory, S.J.: Digital signal processing for coherent transceivers employing multilevel formats. J. Lightwave Technol. **35**(5), 1125–1141 (2017). issn: 0733-8724, 1558-2213
6. Hajomer, A.A.E., et al.: Long-distance continuous-variable quantum key distribution over 100-km fiber with local local oscillator. Sci. Adv. **10**(1), eadi9474 (2024)
7. Jouguet, P., Kunz-Jacques, S., Diamanti, E.: Preventing calibration attacks on the local oscillator in continuous-variable quantum key distribution. Phys. Rev. A **87**(6), 062313 (2013)
8. Kikuchi, K.: Fundamentals of coherent optical fiber communications. J. Lightwave Technol. **34**(1), 157– 179 (2016). issn: 0733-8724, 1558-2213
9. Kumar, R., Qin, H., Alléaume, R.: Coexistence of continuous variable QKD with intense DWDM classical channels. New J. Phys. **17**(4), 043027 (2015). issn: 1367-2630
10. Laudenbach, F., et al.: Pilot-assisted intradyne reception for high-speed continuous-variable quantum key distribution with true local oscillator. Quantum **3**, 193 (2019). arXiv:1712.10242 [quant-ph]. issn: 2521-327X
11. Lucamarini, M., et al.: Implementation Security of Quantum Cryptography Introduction, challenges, solutions| ETSI White Paper No. 27 (2018)
12. Qi, B., et al.: Generating the local oscillator "locally" in continuousvariable quantum key distribution based on coherent detection. Phys. Rev. X **5**(4), 041009 (2015). arXiv:1503.00662 [quant-ph]. issn: 2160-3308
13. Usenko, V.C., Filip, R.: Feasibility of continuous-variable quantum key distribution with noisy coherent states. Phys. Rev. A **81**(2), 022318 (2010). arXiv:0904.1694 [quant-ph]. issn: 1050-2947, 1094-1622
14. Williams, B.P., et al.: Field test of continuous-variable quantum key distribution with a true local oscillator. Phys. Rev. Appl. **21**(1), 014056 (2024)

15. Xu, S., et al.: Counteracting a saturation attack in continuous-variable quantum key distribution using an adjustable optical filter embedded in homodyne detector. Entropy **24**(3), 383 (2022)
16. Xu, Y., et al.: Simultaneous continuous-variable quantum key distribution and classical optical communication over a shared infrastructure. Appl. Phys. Lett. **123**(15) (2023)
17. Zhong, H., et al.: Continuous-variable quantum key distribution coexisting with classical signals on few-mode fiber. Optics Exp. **29**(10), 14486–14504 (2021). issn: 1094-4087

Session 3 Quantum Communications – Towards a Quantum Internet Infrastructure

Crystalline Waveguides for Efficient Integrated Quantum Memories

M. Chan[1,2,3(✉)], A. Ferrier[4,5], A. Talneau[3], P. Berger[2], L. Morvan[2], and S. Welinski[2]

[1] Thales SIX GTS France, 92230 Gennevilliers, France
mickael.chan@thalesgroup.com
[2] Thales Research and Technology, 91767 Palaiseau, France
[3] Centre de Nanosciences et de Nanotechnologies, Université Paris-Saclay, CNRS, 91120 Palaiseau, France
[4] Institut de Recherche de Chimie, Chimie ParisTech, PSL University, CNRS, 75005 Paris, France
[5] Faculté des Sciences et Ingénierie, Sorbonne Université, UFR933, 75005 Paris, France

Abstract. This paper presents progress in the development of crystalline waveguides to enable efficient and integrated quantum memories for long-distance quantum communication. The primary aim is to tackle the difficulties presented by the weak light-matter interaction strength observed in rare-earth ion (REI) doped crystals, specifically in Yttrium Orthosilicate (YSO), which limits optical storage efficiencies. In contrast to other approaches for photonic integration, our method utilizes traditional lithography and dry-etching techniques. This approach maintains the beneficial properties of bulk crystals while offering design adaptability. We refined processes such as bonding, lapping, polishing, and dry-etching to achieve the optimal YSO waveguide structure, as shown in our simulations. This study paves the way for completely integrated quantum repeaters and also improves the integration of other quantum technologies based on REI-doped crystals, including quantum computing, single-photon generation, RF spectral analysis, and microwave-to-optical quantum transduction.

Keywords: Quantum communications · quantum memories · rare-earth-doped crystals · integrated photonics

1 Introduction

Quantum communications aims at the development of the Quantum Internet [1], establishing the secure transfer of quantum information (QI) through a global network composed of quantum computers, sensors, processors and end-users. In this architecture, photons serve as the primary carriers of information, functioning as "flying qubits" that experience minimal interference from their environment. Although fiber optics represent a natural medium for these networks, the transmission of quantum information over long distances is hampered by scattering and absorption losses inherent in fibers. Furthermore, due to the non-cloning theorem [2], the application of classical amplification methods is barred, necessitating the use of a specific protocol and system called

© The Author(s), under exclusive license to Springer Nature Switzerland AG 2026
F. Barbaresco and G. François (Eds.): QUEST-IS 2025, CCIS 2743, pp. 159–169, 2026.
https://doi.org/10.1007/978-3-032-13852-1_17

Quantum Repeaters (QRs) to achieve long-range communication [3]. The QR protocol employs a segmented approach, connecting shorter distance links through intermediate nodes that utilize entanglement swapping to establish communication [4]. Integral to this approach is Quantum Memory (QM), which allows the storage of entangled photonic states at these nodes while waiting to successfully establish entanglement with adjacent nodes. Research groups are increasingly focused on optimizing QM systems for various quantum applications, including distributed quantum computing and quantum key distribution. A critical parameter for effective QM [5] is the storage time, which must exceed the duration required for the entanglement swapping process. Promising candidates for QM include atomic vapors [6] and color centers in diamonds [7], but this paper's focus is on rare-earth ion (REI) doped crystals, which demonstrate significant potential for long-distance quantum repeater applications [8]. In magnetically quiet host materials like Yttrium Orthosilicate (YSO), certain REIs exhibit long optical and hyperfine coherence times at cryogenic temperatures [9]. Another distinguishing feature of REI-doped crystals for network applications is multiplexing, in particular frequency multiplexing where the same device can store multiple photons at different frequencies. This is made possible by the large inhomogeneous linewidths (typically GHz) compared to homogeneous linewidths (typically kHz), enabling the establishment of multiple memory channels within a single crystal absorption line.

Recent advancements have successfully demonstrated the optical storage of single photons in REI-doped crystals [10, 11] alongside several promising network applications [12, 13]. Despite these successes, many existing systems struggle with moderate storage efficiencies due to various challenges such as low optical depth and suboptimal light-matter interactions. The weak oscillator strength of REIs limits absorption, which cannot be enhanced simply by increasing ion concentration or crystal length. Indeed, constraints imposed by spin-spin interaction processes necessitate maintaining low ion concentrations—often in the parts per million (ppm) range—to preserve coherence times. Additionally, the compromise between beam divergence and laser power density restricts overall device length. To address these challenges, integrated photonics offers a pathway to enhance light-REI interactions and push towards fully integrated quantum repeater devices [14]. Various approaches are being explored, including the utilization of $LiNbO_3$-based waveguides [15], silicon photonics [16], and femtosecond laser micromachining [17]. However, many options demonstrate degraded coherence performance or struggle to integrate complex functionalities like couplers, curvature, or cavities effectively. In this work, we introduce a novel fabrication methodology (Fig. 1) for single-crystalline YSO waveguides, created from bulk crystals through conventional lithography and etching techniques. This approach aims to maintain the intrinsic properties of the bulk material while allowing for significant design flexibility. The fabrication process leverages our advancements in crystal bonding to a dedicated substrate, involves thinning YSO layers to micrometric dimensions, and employs dry etching of YSO membranes. Finite Difference Eigenmode (FDE) simulations have been carried out to identify the optimal geometries and dimensions for the REI waveguide structures.

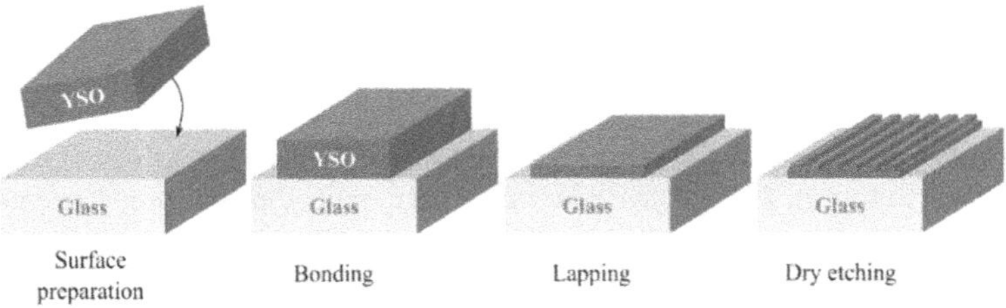

Fig. 1. Fabrication process of crystalline waveguides based on dry etching and UV lithography. Glass substrate is used as a low refractive index media.

2 Simulation Considerations

The preliminary phase of waveguide simulation involves selecting an appropriate fiber-to-structure coupling method suitable for cryogenic conditions. A practical technique is direct edge coupling, utilizing UV-cured polymers to bond the optical fiber to the waveguides. Our optimization focuses on two primary criteria: modal overlap between the fundamental modes of the fiber and waveguide, and the effective indices of both. For convenience, we will denote the product of these two parameters as coupling efficiency in the subsequent analysis. Furthermore, we'll examine the spatial mode distribution and transmission losses attributed to curvature, which will provide insights into the design versatility of our fabrication approach. The device architecture is predicated on YSO bonded to borosilicate SiO_2 glass, designed in either a ridge or shallow ridge configuration. This specific glass was selected due to its thermal expansion coefficient, which is closely aligned with that of YSO. All simulations are conducted at a single operational wavelength of 1536 nm, corresponding to the $(^4I_{15/2}\text{-}^4I_{13/2})$ transition in site 1 of Er^{3+}: YSO. At this wavelength, considering YSO's birefringence, the refractive indices of YSO and borosilicate glass are recorded at 1.7892 and 1.4514, respectively [18]. The waveguide's cross-section is examined within the (b, D_2) plane, while the D_1 axis is oriented parallel to the light propagation direction. This particular choice of axis is constrained by the crystal growth method. The optical fiber used in the simulation is a standard SMF28 Corning fiber with a core/cladding radius of 8/125 μm.

2.1 Ridge Case

YSO/SiO_2 ridges were examined first as the most straightforward approach to our objectives. As detailed in Sect. 3, we consider a maximum achievable etch depth of 1.5 μm. Preliminary simulations reveal that ridges with a vertical thickness (e) exceeding 1 μm always demonstrate multimode behavior regardless of width, while no viable modes are present when e is below 700 nm. Consequently, we have fixed the thickness at 1 μm to ensure single-mode operation. Figure 2a illustrates the coupling efficiency across ridge widths ranging from 1 μm to 30 μm. Given the waveguide's reduced dimensions, an additional SiO_2 top layer is added to improve modal overlap with the optical fiber. The optimal coupling efficiency was determined to be 23% at a width of 14 μm, coupled with a top SiO_2 layer thickness of 1 μm. At these dimensions, the waveguide also exhibits multimode behavior, necessitating modal filtering through a tapered geometry to achieve single-mode conditions. This approach involves an adiabatic reduction of the width from 14 μm down to 1 μm, as explained in [19]. Despite facing challenges related to the relatively low coupling efficiency and the mechanical difficulty of thinning a YSO membrane to a thickness of 1 μm, the structure demonstrates considerable resistance to curvature. As shown in Fig. 2b, with bending radii exceeding 60 μm, losses attributed solely to bending were calculated to remain below 3 dB/cm. This resistance to curvature can be easily understood considering the high differences in refractive indices.

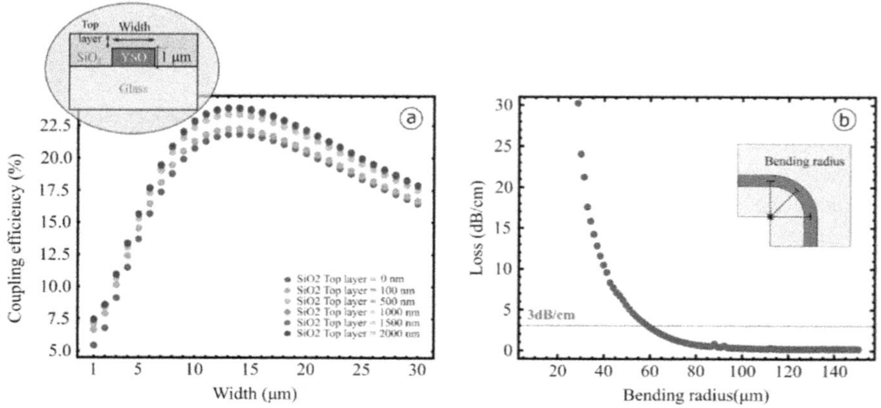

Fig. 2. a) Coupling efficiency (Modal overlap × ratio of effective indices) to ridge width in the case of YSO-SiO$_2$ 1μm-thick ridge, with different SiO$_2$ top layer thicknesses with single mode fiber; b) Bending loss in the case of YSO-SiO$_2$ 1x1 μm ridge

2.2 Shallow Ridge Case

Utilizing shallow ridge structures significantly lightens the manufacturing constraints associated with membrane fabrication. Our analysis indicates that maintaining a total membrane thickness of less than 10 μm, coupled with a shallow etch depth of 1.5 μm, is sufficient for the propagation of eigenmodes. As illustrated in Fig. 3a, the 2D map displays coupling efficiencies with the x-axis representing total membrane thickness (inclusive of the 1.5 μm shallow ridge depth) and the y-axis showing ridge widths ranging from 1 to 30 μm. An optimal coupling efficiency of 75% is achieved at a ridge width of 11 μm, corresponding to a total YSO layer thickness of 8 μm, which is significantly more feasible compared to previous ridge configurations. This improvement is primarily due to the enhanced total thickness, which contributes to a more symmetrical waveguide structure and increases modal overlap. Moreover, Fig. 3b confirms the waveguide's spatially monomodal nature for these dimensions, supporting only the TE00 and TM00 modes. This monomode behavior arises from the minimal effective index differences associated with shallow etch depths, enabling the confinement and propagation of fundamental modes while effectively suppressing the existence of higher-order modes.

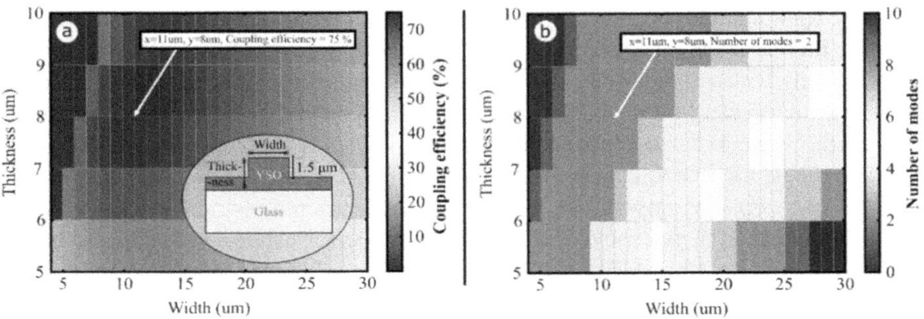

Fig. 3. a) Coupling efficiency for YSO-SiO$_2$ Shallow ridge; b) Number of modes (TE & TM)

The shallow ridge design of the waveguides facilitates fabrication but significantly constrains their resistance to curvature. This limitation arises because the membrane can support membrane modes with effective indices that closely match those of the modes in the ridge section.. To address this, we opted to initially simulate the transmission characteristics of an S-bend, a commonly used configuration in integrated photonics that allows for efficient spatial separation between input and output and simplifies the injection process. Of the various S-bend designs, the Euler bend is the most prevalent, primarily due to its effectiveness in minimizing curvature losses [20].

For simulation efficiency, we developed an algorithm leveraging the commercial ANSYS Lumerical EME (Eigenmode expansion) solver. This algorithm calculates the curvature radius along the length of the waveguide and segments the device into sections where the lengths is chosen based on the rate of curvature change. The greater the curvature variation, the shorter the section. The number of segments was determined through convergence testing. Key parameters for the simulation include input-output spatial offset, etch depth, and total membrane thickness, while maintaining a waveguide width of 11 µm based on previous findings regarding coupling efficiency (Fig. 4).

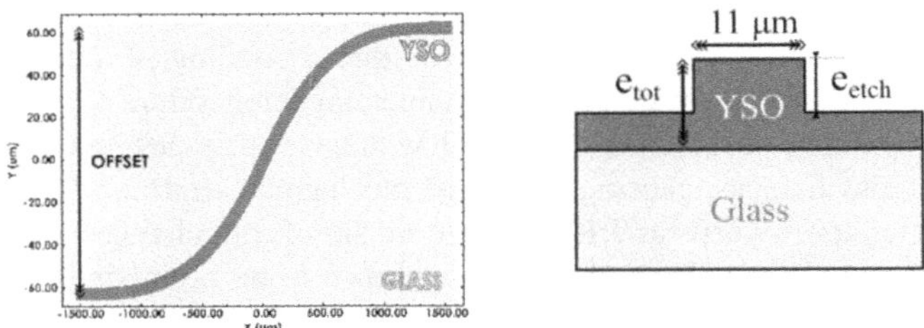

Fig. 4. Simulation parameters for Euler S-bend.

Figure 5 illustrates the transmission characteristics of the S-bend as a function of input-output offset across various etch depths. We conducted simulations for total membrane thicknesses ranging from 1 to 8 µm. For a typical offset of 125 µm with an etch depth limited to 1.5 µm, the ideal waveguide for mode matching presented in Fig. 3 (8 µm total membrane thickness, 11 µm width) exhibits a transmission efficiency of less than

1%. This significant loss can be attributed to the minimal difference in effective indices between membrane modes and ridge modes, where the curvature allows mode coupling not only with waveguide modes but also with multiple membrane modes, resulting in considerable bend losses. Notably, deeper etching enhances transmission efficiency, causing the waveguide to approximate the behavior of a ridge waveguide. Similarly, reducing the membrane thickness yields similar results. Thus, a technological optimization decision becomes imperative. If fabrication constraints restrict etch depth, as we will discuss further, then achieving effective S-bend performance necessitates additional membrane thinning. Conversely, if further thinning is unfeasible, prioritizing etch depth and recipe optimization is crucial for improving transmission performance.

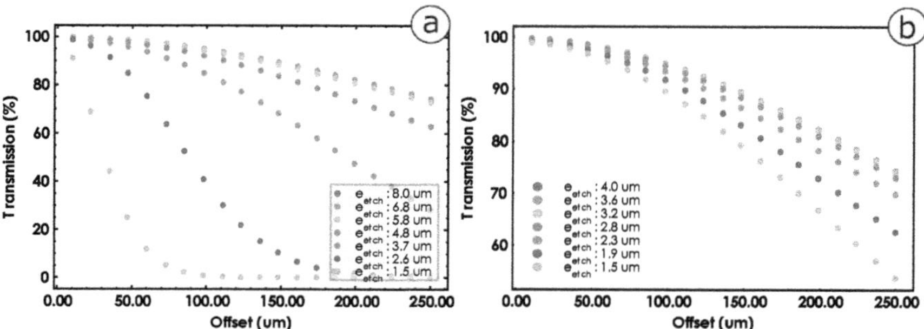

Fig. 5. Euler S-bend transmission for two total membrane thicknesses (a: 8 μm; b: 4 μm) as a function of input-output offset, with varying etch depth

3 Fabrication Process: Bonding, Thinning, Etching

3.1 Bonding

The following paragraph addresses the bonding aspects of YSO. The initial technique explored is direct molecular bonding, as it is supposed to maximize light interaction with the REI-doped crystal and is compatible with cryogenic conditions. As demonstrated in [21], we have successfully achieved direct bonding of a commercial 10 × 5 mm YSO crystal to borosilicate glass at an annealing temperature of 520 °C. However, cracks (Fig. 6a) can be observed in the crystal, which we attribute to the slight difference in YSO and glass flatness, causing increased mechanical strain on the bonded stack. Multiple temperature sweeps to 7 K indicated no significant changes in the assembly's structural integrity. While this technique has shown to be promising, it presents substantial fabrication challenges that must be addressed to ensure reliable outcomes. Key constraints include surface cleanliness, roughness, and flatness. Specifically, achieving superior surface quality on YSO has been a persistent challenge due to its hard and brittle nature, compounded by the absence of established polishing protocols in the literature. To address this issue, we have utilized an intermediate layer of benzocyclobutene (BCB) as an intermediate bonding medium. BCB is a widely used adhesive in semiconductor applications and provides a more manageable approach. However, this intermediate layer must remain minimal in thickness (< 300 nm) to preserve the confining properties of the waveguide. Although this solution addresses the roughness concerns, it still necessitates good surface flatness, requiring at least $\lambda/3$ flatness for an 8×8 mm square (taking the 633 nm of HeNe laser as λ reference). This standard, while less stringent than the $\lambda/10$ flatness required for molecular bonding, still presents significant difficulty for fabrication. An example of such bonding is displayed in Fig. 6b).

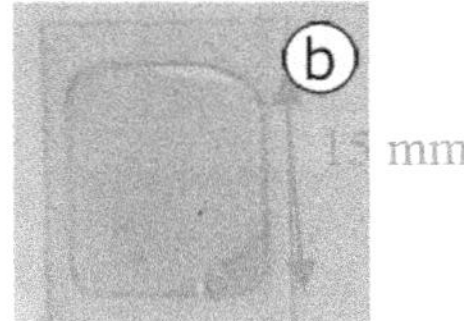

Fig. 6. a) Direct bonding of YSO on Glass. b) BCB bonding of YSO on glass. Interference patterns observed on the lower right-hand corner suggest air pockets that lead to poor bonding. We attribute this effect to flatness differences between the YSO slab and glass

3.2 Lapping and Polishing

Results were obtained with a Logitech PM5 lapping machine, with sample flatness measured via a Fizeau interferometry system (Mahr Type FI 1040 Z). This setup enables direct monitoring of samples on the fixed polishing head, mounted on a jig. In accordance with previously established requirements, the initial polishing phase is focused on achieving optimal flatness for effective bonding. A flat, slightly convex profile is critical, particularly with 80% of the central surface exhibiting a Peak-To-Valley (PV) value below 300 nm. During the polishing process, we continuously monitored sample shape and plateau condition, adjusting parameters such as pressure, plateau spinning speed, and jig positioning every 2 to 3 min. As a result, we achieved a surface flatness of $\lambda/6$, as depicted in Fig. 7a. It is noteworthy that this flatness was recorded while the samples were still on the jig. After removal, an increase in the PV value was observed, likely attributed to thermal effects from the gluing process, which may have induced significant deformation in the samples' profiles (see Fig. 7b).

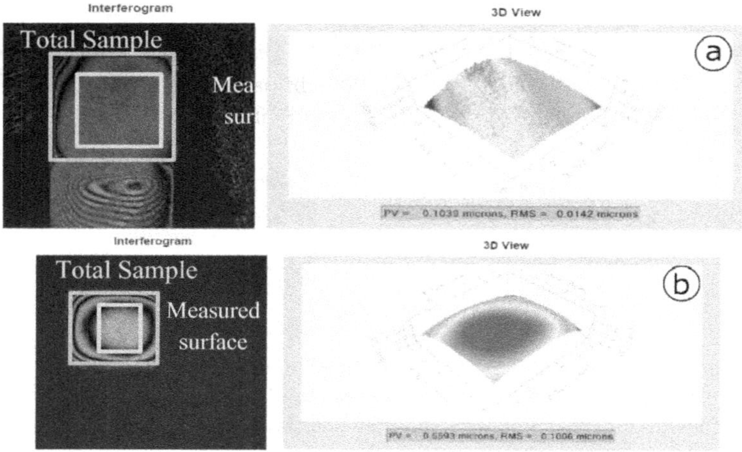

Fig. 7. Fizeau interferometry measurements of a commercial YSO sample under surface preparation polishing. a) Before jig take off PV $= 0.104$ µm. b) after JIG take off PV $= 0.559$ µm. The significant PV value difference is attributed to the heating and gluing process that could have caused material deformation. Peak-to-Valley (PV)

Following the bonding procedure, a lapping process is undertaken to achieve micrometer-level thickness in the crystal membrane. The development of this protocol is beneficial not only for applications involving ridges and shallow ridges of YSO but also essential for single rare-earth ion detection, where a micrometer-thin membrane of YSO is utilized within a Fabry-Perot cavity [22]. The lapping process is performed on a cast iron plateau using progressively finer Al_2O_3 grain size solutions. Throughout this procedure, the thickness is continuously monitored using a precision indicator, with final verifications conducted through a Thetametrics white light reflectance spectrometer for micrometer thickness samples and a Dektak XT. Stylus mechanical profilometer for slightly thicker ones. As illustrated in Fig. 8, we achieved an 8 µm-thick membrane that maintained commendable surface homogeneity. Notably, the membrane exhibited no cracks or displacements during the lapping process, with the exception of initially unbonded regions, as shown by the preliminary interference pattern (Fig. 8b). This highlights the critical role of the bonding process in the overall fabrication protocol, as it has a direct impact on the final flatness and effective surface area of the membrane.

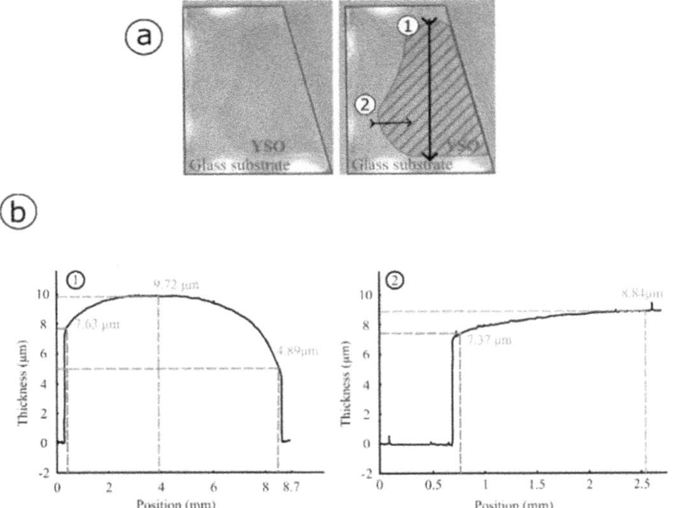

Fig. 8. a) Lapped YSO membrane from the bonded assembly in Fig. 4b) with a delimitation of YSO and glass on the right picture; b) Two stylus profilometry following paths 1 and 2 are shown as an example, indicating a convex shape and a mean thickness of around 8 μm.

3.3 Etching

After completing the polishing steps, we will present our findings regarding the ICP (Inductively Coupled Plasma) Reactive Ion Etching (RIE) of YSO, with further details available in reference [23]. The etching mechanism involves both physical etching, driven by accelerated Argon (Ar) ions, and chemical etching relying on the reactive plasma environment to enhance material displacement and etch rates. The interplay between these two contributions plays a critical role in defining the quality, morphology of the etched surface, etch rate, and selectivity. ICP-RIE is a well-established etching method utilized extensively for various oxides and glasses. Notably, JB Bradley's research [24] evaluated a spectrum of chemistries, including Fluorine, Chlorine, and Bromide-based systems, particularly focusing on Al_2O_3 and Y_2O_3. This study highlighted Chlorine-based etching as superior, achieving the highest etch rates for Y_2O_3, thereby establishing a solid foundation for defining etching parameters for YSO. Several ICP chlorine-based etching processes have been investigated to fabricate shallow ridge waveguides in YSO. The plasma compositions explored include Cl_2/Ar, Cl_2/N_2, and Ar-only configurations. This variation allows for a thorough examination of the underlying etching mechanisms. Dielectric masks, such as SiO_2 or SiN, were chosen based on their compatibility with the respective plasma compositions. Experimental results indicate that the etching process is predominantly ion-driven. The Ar-only plasma configuration produced the highest YSO etch rate at 80 nm/min, with a selectivity ratio of 0.5 using a SiN mask under high ICP power (800 W) and a considerable RF bias (350 V); however, this setup led to trenching issues. A decrease in RF bias to −100 V reduced trenching significantly, although it also resulted in a lower etch rate, but increased selectivity, both of which can be attributed to the reduced energy of the Ar ions. In contrast, both Cl_2/Ar and

Cl_2/N_2 plasma configurations, utilizing chlorine ions, yielded lower etching rates and selectivities against a SiO_2 mask but resulted in smoother sidewalls and eliminated trenching. SEM images of the etched YSO under these conditions are shown in Fig. 9. Practically, the etch depth achievable for YSO is constrained by the thickness of the dielectric mask; given the selectivity values obtained, a 3 μm mask thickness is required to etch down to 1.5 μm of YSO. Utilizing higher dielectric thicknesses would necessitate stricter lithography requirements to sustain vertical etch walls. With the current etching parameters, achieving a 1.5 μm YSO etch depth with vertical walls for an 11-μm wide shallow ridge geometry, as discussed in Sect. 2, remains feasible.

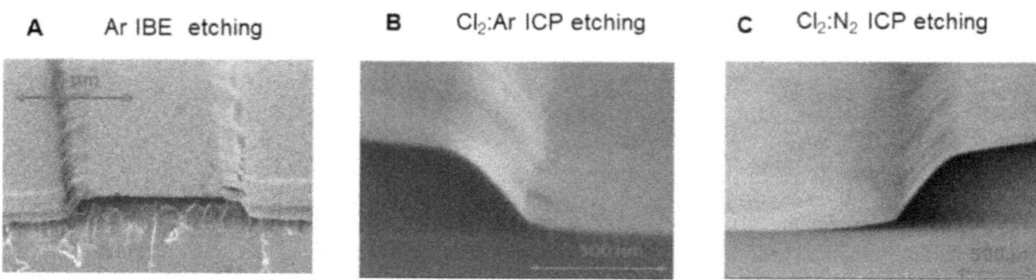

Fig. 9. SEM images of etched YSO with different plasma compositions. Data and discussions will be presented in detail in [23]

4 Conclusion

REI-doped materials are emerging as strong candidates for quantum repeaters and memory applications due to their long optical and spin coherence times, especially when doped into low magnetic noise matrices, such as YSO. Nevertheless, the intrinsic weak light-matter interaction in bulk samples significantly limits optical storage efficiency. This can be addressed through integrated photonics, which allows for the confinement of light within a reduced modal volume. In this study, we detailed improvements to our dry etching fabrication process, highlighting key procedures such as bonding, lapping, and etching. These advancements pave the way for the reproducible and scalable fabrication of YSO waveguides directly from bulk substrates, ensuring the retention of their unique spectroscopic features while offering design versatility. Future investigations will focus on performing waveguide characterization under cryogenic conditions. Additionally, we will undertake spectroscopic measurements to determine any potential degradation in ion characteristics and investigate light-matter interactions, possibly through Rabi frequency measurements, to evaluate enhancements in efficiency and overall performance.

References

1. Kimble, H.J.: The quantum internet. Nature **453**(7198), 1023–1030 (2008)
2. Dieks, D.: Communication by EPR devices. Phys. Lett. A **92**(6), 271–272 (1982)
3. Sangouard, N., Simon, C., de Riedmatten, H., Gisin, N.: Quantum repeaters based on atomic ensembles and linear optics. Rev. Mod. Phys. **83**(1), 33–80 (2011)

4. Briegel, H.-J., Dür, W., Cirac, J.I., Zoller, P.: Quantum repeaters: the role of imperfect local operations in quantum communication. Phys. Rev. Lett. **81**(26), 5932–5935 (1998)

5. Goldner, P., Ferrier, A., Guillot-Noël, O.: Rare Earth-Doped Crystals for Quantum Information Processing, in Handbook on the Physics and Chemistry of Rare Earths, vol. 46, Elsevier (2015)

6. Hosseini, M., Sparkes, B.M., Campbell, G., Lam, P.K., Buchler, B.C.: High efficiency coherent optical memory with warm rubidium vapour. Nat. Commun. **2**(1), 174 (2011)

7. Togan, E., et al.: Quantum entanglement between an optical photon and a solid-state spin qubit. Nature **466**(7307), 730–734 (2010)

8. Böttger, T., Thiel, C.W., Cone, R.L., Sun, Y.: Effects of magnetic field orientation on optical decoherence in Y2SiO5. Phys. Rev. B **79**(11), 115104 (2009)

9. Zhong, M., et al.: Optically addressable nuclear spins in a solid with a six-hour coherence time. Nature **517**(7533), 177–180 (2015)

10. Ortu, A., Holzäpfel, A., Etesse, J., Afzelius, M.: Storage of photonic time-bin qubits for up to 20 ms in a rare-earth doped crystal. npj Quantum Inf. **8**(1), 29 (2022)

11. Rakonjac, J.V., Lago-Rivera, D., Seri, A., Mazzera, M., Grandi, S., de Riedmatten, H.: Entanglement between a telecom photon and an on-demand multimode solid-state quantum memory. Phys. Rev. Lett. **127**(21), 210502 (2021)

12. Rakonjac, J.V., Grandi, S., Wengerowsky, S., Lago-Rivera, D., Appas, F., de Riedmatten, H.: Transmission of light-matter entanglement over a metropolitan network. Optica Quantum **1**(2), 94–102 (2023)

13. Zhu, T.-X., et al.: A metropolitan-scale multiplexed quantum repeater with bell nonlocality. arXiv preprint arXiv:2508.17940 (2025)

14. Saglamyurek, E., et al.: A multiplexed light-matter interface for fibre-based quantum networks. Nat. Commun. **7**(1), 11202 (2016)

15. Chakraborty, T., et al.: Towards a spectrally multiplexed quantum repeater. npj Quantum Inf. **11**(1), 3 (2025)

16. Craiciu, I., Lei, M., Rochman, J., Bartholomew, J.G., Faraon, A.: Multifunctional on-chip storage at telecommunication wavelength for quantum networks. Optica **8**(1), 114–121 (2021)

17. Zhu, T.-X., et al.: Integrated spin-wave quantum memory. Nat. Sci. Rev. **11**(11), nwae161 (2024)

18. Böttger, T., Sun, Y., Thiel, C.W., Cone, R.L.: Spectroscopy and dynamics of Y2SiO5 at 1.5µm. Phys. Rev. B, **74**(7), 075107 (2006)

19. Van Asch, J., et al.: A methodical approach to design adiabatic waveguide couplers for heterogeneous integrated photonics. J. Phys. Photonics **6**, 045013 (2024)

20. Zhang, L., et al.: Low-loss, ultracompact n-adjustable waveguide bends for photonic integrated circuits. Opt. Express **31**(2), 2792–2806 (2023)

21. Talneau, A., Welinski, S., Ferrier, A.: Rare-earth-doped Y2SiO5 crystal directly bonded on glass for efficient optical quantum technologies platform. physica status solidi (a), **221**, 2300718 (2024)

22. Ulanowski, A., Früh, J., Salamon, F., Holzäpfel, A., Reiserer, A.: Spectral multiplexing of rare-earth emitters in a co-doped crystalline membrane. Adv. Opt. Mater. **12**(15), 2302897 (2024)

23. Talneau, A., et al.: High surface quality Y2SiO5 silicate-crystal waveguides etched by chlorine-based inductive coupled plasma reactive ion etching. J. Vac. Sci. Technol. A **43**(4), 043002 (2025)

24. Bradley, J., Ay, F., Worhoff, K., Pollnau, M.: Fabrication of low-loss channel waveguides in Al2O3 and Y2O3 layers by inductively coupled plasma reactive ion etching. Appl. Phys. B **89**, 311–318 (2007)

Entanglement Routing in Large-Scale Quantum Networks with Ground-Space Interconnection

Agathe Blaise[1(✉)], Luca Paccard[2], Fabrice Arnal[2], Laurent de Forges de Parny[2], and Stéphanie Molin[1]

[1] Thales SIX GTS, 4 avenue des Louvresses, 92230 Gennevilliers, France
`agathe.blaise@thalesgroup.com`
[2] Thales Alenia Space, 26 avenue Jean-François Champollion - BP 33787, 31037 Toulouse-Cedex 1, France

Abstract. Quantum Information networks (QIN) aim to interconnect quantum systems via quantum repeaters through end-to-end entanglement of qubits. The integration of both satellites and ground segments is essential to provide global connectivity, even over large distances. In this paper, we study the problem of entanglement routing in large-scale quantum networks – i.e., with a large number of nodes, with a hybrid space-ground architecture. We decompose the routing problem into two subcomponents: request scheduling and path selection, and structure our approach around two phases: a proactive and a reactive step. We design and implement an entanglement routing strategy in an emulated European quantum network, first in a custom Python-based simulator for performance forecasting, and then into the NetSquid quantum network simulator. Our work thus provides a tool allowing for a comparison between terrestrial and satellite segments in terms of successful end-to-end entanglement in a large-scale network and shows that both segments are required for optimal service. Our results indicate that, within the emulated European topology, terrestrial networks can support end-to-end entanglement for approximately 35% of user pairs – the remaining pairs are too distant and suffer from prohibitive losses. In contrast, the satellite-based network achieves connectivity for up to 91% of user pairs, although some pairs remain unreachable due to a lack of simultaneous satellite visibility. However, while satellites can outperform terrestrial networks in total photon distribution, they may not be suitable for low-latency requests if out of visibility, making ground networks essential. This highlights the complementary roles of space and terrestrial segments in quantum networking.

Keywords: QIN · Quantum repeaters · Entanglement routing

A. Blaise and L. Paccard—Contributed equally to this work.

1   Introduction

The future Quantum Internet will enable quantum communication between local quantum processors located anywhere on Earth. Such an infrastructure will support a wide range of applications, including secure communications, distributed quantum computing, and high-precision clock synchronization.

The fundamental resource of a QIN is *entanglement distribution*. One of its key building blocks is quantum repeater, which allows the transfer of quantum states over long distances through *entanglement swapping* – as optical amplification is prohibited in QIN. We consider the establishment of entanglement between two distant quantum nodes – A and C – via an intermediate repeater node, B. Quantum sources generate entangled Einstein-Podolsky-Rosen (EPR) pairs locally and send them to both the repeater and the endpoint nodes. Upon receiving one entangled particle from each source, B performs a projective measurement and transmits appropriate instructions to C over a conventional communication channel. C can then apply a corrective operation on her qubit, which creates an end-to-end (E2E) entanglement with the qubit received by A. The entanglement done locally at every repeater for every link toward an adjacent node can fail with a non-negligible probability, but it is heralded by some technology means such that the outcome is known to the repeater itself. We then assume that there is a network element that collects all these outcomes and, therefore, has perfect knowledge of which links between quantum computers can be used to transfer the status of qubits. Thus, the controller tries to satisfy a set of demands from the upper layer applications by performing an ephemeral **routing** of multiple requests, which is only valid until the next round of local link entanglements.

The integration of **satellite segments** with terrestrial networks aims to provide global connectivity, even over large distances. A global quantum Internet will therefore consist of both space segments, with satellites serving ground stations, and terrestrial segments. Satellites are expected to operate in Low or Medium Earth Orbit (LEO or MEO), as these altitudes allow for a better photon reception rate on the ground.

These systems would be composed of multiple satellites organized in constellations, to enable the deployment of terminals anywhere within the coverage area and move toward near-real-time service.

In this work, we focus on the problem of **entanglement routing** in large-scale quantum networks that include both satellite and terrestrial components. We provide a tool allowing for a comparison between terrestrial and satellite segments in terms of successful E2E entanglement and show that both segments are required for optimal service. We also propose and evaluate several strategies that address request scheduling and path selection. The remainder of the paper is organized as follows. Section 2 introduces the related work on entanglement routing and large-scale quantum networks. In Sect. 3, we present the architecture of quantum networks and detail our methodologies for entanglement routing in large-scale quantum networks. Section 4 presents our results obtained via extensive simulations. Finally, Sect. 5 concludes the paper.

2 Related Work

Many studies have addressed the problem of entanglement routing in QIN. Some challenges are similar to those found in classical networks, such as handling multiple simultaneous connection requests between different user pairs, request scheduling and resource allocation, multi-path management, network reconfiguration, and link capacity estimation. However, there are also several challenges arise due to the specific to the nature of quantum networks, such as: 1) the stochastic nature of links, due to the low success probability of entanglement distribution, 2) the very short lifetime of quantum memories, 3) the handling of concurrent entanglement requests, which can lead to congestion on quantum links as each link can only handle one request at a time unlike packet switching in classical networks. Several major strategies for entanglement routing have emerged in the literature.

The first major distinction is between **proactive** (continuous) and **reactive** (on-demand) routing strategies. Proactive strategies involve continuously generating entangled pairs between neighboring nodes, allowing fast E2E entanglement establishment through entanglement swapping once a request arrives. Reactive strategies, by contrast, initiate entanglement generation only upon receiving a request. They begin by selecting an appropriate path, then sequentially establish entanglement between neighboring nodes and perform entanglement swapping as needed – reducing idle occupancy and decoherence risks.

A second distinction involves **centralized** vs. **distributed** routing approaches. Centralized approaches [1,5,8,10,13] rely on global knowledge of the network and use classical algorithms like Dijkstra to compute paths based on physical characteristics (link transmittance, EPR pair generation rate). These assume either each node or a controller maintains up-to-date knowledge of the entire network state and produces globally optimal routing decisions, maximizing network throughput and minimizing latency. On the other hand, distributed approaches [2,11] operate using only local information shared among neighboring nodes, making them more suitable for dynamic and large-scale networks.

Another critical aspect of routing is the management of **concurrent** entanglement requests, which can compete for limited quantum resources (e.g., memories or channels). For instance, [7] proposes a centralized scheduler that determines paths, orchestrates request timing, and disseminates decisions across the network. Finally, **multi-path** routing [6,8,11] supports robustness and load balancing by precomputing multiple routes for a single user pair. If one route becomes unavailable, due to an entanglement swap failure or expired links, alternative paths can be quickly utilized.

New challenges to the entanglement routing problem arise with the integration of satellites into large-scale QIN: time-varying topologies due to satellite motion, atmospheric losses and weather sensitivity affecting link reliability, and short-time communication windows. The work in [4] defines a high-level architecture of a generic QIN, then focuses on the architecture of the space segment.

Specific characteristics must be considered when selecting a routing algorithm. Proactive strategies can be particularly useful for atmospheric links, which are unstable and affected by external factors such as weather. Efficient routing in networks combining terrestrial and satellite components requires hybrid-aware algorithms that can exploit satellites for long-distance connectivity. This entails dynamic path planning based on factors such as inter-city traffic demand, meteorological conditions, and dynamic satellite visibility maps. Leveraging satellite links in this way can relieve congestion in terrestrial networks and enable global entanglement distribution.

3 Large-Scale Entanglement Routing

3.1 Quantum Network Architecture

Figure 1 illustrates the reference architecture of the global quantum Internet network under study, composed of multiple metropolitan networks interconnected via both space segments (satellites serving ground stations) and terrestrial segments (when the physical distance between access stations is reasonable, i.e., less than a hundred kilometers). The satellites are equipped with an entangled photon source and handle the distribution of the photon pairs to the ground – we do not consider Inter-Satellite Links in this scenario.

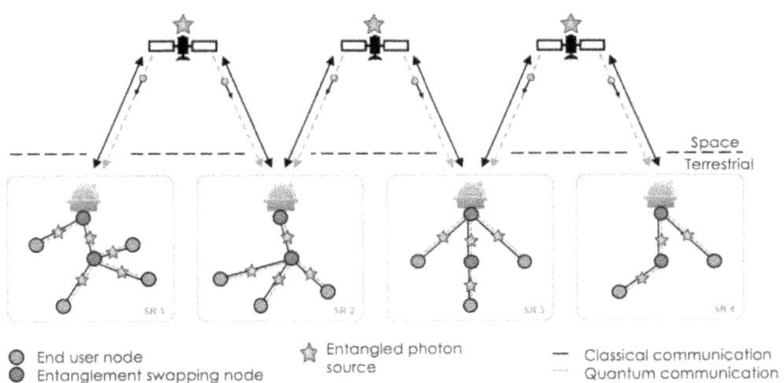

Fig. 1. Reference architecture.

3.2 Problem Definition

Let $\mathcal{G}(\mathcal{V}, \mathcal{E})$ be a static graph physical defining the network so that:

- Each node in $\mathcal{V}$ (with $N = |\mathcal{V}|$) hosts a quantum repeater;
- Each edge in $\mathcal{E}$ (with $M = |\mathcal{E}|$) is a lossy optical channel of length L_e (km) and power transmittivity $\eta_e = e^{-\alpha L_e}$, where α depends on the propagation media of the quantum channel. Each supports $S(e) \in \mathbb{Z}^+$ parallel channels(e.g., spatial, spectral, polarization modes);
- The network serves K source-destination pairs (A_j, B_j), for $1 \leq j \leq K$, with $A_j, B_j \in \mathcal{V}$.

The time is slotted, with the assumption that each memory can hold a qubit perfectly for $T \geq 1$ time slots. At the link level, entanglement generation is attempted repeatedly over the $S(e)$ parallel channels available on each edge e. Each attempt succeeds with probability $p_0(e) = e^{-\alpha L_e}$.

Two probabilities are defined throughout the whole process:

- Probability p (related to the external phase) that is the success probability of local entanglement between two quantum repeaters;
- Probability q (related to the internal phase) that is the success probability of entanglement swapping (internal-link attempts).

4 Simulations and Experiments

In this section, we describe the continental European quantum network we simulated and the results of our forecast and operational simulations. We note that it sometimes rely on simplifying assumptions and parameters, like the lifetime and heralding capabilities of quantum memories. Additionally, our geographic model is limited to continental Europe: we do not include overseas territories not to introduce too much complexity to the problem.

In this work, we focus on polarization encoding, which remains one of the most widely used schemes in free-space quantum communication. Polarization qubits are straightforward to generate and measure, and they have already been successfully employed in satellite-based experiments. While alternative encodings such as time-bin or orbital angular momentum (OAM) may offer advantages in certain scenarios, polarization is currently the most practical choice for long-distance free-space links. We note, however, that polarization encoding is also subject to logical errors due to atmospheric turbulence and misalignment, which must be compensated in a full implementation.

4.1 Emulated Network

We define a hypothetical European network composed of 84 cities, including capitals and big cities of Western European countries. Each city is connected to its three nearest neighbors. This results in the interconnection graph shown in Fig. 2. Two approaches are considered to connect two end users located in different cities:

1. **Terrestrial**: Entanglement is distributed between the end users by relaying through other cities, using a hop-by-hop approach. The selection of intermediate nodes is made using Dijkstra to determine the shortest path, taking into account network constraints.
2. **Satellite**: Another means of connecting users is through a satellite. We focus on low Earth orbit (LEO) satellites, as they offer shorter propagation distances compared to medium (MEO) and geostationary (GEO) satellites, reducing transmission losses and increasing the photon detection probability.

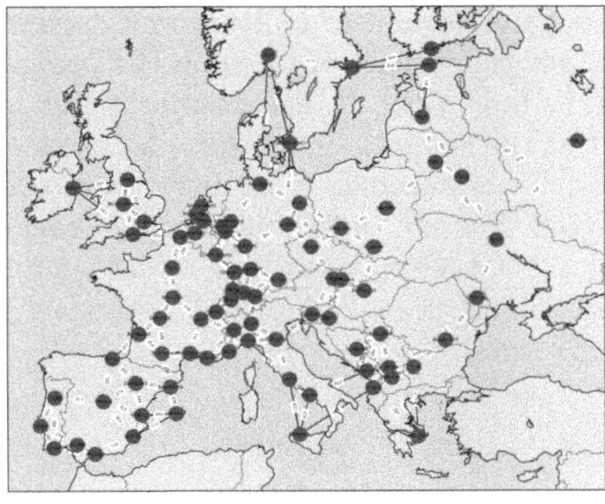

Fig. 2. Interconnection graph of a possible European quantum network.

4.2   Forecast Performance Simulation

The objective is to study and compare the predictive performance of entanglement distribution on a continental Europe scale using a terrestrial network or a satellite link. The aim is to determine macroscopically the most efficient path to connect two users located in Europe.

Elementary Link. We model the quantum network as a chain of elementary links. An elementary link is composed of an entangled photon source (EPS), two propagation channels, and two quantum memories with N storage slots (or modes) each. Quantum memories are assumed to be heralded, meaning that writing a quantum state onto the memory outputs a signal indicating that the memory is storing a quantum state. We assume multimode quantum memories with a capacity of up to 50 *temporal* storage modes. Temporal multiplexing is particularly relevant for quantum repeater architectures, as it allows multiple entanglement attempts within the coherence time of the memory. Other types of multiplexing, such as spatial or orbital angular momentum modes, are more vulnerable to degradation during long-distance free-space propagation and are therefore not considered in this work.

The elementary links are connected through a Bell State Measurement (BSM) module as highlighted in Fig. 3.

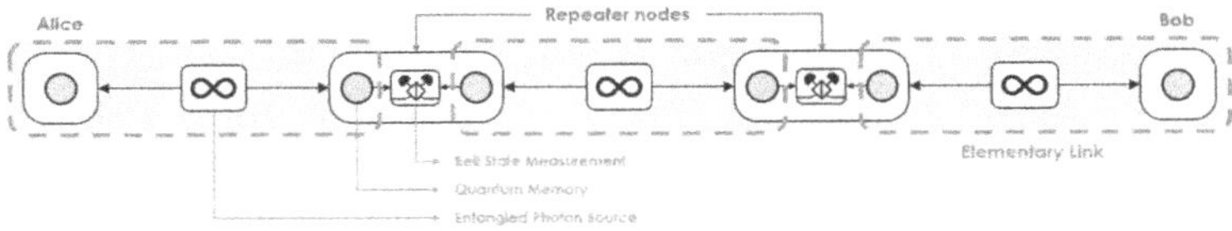

Fig. 3. Chain of quantum repeaters.

We aim to take advantage of the multi-modal capability of the quantum memories, as they reduce the impact of losses in the propagation channels. In a chain of two elementary links (Fig. 4), if a photon is lost between a source and a node on link A, but at the same time another photon is received from another pair, then the latter can be stored in a memory until a photon from another pair of link B is received. We can individually model each elementary link and connect them through a BSM. Our objective is then to determine the probability of successfully transmitting at least one pair of entangled photons between the two nodes after N time slots.

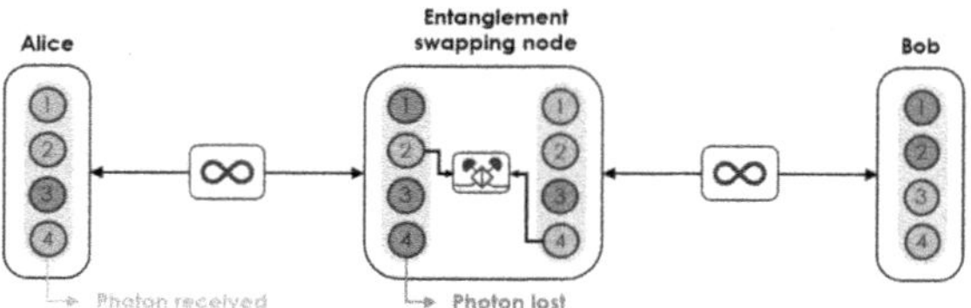

Fig. 4. Two elementary links of a chain of quantum repeaters.

We find that the probability of successfully transmitting at least one EPR pair between the two end-user nodes, on a chain of M quantum repeaters is:

$$p^{chain}_{(success \geq 1)} = \frac{\left[1 - \prod_{t=1}^{N}(1 - \eta_{src} \cdot \eta_{ch}^2 \cdot \eta_{QM}^2)\right]^{M+1} \times \left[\eta_{BSM} \cdot \eta_{det}^2\right]^{M}}{N} \tag{1}$$

The development of the model can be found in Sect. A.

Performances of Terrestrial and Space Paths. We compare the performance of the entanglement distribution between end-users with a terrestrial path and a satellite path. For the terrestrial path, we consider the network proposed in Fig. 2 and connect the end-users with the different nodes in the network using a shortest path algorithm. For the space path, we consider a LEO satellite with an altitude of 800 km and an inclination of 60°. We run a 4-month simulation using STK (Systems Tool Kit) and compute the dual link budget between the satellite and the pairs of ground users. The total dual visibility is also considered to take into account the availability of the satellite during the simulation period. Details of these data can be found in Sect. B.

Figure 5 illustrates the number of EPR pairs distributed over the simulation period of 4 months. Each point corresponds to a pair of users. For the ground network, the link budget between end-users is computed using Eq. 7. To obtain the number of pairs distributed over 4 months, we consider that the network is running 24 h a day and 7 days a week during the period.

The satellite link is considered as an elementary link with a source (the satellite) and two quantum memories (the end-users). Its performance is thus computed using Eq. 5. We observe that a satellite link becomes more interesting than the ground network when the distance between the end-users exceeds 400 km. For lower distances, the ground network is naturally more effective due to lower losses and greater availability of the service. However, even when considering its limited availability, the satellite allows the distribution of a greater number of EPR pairs than the ground network. Beyond a limit distance of around 2200 km, we observe a cut-off in the number of entangled photons pairs distributed by the satellite because, beyond this distance, two nodes cannot be simultaneously visible to the satellite according to the parameters chosen for the simulation (altitude, minimum elevation, etc.).

We note that these results depend on the assumptions and parameters used in the simulation. Our simulations are designed to capture the global behavior of entanglement distribution over the two network segments, enabling qualitative analysis of routing strategies and architectural trade-offs.

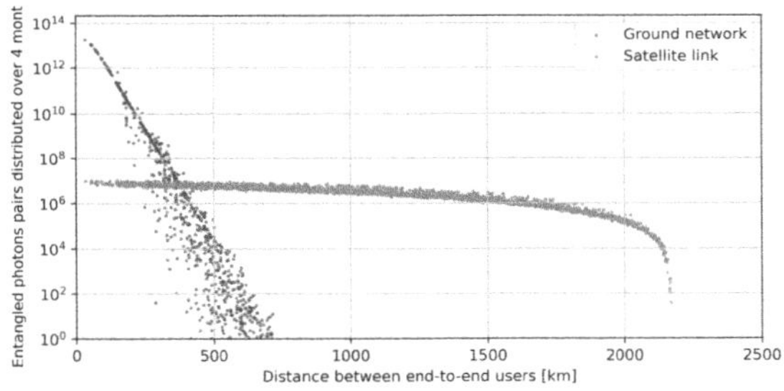

Fig. 5. Number of EPR pairs distributed over 4 months.

Satellite, while outperforming ground networks for distant users, has a significant drawback in terms of availability. Users needing low-latency entanglement resources might find the satellite on the opposite side of the planet, making terrestrial networks the only viable option.

For satellite-based quantum communication, additional effects such as Doppler shifts and imperfect mode matching must be taken into account. Doppler shifts can lead to temporal misalignment between entangled photons, while mode mismatch reduces interference visibility and therefore the overall success probability of entanglement swapping. Both effects can be modeled as an effective reduction in detection efficiency, meaning that the optimistic 99% efficiency assumed in our baseline analysis should be interpreted as an upper bound.

4.3 Operational Performance Simulation

The objective is to evaluate the routing performance as implemented in the NetSquid simulator. NetSquid [9] is a Python-based library designed to simulate a wide range of components in a quantum network, with a strong emphasis on real-time network behavior simulation.

Implementation. We developed a simulation of E2E entanglement routing in a quantum network using NetSquid. Our work builds upon a source code implementing routing [3], extends it with space (satellites, ground stations) and ground (metropolitan nodes) components, and implementing the topology of a European network. Quantum links encompass both fiber and free-space links. Each quantum node also contains a quantum processor (e.g., for entanglement swapping) and a quantum memory. More details can be found in Sect. C.

The selected path for connecting two stations is computed with the shortest path that maximizes the E2E available capacity, as described in the previous simulation. After finding the shortest path for each pair, we must determine which pairs will be processed and in what order, in cases where not all pairs can be handled at the current time slot. We implemented and compared several scheduling policies:

- **FIFO**: Pairs are processed in order of arrival without considering path costs, ensuring strong fairness.
- **Best Efficiency**: A path is skipped if its cost exceeds the average cost of all candidate paths, ensuring overall efficiency.
- **Weighted FIFO**: A path is skipped with a probability $P_{skip} = max\{0, 1 - \frac{\sigma}{E[c]}\}$, with σ the path cost and $E[c]$ the average path cost over all paths.

For each simulation, we select random pairs of cities to connect over T time slots. The tuning of parameters in the simulation is described in Sect. D.

Figure 6a illustrates the cumulative distribution function (CDF) of the average fidelities for the three selected scheduling policies. On average, higher fidelity is achieved with Best Efficiency and Weighted FIFO. This is because, by avoiding the processing of requests whose path cost is significantly above the average, network resources are used more efficiently, favoring entangled pairs whose nodes are not too distant. Figure 6b shows the CDF of the average delays for the three scheduling strategies. Both Weighted FIFO and Best Efficiency outperform FIFO, as they achieve lower average E2E entanglement distribution latency. However, we note that Best Efficiency and Weighted FIFO tend to favor shorter paths, which penalize other requests systematically.

We compare the average fidelity depending on the number of intermediate swaps in Fig. 7a, to evaluate the performance of the space and ground segments. We see that the configuration with two metropolitan nodes connected by a fiber link yields the highest fidelity. When a single intermediate swap is required, using either a satellite or a ground station results in roughly equivalent fidelity. However, beyond one swap, the fidelity is consistently higher using the satellite. Figure 7b further shows the fidelity as a function of the total distance traversed. We observe that the distribution of photons via the space segment results in higher fidelity, even though the physical distance is significantly greater.

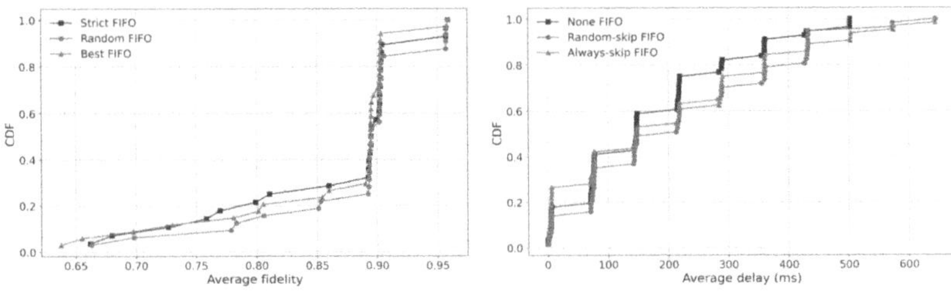

(a) CDF of the average fidelity.   (b) CDF of the average delay.

Fig. 6. CDF for E2E routing between 2 nodes for different scheduling strategies.

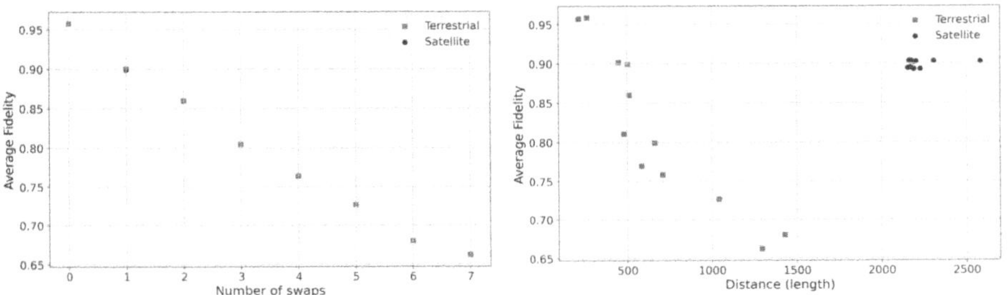

(a) Depending on the number of swaps.   (b) Depending on the distance (kms).

Fig. 7. Fidelity values for the terrestrial and satellite scenarios.

5 Conclusion

This paper investigated the challenges and potential of a QIN composed of both terrestrial and satellite-based components. We focused on the problem of establishing E2E entanglement between distant quantum nodes. We modeled an emulated European-scale quantum network, incorporating optimistic physical parameters for quantum memory, decoherence, link loss, and processing delays. Our simulations explored both forecast and real-time operational scenarios, allowing us to evaluate the performance of different routing mechanisms under various network conditions. We implemented our routing strategies within a simulation framework built on NetSquid, extended with a space-ground model including both fiber and free-space quantum links. Our results highlight the complementary roles of space and terrestrial segments in quantum networking.

A Model for forecast performance simulation

In this appendix, we will develop the model for the forecast performance simulations presented in Sect. 4.2.

The parameters involved are the following:

- η_{src}: efficiency of the EPS;
- η_{ch}: efficiency of the propagation channel (either free-space or optical fiber);
- η_{QM}: efficiency of the quantum memory (we include both writing and reading efficiencies in this parameter);
- N: number of storage modes of the quantum memory (i.e., multiplexing capabilities in the quantum memory).

The probability of failing to transmit entanglement between two nodes at time t is:

$$p_{fail}(t) = 1 - \eta_{src} \cdot \eta_{ch}^2 \cdot \eta_{QM}^2 \tag{2}$$

For the description of the model, the channel efficiencies will be considered equal, but could be tuned to reflect the real losses of the network.

The probability of failing to transmit entanglement between two nodes after N time slots is then:

$$p_{fail}^{all} = \prod_{t=1}^{N}(1 - \eta_{src} \cdot \eta_{ch}^2 \cdot \eta_{QM}^2) \tag{3}$$

Thus, the probability of transmitting at least one pair of entangled photons on an elementary link after N time slots is:

$$p_{(success \geq 1)}^{N} = 1 - \prod_{t=1}^{N}(1 - \eta_{src} \cdot \eta_{ch}^2 \cdot \eta_{QM}^2) \tag{4}$$

Generally speaking, when we consider a link budget, we look at the losses or reception probabilities for a photon (or a pair of photons in our case) on a single time slot. Here we need to divide the probability in Eq. 4 by the number of time slots considered to be able to compare the performances of links with quantum memories of various storage modes:

$$p_{(success \geq 1)} = \frac{1 - \prod_{t=1}^{N}(1 - \eta_{src} \cdot \eta_{ch}^2 \cdot \eta_{QM}^2)}{N} \tag{5}$$

A chain of quantum repeaters is simply a succession of elementary links, to which BSM modules are added as shown in Fig. 3. A BSM is characterized by two elements:

- η_{BSM}: intricate probability of success of the BSM (i.e., the projected state is either $|\psi^+\rangle$ or $|\psi^-\rangle$);
- η_{det}: efficiency of the single photon detectors in the BSM module (two detectors must click during a successful BSM).

Thus, entanglement swapping has the following efficiency:

$$\eta_{swapping} = \eta_{BSM} \cdot \eta_{det}^2 \tag{6}$$

From Fig. 3, we see that in a repeater chain composed of two repeater nodes (i.e., nodes performing the BSM), there are three elementary links. By generalizing to M repeater nodes, we count:

- M repeater nodes;
- $M + 1$ elementary links

From Eq. 5 and Eq. 6, the probability of successfully transmitting at least one pair of entangled photons between the two end-user nodes, on a chain of M quantum repeaters is:

$$p^{chain}_{(success \geq 1)} = \frac{\left[1 - \prod_{t=1}^{N}(1 - \eta_{src} \cdot \eta_{ch}^2 \cdot \eta_{QM}^2)\right]^{M+1} \times \left[\eta_{BSM} \cdot \eta_{det}^2\right]^{M}}{N} \tag{7}$$

B   Parameters for forecast performance simulations

Table 1 presents the parameters of the forecast performance simulation. Some are based on optimistic assumptions, as they reflect the expected performance of future hardware rather than the current state of the art. Recent advances in superconducting nanowire single-photon detectors (SNSPDs) have demonstrated system detection efficiencies as high as 99.5% at telecom wavelengths [12]. While such performance is not yet standard in laboratory environments, we use this value as an upper-bound benchmark to assess the ultimate performance of satellite-based quantum communication scenarios.

Table 1. List of the parameters of the forecast performance simulation.

Parameter	Symbol	Value
Wavelength	λ	1550 nm
Source rate	R_{src}	100 MHz
Source efficiency	η_{src}	0.25
Quantum memory efficiency	η_{QM}	0.9
Number of storage modes per quantum memory	N	50
BSM efficiency	η_{BSM}	0.5
Detector efficiency	η_{det}	0.99
Fiber attenuation factor	α	0.2 dB/km
Satellite telescopes diameter	D_{TX}	50 cm
Ground station telescope diameter	D_{RX}	80 cm

Table 2 illustrates the satellite data for the pairs of users Paris-Toulouse and Madrid-Amsterdam. The number of EPR pairs distributed is lower for Madrid-Amsterdam because the two users are further apart, hence reducing the mean elevation angle and increasing the mean slant range, as well as reducing the total duration of dual visibility.

Table 2. Example of satellite simulation data for two pairs of users.

User pair	Distance (km)	Mean slant range (km)	Mean elevation angle (°)	Dual visibility [hh:mm:ss]
Paris	588	1097	46.9	18:56:15
Toulouse		1063	49.2	
Madrid	1481	1187	39.6	05:49:28
Amsterdam		1199	39.0	

C   Details of the implementation within NetSquid

As highlighted in Fig. 8, the three nodes Alice, Bob, and Carol are interconnected by both quantum and classical links, each of equal length. Two types of nodes were implemented: a `TerrestrialNode` class for metropolitan nodes, which also serve as ground stations, and a `Satellite` class for the satellites. Both classes take as input the node's position (latitude, longitude, and altitude). Similarly, two types of quantum links are distinguished: fiber links on the terrestrial segment connecting two `TerrestrialNode` instances, and free-space links between a `TerrestrialNode` and a `Satellite` node. A quantum source is placed at the midpoint of each fiber link and co-located with each satellite, emitting EPR pairs toward the connected nodes.

The quantum channel introduces a constant propagation delay $\delta = d/c'$ through a link of length d (in kilometers), where c' is the propagation speed (set to 200,000 km/s for fiber, and 318,000 km/s in free space in the simulations). Additionally, the quantum channel may cause the loss of a qubit with a certain probability. Different loss models are used depending on the type of link. NetSquid includes a fiber loss model, `FibreLossModel`, where the probability of photon loss is given by: $p_{loss} = 1 - (1 - p_{init}) \cdot 10^{\frac{-\eta d}{10}}$, where p_{init} is the probability of qubit loss immediately after generation, and η is the attenuation factor along the fiber (in dB/km).

Figure 8 also illustrates the structure of a quantum node as implemented in the NetSquid simulations. At each node, every flying qubit received from a quantum source is immediately absorbed into its designated slot in the quantum memory. The node then determines whether the local entanglement attempt with the paired qubit was successful. Qubits stored in memory undergo decoherence due to dephasing noise, modeled by stochastically applying a Pauli-Z gate with probability: $p_{dephasing} = 1 - e^{-\Delta t \cdot R_{dephasing}}$, where $R_{dephasing}$ is the dephasing rate (in Hz), and Δt is the time elapsed since the qubit was absorbed into memory.

The quantum processor acts on the stored qubits to perform entanglement swaps, implemented via BSM on pairs of qubits, and applies correction operations using X and Z gates (based on received correction bits). Each operation requires a fixed time of 10 ns. Moreover, the X and Z gate instructions introduce noise modeled as depolarizing noise, by stochastically applying Pauli X, Y, and Z gates with probability: $p_{depol} = 1 - e^{-\Delta T \cdot R_{depol}}$, where R_{depol} is the depolarization rate (in Hz), and ΔT is the time required to perform the operation.

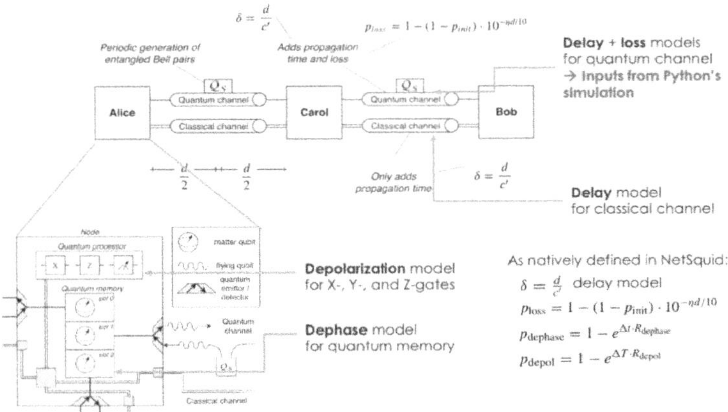

Fig. 8. Illustration of the nodes and channels as implemented in NetSquid.

D   Tuning of the parameters for the NetSquid simulation

The first step is to determine suitable parameters for the simulation. We select values for the depolarization rate $R_{depol} \in \{0.1, 1, 10, 100\}$ Hz and $R_{dephase} \in \{0.01, 1, 100, 10000\}$ Mhz. Figure 9a shows the fidelity of the distributed quantum state between two node pairs as a function of the number of intermediate quantum repeaters, for various depolarization rates. For $R_{depol} \leq 1$ Hz, the number of entanglement swapping operations has little influence on the E2E fidelity, which appears unrealistic. For $R_{depol} = 100$ Hz, the fidelity drops below 0.3 after only two entanglement swaps, which is too low. We then select an intermediate and realistic value: $R_{depol} = 10$ Hz. Figure 9b shows the fidelity of the distributed quantum state as a function of the number of repeaters, for various $R_{dephase}$ values, with $R_{depol} = 10$ Hz. We observe that for $R_{dephase} = 1$ MHz, the curves are nearly identical, suggesting that dephasing has no significant impact in that range. Starting from $R_{dephase} = 100$ MHz, a clear influence of the number of swapping operations is visible. We thus select the combination $R_{depol} = 10$ Hz and $R_{dephase} = 100$ MHz for our simulations.

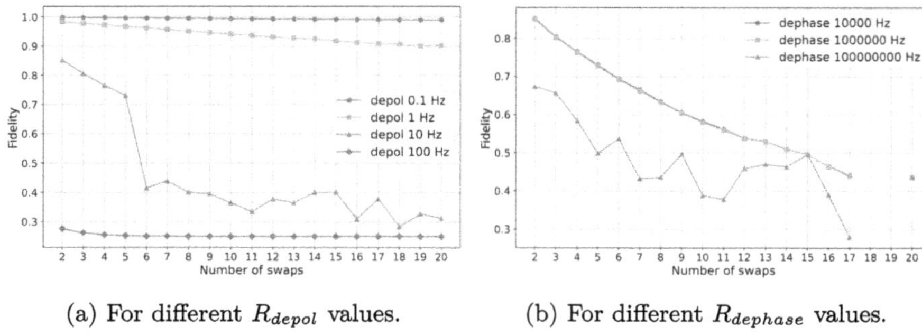

(a) For different R_{depol} values.

(b) For different $R_{dephase}$ values.

Fig. 9. Fidelity depending on the number of swaps.

References

1. Caleffi, M.: Optimal routing for quantum networks. IEEE Access **5** (2017)
2. Chakraborty, K., et al.: Distributed routing in a quantum internet (2019)
3. Cicconetti, C.: Quantum networking experiments with NetSquid (2025). https:// github.com/ccicconetti/netsquid
4. de Forges, L., et al.: Satellite-based quantum information networks: use cases, architecture, and roadmap (2022)
5. Gyongyosi, L., Imre, S.: Entanglement-gradient routing for quantum networks. Sci. Rep. **7** (2017)
6. Kozlowski, W., et al.: Designing a quantum network protocol. In: CoNEXT. ACM (2020)
7. Li, C., et al.: Effective routing design for remote entanglement generation on quantum networks. NPJ Quantum Inf. **7** (2021)
8. Pant, M., et al.: Routing entanglement in the quantum internet. NPJ Quantum Inf. **5** (2017)
9. QuTech: NetSquid: The Network Simulator for Quantum Information using Discrete Events (2025). https://netsquid.org/
10. Schoute, E., et al.: Shortcuts to quantum network routing (2016)
11. Shi, S., et al.: Concurrent entanglement routing for quantum networks: model and designs. IEEE/ACM Trans. Netw. (2024)
12. Shibata, H., et al.: Detecting telecom single photons with 99.5% system detection efficiency. APL Photonics, **6**(3), 036114 (2021)
13. Van Meter, R., et al.: Path selection for quantum repeater networks. Networking Sci. **3** (2012)

Paving the Road Towards Quantum-Safe Communications for the Alliance

Joanna Sliwa[✉], Konrad Wrona, and Mikolaj Pietruczuk

NATO Communications and Information Agency, The Hague, Netherlands
{joanna.sliwa,konrad.wrona,mikolaj.pietruczuk}@ncia.nato.int

Abstract. This paper presents an overview of experimental studies focusing on quantum-safe communications in military applications, conducted within the NATO Innovation Continuum. The experiments explore a range of topics, including cryptographic agility, secure software supply chains, and quantum-resilient friend-or-foe identification for unmanned aerial vehicles (UAVs). Collectively, these studies highlight challenges and design considerations for military communication and information systems (CIS) architectures that prioritize resilience to emerging quantum threats.

Keywords: cryptographic agility · quantum-safe satellite communications · key management systems · quantum key distribution (QKD) · post-quantum cryptography (PQC)

1 Introduction

Governmental and defense organizations worldwide increasingly recognize the need for cryptographic agility in their systems, required for securing their communications against the growing threat posed by quantum-enabled adversaries. While post-quantum cryptography (PQC) appears to be a promising short-term mitigation against cryptographically relevant quantum computers, it may prove vulnerable to new cryptanalytic attacks. Quantum security technologies, such as quantum key distribution (QKD), offer the promise of enabling secure communication channels that can withstand attackers with virtually unlimited computational resources. However, current QKD solutions also present limitations, including a short effective range and vulnerability to side-channel and availability-based attacks on the quantum channel. In such an uncertain environment, designing systems with built-in cryptographic agility is the most future-resilient approach.

The experiments planned to be conducted within the NATO Innovation Continuum (IC25) in 2025 are aimed at exploring ways to achieve quantum-secure communication channels in key military use cases over various transmission media, as well as to investigate the benefits, limitations, and performance of cryptographically agile systems' architecture. Moreover, as shown by extensive industry and deep tech participation in the experiments, NATO Innovation Continuum fosters cooperation with companies and national research centers developing dual-use solutions for the Alliance. Thanks to this

© The Author(s), under exclusive license to Springer Nature Switzerland AG 2026
F. Barbaresco and G. François (Eds.): QUEST-IS 2025, CCIS 2743, pp. 185–194, 2026.
https://doi.org/10.1007/978-3-032-13852-1_19

approach, we aim to accelerate innovation, reduce costs, and enhance the agility of supply chains in the defense industry.

2   Motivation

Some of the top concerns in modern warfare are staying at the forefront of innovation and delivering capabilities that meet the needs of operators and provide an advantage over potential adversaries. These concerns underscore the importance of investing in innovative activities and planning for the application of emerging and disruptive technologies in military contexts.

The NATO Communications and Information Agency (NCIA) recently published its Technology Strategy Towards 2030 [11], which addresses the need for delivering high-value, modern solutions for the Alliance and thus advancing the Technology, People, and Processes pillars of the NATO Digital Transformation Vision [12]. One of the key themes of the NCIA Technology Strategy is Quantum Technology (QT), broken down into four strategic objectives: mitigating the quantum threat, performing technology watch, innovating, and creating quantum-literate personnel.

Following its Technology Strategy, NCIA is actively engaged in short-term and long-term activities aimed at satisfying the objectives for the QT Theme. The most immediate ones relate to the approaching Q-day[1] and the necessity of implementing secure and resilient communication patterns. These efforts encompass engagement in the NATO STO Research Task Group on Quantum Technologies Vulnerabilities, which explores means to potentially disrupt the operation of quantum sensors, computers, and QKD devices with different attack vectors. The path from the laboratory to the field for QT must demonstrate that any new technology does not introduce additional security gaps in the system, providing a reliable and highly available capability to the warfighter. Another activity addresses long-term solutions that would make the communications infrastructure independent from the cryptographic primitives protecting these information flows, supporting cryptographic agility and resilience of the CIS. For this purpose, NCIA, together with its partners, prepares a series of experiments for the NATO Innovation Continuum exercise in 2025, organized by the NATO Allied Command Transformation (ACT).

The remainder of the paper provides details of these experiments, with a particular focus on the research covered by each of them.

3   Background

In the context of national security and defense, the implementation of cryptographically agile systems is of critical importance, mainly due to the extended operational lifespan of such infrastructures. These systems implement complex and rigid architectures that require fulfilling strict requirements for resilience and availability, making rapid cryptographic transitions particularly challenging. Additionally, the classified information

---

[1] Approaching moment of building Cryptographically Relevant Quantum Computer able to break current encryption schemes.

they store often has a prolonged confidentiality requirement, significantly increasing the threat posed by "Harvest-Now, Decrypt-Later" attacks. Mosca's Theorem [10] highlights this urgency, demonstrating that if the required confidentiality duration of data exceeds the time it will take to break current cryptographic algorithms, then measures towards securing such systems must be taken now. In military systems, where sensitive information may need to remain secure for decades, this timeline often extends well beyond current estimates for the introduction of Cryptographically Relevant Quantum Computers. The combination of long-term data sensitivity and low cryptographic agility renders such systems particularly vulnerable to emerging quantum threats.

Simultaneously, mobile communication platforms such as unmanned aerial vehicles (UAVs) and unmanned ground vehicles (UGVs) are playing an increasingly important role in tactical environments due to their agility and real-time responsiveness on the battlefield. However, they also introduce new risks, and if compromised, the consequences may be even loss of life. This potentially very high threat impact highlights the urgency of developing quantum-resilient communication mechanisms customized to mobile tactical platforms. Post-quantum cryptography (PQC) offers a practical and scalable solution for several key applications, including platform authentication and secure key exchange for encrypting real-time data, such as video streams transmitted over radio frequencies. Its compatibility with existing infrastructure makes it a more immediately viable approach. Nonetheless, quantum key distribution (QKD) performed via UAVs remains a valuable fallback, potentially useful in establishing encrypted ad-hoc networks in contested or infrastructure-less environments while reducing reliance on centralized key distribution. Exploring both technologies ensures greater resilience for future battlefield communications.

Cryptographic agility is understood as a capability for flexible addition of new cryptographic scheme implementations in a software or hardware system, to reduce the time required to transition and to allow for seamless updates to future cryptographic standards. The latest NIST Cybersecurity White Paper on Considerations for Achieving Crypto Agility [13] defines crypto agility in 3 dimensions, as the ability for machines to select their security algorithms in real time and based on their combined security functions, add new cryptographic features or algorithms to existing hardware or software, resulting in new, stronger security features, and the ability to gracefully retire cryptographic systems that have become either vulnerable or obsolete. Crypto agility can be used on a system, protocol and application level. In our experimental work, we focused on the ability of a system to swiftly and securely transition between cryptographic primitives as threats evolve or standards change over time.

If we consider that certain vulnerabilities can cause cryptographic algorithms be compromised and trigger their update, this approach in a long term also promotes cyber resilience, which is the system's ability to resist, minimize, and mitigate a degradation caused by a successful cyber-attack on a system or network of computing and communicating devices [9]. Cyber resilience by design means that such an ability is intrinsic to the system; the resilience mechanism seamlessly integrates with the primary functional elements and other modules of the system. Cyber resilience by design differs from cyber resilience by intervention, where the intervention of an external actor that "resides" outside of the system is required to provide resilience.

4 State of the Art

The National Institute of Standards and Technology (NIST) has been at the forefront of the effort to standardize post-quantum cryptography. In August 2024, NIST released initial PQC standards, including key encapsulation mechanism FIPS 203 (derived from CRYSTALS-Kyber) and lattice-based digital signature algorithm FIPS 204 (derived from CRYSTALS-Dilithium), crucial for quantum-resistant encryption [14]. Simultaneously, the ITU (International Telecommunication Union) is developing standards for quantum-safe networks (in the FG-QIT4N working group [15]), while ETSI (European Telecommunications Standards Institute) actively researches and promotes quantum-safe cryptography and Quantum Key Distribution (QKD) through dedicated working groups and conferences [16]. Migration plans promote creation of cryptographic inventories, emphasize cryptographic agility and hybrid key approaches [3, 4].

As already emphasized, cryptographic agility can be analysed in scope of the system, application and protocol. In this article we focus on the out of band key distribution and agility in selection of cryptographic key providers. Several interfaces define the state of the art in key supply. These are crucial for designing a cryptographically agile system as they effectively allow for decoupling the key management layer from the underlying transmission layer, which is the consumer of keys. Due to this characteristic, the application that is ingesting cryptographic keys is agnostic to the key establishment mechanism.

ETSI has played a central role in defining how applications receive keys. Especially important are two standards [16]: ETSI GS QKD 014, specifying a simple, widely adopted REST API that allows applications to request keys on demand and ETSI GS QKD 004, session-based interface that supports Quality of Service (QoS) parameters, such as guaranteed key rate and jitter. Based on QKD 004, a dedicated commercial key supplier interface has been proposed by CISCO. SKIP (Secure Key Integration Protocol) can provide Post-Quantum Preshared Keys (PPKs) for quantum-safe encryption in IKEv2 and IPsec VPNs. It enhances the standard IKEv2 protocol [8] by mixing traditional Elliptic Curve Diffie-Hellman (ECDH) keys with PPKs during the session key derivation process. Functioning as a REST API over HTTPS, SKIP forwards these PPKs from an external key source into switching devices (e.g. routers). This source, which can be a QKD device, a Key Management System (KMS), or another service, must implement the SKIP Server and use an out-of-band mechanism to deliver the identical PPK to both communicating peers.

5 Experiments

When seeking technical solutions for implementing cryptographic agility, we can assume that different key providers offer varying levels of key management security. This variation allows for selecting the optimal provider based on the transmission source's requirements (e.g., national security policy regulations) or the data type's classification (e.g., security marking of information). Additionally we focus on less dynamic scenarios: when the cryptographic algorithms implemented become compromised, and having the possibility of changing the key provider without necessity to redesign the system brings time and cost savings.

In our experiments we focus on the out of band key distribution and agility in selection of cryptographic key providers. We defined several experiments that will be carried out in October 2025 during the NATO Innovation Continuum (IC25) exercises, organized by the NATO Allied Command Transformation.

5.1 Design for Cryptographic Agility and Cyber Resilience

The first experiment focuses on the crypto agility by design and is aimed to explore the following research question: How does the decoupling of the key management layer from the transmission layer facilitate the creation of quantum-secure systems through cryptographic agility and cyber resilience in military communication systems? There are 3 objectives for experiment 1: evaluate the performance and integration feasibility of ML-KEM and QKD (MDI-QKD[2], DV QKD and BBM92 QKD) for secure key exchange in military communications.; implement a modular architecture that decouples the key management layer from the transmission layer to support independent cryptographic algorithm updates; demonstrate cryptographic agility and cyber resilience by testing the ability to swap key exchange mechanisms with minimal disruption to the underlying transmission channel.

The scenario assumes that military data centers serve as the backbone for tactical, operational, and strategic communications, storing critical data, running intelligence analytics, and hosting Command and Control (C2) systems. These data centers typically feature a high-throughput network environment with large volumes of data traffic, including encrypted communications between data centers themselves. Current widely used network security protocols, such as IPSec, are not well-suited for the high throughput demands set by data centers. MACSec, being a Layer 2 protocol, should allow for significantly higher throughput.

In our system, the decoupling of key management from the transmission channel allows for the interchangeability of key exchange mechanisms, providing support for both quantum and post-quantum key exchange schemes without altering the transmission layer (see Fig. 1). Such an abstraction enables us to achieve crypto agility, as the transmission layer does not need to know whether the key originates from QKD or PQC, nor does it require significant changes to itself to accommodate a key from a different source. Subsequently, this innate crypto agility equips our system with cyber resilience by design. If an incident occurs, such as a failure of QKD or the compromise of a specific cryptographic primitive, the system can rapidly respond by substituting the affected cryptographic component without requiring a complete system redesign. The architectural decoupling of the key management layer from the transmission layer ensures that such failures are isolated, preventing cascading effects across the system. This, in turn, leads to minimizing the impact of cryptographic failures and maintaining the system's operational continuity.

The system will be tested against crypto agility and cyber resiliency criteria. For the first one, the system's ability to integrate and switch between PQC and QKD mechanisms without disrupting communication. Moreover, the ease of implementation will be tested,

---

[2] MDI – Measurement Device Independent, DV – Discrete Variable, BBM92 – Standard developed in 1992 by Charles H. Bennett, Gilles Brassard, and N. David Mermin.

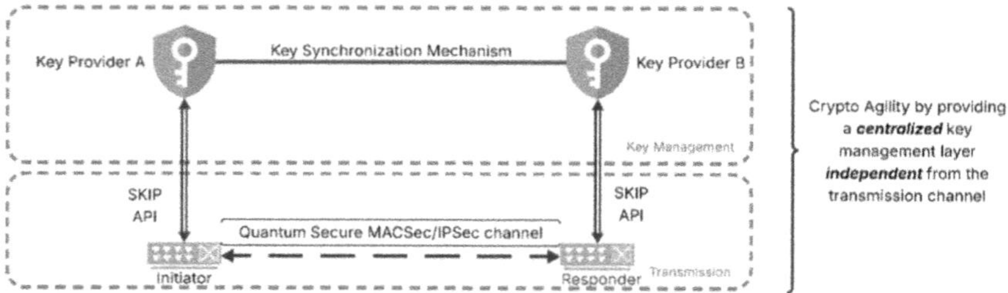

Fig. 1. Architectural decoupling of key management from the underlying transmission

with an emphasis on the scalability of incorporating new cryptographic mechanisms into the key management layer. For the latter, we will focus on the emulation of system's behavior when one of the cryptographic primitives fails (e.g., a QKD device failure, deprecation of one of the key exchange algorithms, or a successful attack on it), with an emphasis on the system's ability and ease of switching to an alternative method. We are interested in measuring the performance metrics, including the key rate, key failure rate, and number of successful key derivations that will be logged and analyzed to evaluate the feasibility of integrating PQC and QKD into systems with real-world data throughput. For the QKD we are also interested in verification of the resilience to atmospheric interference. The key performance metrics will be logged for an open-space quantum channel under non-optimal and potentially disruptive atmospheric conditions. Subsequently, they will be compared against the same metrics under optimal atmospheric conditions to test the resilience of the open-space QKD channel.

5.2   Secure Mobile Platforms

The second set of experiments focus on the performance of the PQC standards in dynamic tactical scenarios. Mobile platforms are an integral part in modern military scenarios, providing real-time intelligence and communication support, both of which ultimately lead to increased situational awareness. Secure Identification Friend or Foe (IFF) can effectively mitigate adversarial spoofing attempts or unauthorized access to friendly networks. PQC offers a viable solution for implementing IFF and securing high-value video streaming. Nonetheless, PQC integration poses challenges, including computational and data overhead, key size constraints and energy limitations on UAVs.

This experiment is aimed at verification to what extent post-quantum cryptography mechanisms can be integrated into mobile platforms for secure authentication and real-time, bandwidth-constrained data transmission, such as UAV identification and low-latency 1080p video over tactical radio. It has 2 main objectives related to 2 scenarios: (1) to implement a system that recognizes friendly UAVs using quantum-secure public key infrastructure systems, evaluate the performance of successful authentication attempts and define limitations of such an approach; (2) to showcase the feasibility of integration and evaluate the performance of post-quantum cryptography in high-bandwidth video encryption over radio frequency.

The relevance of these two scenarios relate to the fact that in highly critical tactical environments, UAVs often operate in close coordination with ground forces and other airborne assets. In such scenarios, the ability to reliably authenticate friendly drones and securely provide live video feed can be vital to the success and protection of friendly forces. Adversarial attempts to spoof or disrupt these systems can compromise operations and endanger human lives. Evaluating PQC under these operational pressures ensures it can meet the demands of real-world deployments.

Two systems will be prepared for (1) performing UAV authentication and (2) encrypting video stream over tactical radio. The former will include three UAVs equipped with a PQC capable FPGA accelerators and directional antennas. The latter will consist of a mobile platform equipped with a camera, a radio transmitter (also capable of encoding video), a radio receiver (capable of decoding the video) and an operation workstation.

Quantum secure IFF scenario will be evaluated based on timing metrics, gathered for both the initiator and the responder. These will include total authentication time, data link latency, data link throughput, computation time for both parties, authentication throughput. After that the protocol disruption will be analyzed based on bit loss ratio, e.g. how many times the protocol should be repeated to perform a successful authentication. Quantum secure video over tactical radio will use multiple metrics collected from the encrypted and unencrypted video feed and then compared against themselves. The metrics will include encoder, encryption and decryption latency.

5.3 Quantum Key Distribution Analysis

There are only a handful of scenarios where QKD offers a promising solution for the use in military applications. The IC25 exercises focus on dynamic operational scenarios that could provide high resiliency of key provision for the deployed forces. As the first step in these considerations, the use of the free-space entanglement - based QKD (BBM92) has been proposed to identify challenges of this technology implementation for key distribution from space (e.g. HAPS[3] platforms, satellites). As an emulation of the flying platform, we assume to use a UAV (drone) that will be stabilized to provide entangled photons to two base stations.

This scenario poses several challenges. Maintaining precise laser alignment between ground stations and the possibly moving platform is difficult due to the motion, vibrations and limited stability of drones in flight, which can be also further degraded by windy conditions. Additionally, free-space optical QKD is highly sensitive to atmospheric conditions such as fog, rain or even intensive sunlight. These factors render reliable key exchange under real-world conditions a complex and demanding task. Nonetheless, QKD is analysed as an important fallback option which, as opposed to PQC, is unlikely to be affected by developments in cryptanalysis or increasing computing resources available to adversaries[4].

The third experiment aims to verify to what extent are Quantum Key Distribution systems viable for mobile secure communication networks, and what are their potential vulnerabilities. It is planned around 3 objectives: Evaluate the performance and atmospheric

---

[3] High-Altitude Platform Stations.

[4] Note that QKD may also be subject to additional Side – Channel Attacks that would significantly reduce its overall security [17].

resilience of an open-space QKD channel to provide quantum secure video call encryption (*Exp3.scenario 1*); perform a successful side-channel attack (SCA) on a UAV-based mobile QKD system, effectively extracting a quantum-generated key (*Exp3.scenario 2*); evaluate mitigations for such SCA (*Exp3. Scenario2*).

Ad-hoc mobile networks (MANETs) secured with QKD can be relevant in environments where conventional infrastructure is unavailable or compromised. By using high altitude flying platform to establish secure optical links between ground nodes[5], encryption keys can be distributed dynamically and on-demand across mobile units. That can aid in tactical flexibility for units in dynamic situations that still require encrypted communications, for example over Radio Frequency (RF).

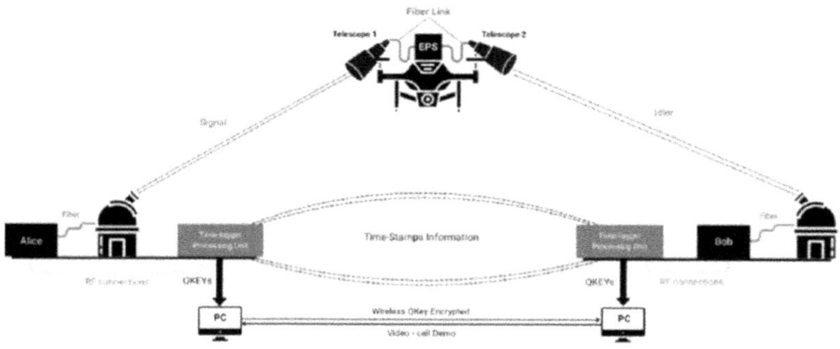

Fig. 2. Proposed system design for experiment 3

The system for *Exp3.scenario1* consists of a drone, hovering in one place, equipped with an entangled photon source (EPS), two ground stations (Alice and Bob), the time tagging processing unit and two clients consuming the keys to encrypt a video channel using AES-256. Additionally it is assumed that communications between the time-tagging units will be encrypted with the FPGA – based PQC (see Fig. 2). For *Exp3.sceanario2* the adversarial setup for performing an SCA consists of n antennas, each dedicated to a single photon detector (SPD), an analog-to-digital converter (ADC) and a reinforcement learning algorithm specifically trained with previous ADC outputs.

It is planned to measure free-space QKD performance with quantum key generation rate, Quantum Bit Error Rate (QBER), and link stability over the duration of the drone's flight. Coincidence windows between Alice and Bob's detection events will be evaluated to optimize temporal filtering and maximize the secure key yield. QBER will be tracked over time to identify any degradation due to optical misalignment, atmospheric turbulence, or external disturbance. The encryption performance will be verified through the continuity and integrity of the real-time video call. In cases of protocol failure (e.g. high QBER or depleted key supply), the time and cause of the failure will be logged.

For *Exp3.scenario 2* we will focus on QKD side channel vulnerability analysis and possibility to infer keys exchanged by the endpoints. RF traces will be analyzed to detect

---

[5] Even though the scenario covers mobile units, it is assumed that ground station is not moving during the key transfer.

repeatable patterns corresponding to detector clicks. Machine learning models trained on labeled RF data will be used to classify pulses and assign them to specific SPDs. The inferred time tags will be then compared against the legitimate QKD data to assess accuracy.

6 Summary

The article focuses on exploring technical implementation of different quantum-resistant solutions supporting communications, with great emphasis on dynamic, tactical scenarios and challenges of both – PQC and QKD. Experimenting with cryptographic agility and out-of-band key distribution is critical for future-proofing security. Agile key provider exchange allows organizations to switch providers based on security needs or policy changes, ensuring optimal key management. Hybrid scenarios (combining classical and PQC) offer a crucial transition bridge, mitigating risks from unknown PQC vulnerabilities. However, deployment of any solution needs to take into account performance issues resulting e.g. from larger key sizes and higher computational overhead (for current PQC algorithms), sensitivity to atmospheric conditions, jamming and vulnerability to side-channel attacks (for QKD). These pose significant challenges and performance problems in resource-constrained tactical environments.

The results of experiments will be used to inform NATO ACT about the efficiency of evaluated technologies and prepare recommendations for the NATO CIS quantum threat mitigation plan.

References

1. Shim, K.-S., et al.: Design and validation of quantum key management system for construction of KREONET quantum cryptography communication. J. Web Eng. (2022). https://doi.org/10.13052/jwe1540-9589.2151
2. James, P., Laschet, S., Ramacher, S., Torresetti, L.: Key Management Systems for Large-Scale Quantum Key Distribution Networks (2023). https://doi.org/10.1145/3600160.3605050
3. Barker, E., et al.: Considerations for Achieving Crypto Agility. NIST CSWP 39. https://doi.org/10.6028/NIST.CSWP.39.ipd
4. Timelines for migration to post-quantum cryptography, UK NCSC (2025). https://www.ncsc.gov.uk/guidance/pqc-migration-timelines
5. ITU-T Y.3800 (2019) Corrigendum 1. Overview on networks supporting quantum key distribution, April 2020
6. ITU-T Y.3801. Functional requirements for quantum key distribution networks (2020)
7. ITU-T Y.3803. Quantum key distribution networks – Key management (2020)
8. Fluhrer, S., Kampanakis, P., McGrew, D., Smyslov, V.: Mixing Preshared Keys in the Internet Key Exchange Protocol Version 2 (IKEv2) for Post-quantum Security. IETF (2020). https://datatracker.ietf.org/doc/html/rfc8784
9. Kott, A., Golan, M.S., Trump, B.D., Linkov, I.: Cyber resilience: by design or by intervention? arXiv.org. https://arxiv.org/abs/2201.11152
10. Mosca, M.: Cybersecurity in an era with quantum computers: will we be ready? IEEE Secur. Priv. **16**(5), 38–41 (2018). https://doi.org/10.1109/msp.2018.3761723
11. NATO Communications and Information Agency Technology Strategy Towards 2030, NU, Brussels (2024)

12. Soare, S.R.: Digitalisation of Defence in NATO and the EU: Making European Defence Fit for the Digital Age, Aug 2023, The International Institute for Strategic Studies
13. Barker, E., et al.: Considerations for Achieving Crypto Agility. NIST CSWP 39. https://doi.org/10.6028/NIST.CSWP.39.ipd10.6028/NIST.CSWP.39.ipd
14. NIST Post Quantum Cryptography standardization project. https://csrc.nist.gov/projects/post-quantum-cryptography
15. ITU-T Study Group 17 (SG17) – Security. https://www.itu.int/en/ITU-T/studygroups/2025-2028/17/Pages/default.aspx
16. Industry Specification Group (ISG) on Quantum Key Distribution (QKD). https://www.etsi.org/committee/qkd
17. BSI. Implementation Attacks against QKD Systems v. 1.0 (2023). https://www.bsi.bund.de

On the Road to Satellite-Based Quantum Information Networks: Hong-Ou-Mandel Effect with Pulsed Coherent States

Mathieu Bertrand[1]([✉]), Luca Paccard[1], Pedro Roldan Gomez[1],
Laurent de Forges de Parny[1], Michel Sotom[1], Etienne Samain[2], Olivier Alibart[3],
Matteo Schiavon[4], Sacha Gressani[5], Erik Kerstel[5], Thierry Lanz[6], Eleni Diamanti[4],
Sébastien Tanzilli[3], and Mathias van den Bossche[1]

[1] Thales Alenia Space, 26 Avenue Jean-François Champollion - BP 33787, 31037
Toulouse-Cedex 1, France
`mathieu.bertrand@thalesaleniaspace.com`
[2] SigmaWorks, 8, Allée Bellevue, 06460 Saint Vallier de Thiey, France
[3] Université Côte d'Azur, CNRS, Institut de Physique de Nice (INPHYNI), UMR 7010,
Parc Valrose, 06108 Nice Cedex 2, France
[4] Sorbonne Université, CNRS, LIP6, 4 Place Jussieu, 75005 Paris, France
[5] Centre Spatial Universitaire de Grenoble, 120 Rue de La Piscine, 38400
Saint-Martin-d'Hères, France
[6] OGS Technologies, 7 Chemin Des Pins, 06000 Nice, France

Abstract. The so-called Quantum Information Networks (QIN) promise to revolutionize the world with new applications based on the interconnection of quantum devices such as quantum computers, quantum sensors and physically secured cryptographic receivers. Such networks employ photons as a propagation mean of quantum information in order to create entanglement between the end-users' devices. Over long distances, satellites will become mandatory in the network as they offer a better optical losses scaling than fibres. In this paper, we focus on the physical principle at the heart of the two-photon Bell-State Measurement devices used in the QIN to swap entanglement between and inside the network nodes: the Hong-Ou-Mandel effect.

Keywords: Quantum Information Networks · Satellite · Bell-State Measurement · Hong-Ou-Mandel

1 Satellite-Based QIN Architecture

1.1 Rationale

QIN architectures are a thriving topic in the quantum information community [1, 2]. Generally speaking, QIN are made up of nodes (often called Quantum Repeater Nodes, QRN) combining Entangled Photon Sources (EPS), Bell-State Measurement (BSM) devices and Quantum Memories (QM) [3, 4]. Long-distance connections on ground are achieved using optical fibres in the telecom bands, with typical losses of $0.2dB/km$.

© The Author(s), under exclusive license to Springer Nature Switzerland AG 2026
F. Barbaresco and G. François (Eds.): QUEST-IS 2025, CCIS 2743, pp. 195–203, 2026.
https://doi.org/10.1007/978-3-032-13852-1_20

On the other hand, the link budget of satellite Free-Space Optical (FSO) connections scales as

$$\frac{P_{R_x}}{P_{T_x}} \propto \left(\frac{\pi D_{R_x} D_{T_x}}{4\lambda L} \right)^2, \tag{1}$$

with the subscript T_x (R_x) corresponding to the emitter (receiver), P the optical power, D the telescope diameter, λ the optical wavelength and L the distance between the satellite and the Optical Ground Station (OGS) [5]. For end-users' distances typically above 100km, direct satellite connections thus provide higher transmission rates than optical fibres. When designing the satellite segment of the QIN, two main constraints have to be taken into account.

The first one is linked to the limitations in Size, Weight and Power (SWaP) of the equipment a satellite payload can embark. This constraint rules out architectures where the BSM devices of the QIN are placed on-board the satellite. Indeed, these devices rely on Single Photon Detectors (SPD). Today, the most efficient type of SPD are Superconducting Nanowire SPD (SNSPD), which require cryogenic cooling in the 1K temperature range and are thus very challenging to embark. Moreover, efficient QM technologies are either based on cryostats for Rare-Earth doped ions, or on complex and high-SWaP apparatuses for cold atoms and trapped ions, ruling them out too for integration into the satellite payload [6].

The second constraint is due to the fact that uplink Free-Space Optical (FSO) connections are more prone to losses than downlinks, typically by a factor of 10 in the optical link power budget. The reason is that the turbulence induced by the atmosphere occurs in the first few tens of kilometres above the Earth surface, while the rest of the propagation channel is basically vacuum. In uplink, these turbulences induce a beam wandering, with a strong negative impact on the link budget. In downlink, the atmosphere disturbances only induce wave front deformations, which can be corrected by an Adaptive-Optics (AO) assembly.

Combining these two constraints leads to defining satellite-based QIN nodes with EPS on-board, and to placing BSM and QM on the ground segment. A simplified architecture of such a network is represented in Fig. 1.

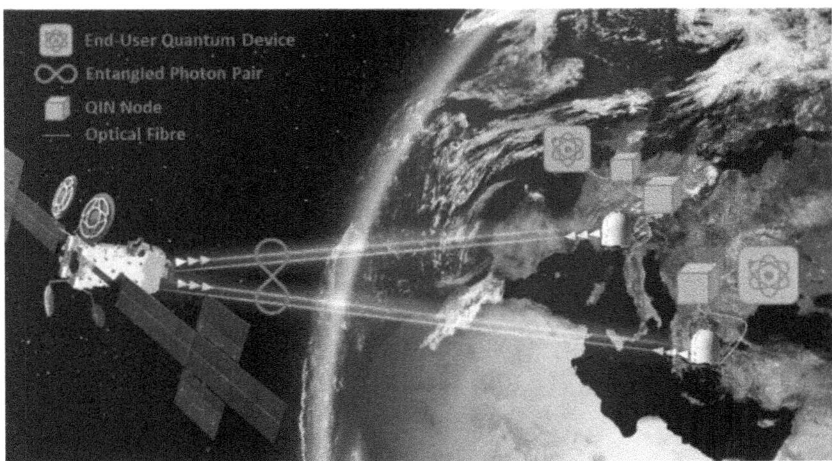

Fig. 1. Simplified architecture of a QIN including space and ground segments. Entanglement generated by EPS is propagated through satellite FSO and ground optical fibre links. QIN nodes include EPS, BSM and QM. Red lines correspond to qubits, while blue ones include all the auxiliary channels such as synchronization signals, polarisation references, etc.

1.2 Scope of the Work

As part of this work, we develop an embryonic QIN space segment demonstrator. To end up with a fully fibered optical testbed at telecom wavelength, we greatly simplify the above architecture by making several assumptions, considered valid for a Proof-Of-Concept (POC) demonstration of such a satellite system. First, we remove the QM, as their maturity is still quite low and their cost is too high to be included in a mid-term project. This simplification leads to the schematic of the POC sketched in Fig. 2.

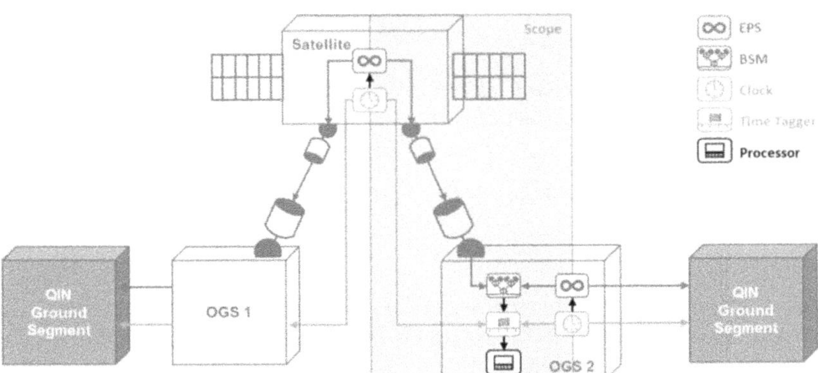

Fig. 2. POC demonstrator without QM, including a satellite containing an EPS and a clock and two OGS with dedicated quantum receivers.

In this picture, the BSM device is assumed to be a passive 2-photon polarisation BSM apparatus. The simplest way of implementing this element is to use a combination of Beam-Splitters (BS) and Polarising Beam-Splitters (PBS). The physical principle behind the projection onto the Bell-State basis is based on the so-called Hong-Ou-Mandel effect [7]. If all the degrees of freedom of the two incoming photons are identical, they will bunch on the BS if they have similar polarisations. Post-selecting the right correlation events at the BS output then allows to project the incoming photons on one or two of the four Bell states [8, 9]. The second simplification thus consists in reducing the BSM device to a simple BS.

Finally, the third and final simplification consists in considering only the two photons interacting on the BS. In our experiment, we then forget about the two other photons in Fig. 2, directed towards the OGS 1 and QIN ground segment. This last assumption allows us to work with attenuated laser sources instead of EPS to emulate the on-board and ground sources. The resulting HOM testbed is represented in Fig. 3.

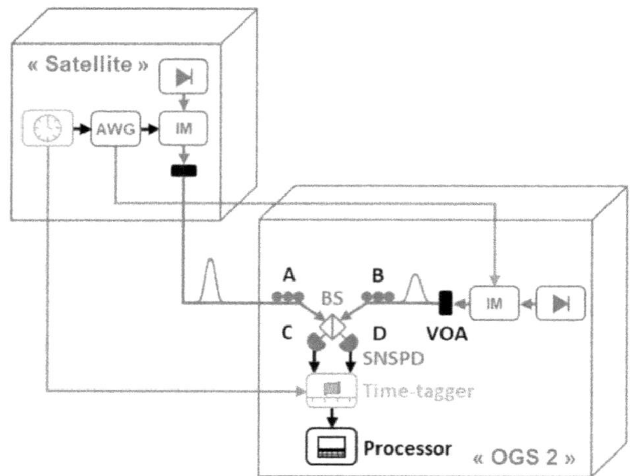

Fig. 3. Optical layout of the HOM testbed. AWG: Arbitrary Wave Generator, IM: Intensity Modulator, VOA: Variable Optical Attenuator.

2   Hong-Ou-Mandel Effect with Pulsed Coherent States

2.1   Experimental Setup

First demonstrated in 1987 by the 3 physicists Chung Ki Hong, Zheyu Jeff Ou and Leonard Mandel, the HOM effect is a quantum interference between two photons imping-ing on each input port of a 4-way BS [7]. As the photons are bosons, if all their degrees-of-freedom are strictly identical, the two-photon wave function symmetrisation implies that they will go out in the same output port, even though this port is chosen randomly. Studying the correlation between the detections in the BS output ports with SPD thus reveals the existence of the effect, as depicted in Fig. 3.

In a satellite-based QIN, the interference between downlink and ground photons can be affected by 4 types of impairments: timing jitters due to synchronisation imperfections, frequency mismatch due to laser drifts and Doppler effect, polarisation alignment issues and link budget variations. In this work, we have studied the dependency of the so-called HOM dip, *i.e.* the reduction of the coincidences between the output ports, as a function of the arrival time delay and frequency offset between two attenuated optical pulses, as well as their mean photon number ratio.

To generate the pulses, we use two telecom C band lasers at 1545 nm: one temperature/current controlled Distributed Feedback Laser (DFB) and a tuneable external cavity laser. The frequency offset between the lasers is preliminary matched and then let free-running. This offset Δ can be tuned by changing the optical frequency of the external cavity laser. The intensity of each laser is then carved by a Mach-Zehnder interferometer modulated by one output of a common Arbitrary Wave Generator (AWG). This setup allows to control the arrival time delay δ between the two coherent states. Moreover, the internal clock of the AWG serves as a time reference for the time-tagger. Figure 4 shows the intensities of the optical signals obtained with this testbed for a repetition period of 30ns and a Full Width at Half Maximum (FWHM) of 2.5 ns.

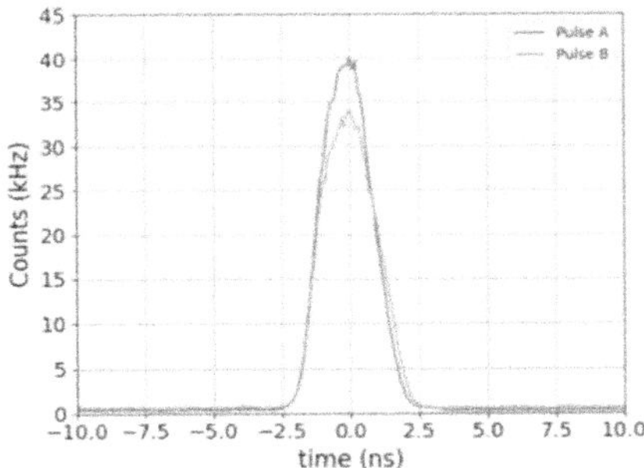

Fig. 4. Optical intensities of the two pulses measured with SNSPDs.

2.2 HOM Effect Theory

Taking the normalised square root of the optical pulses shown in Fig. 4 allows to define the wave function amplitudes $\epsilon_A(t)$ and $\epsilon_B(t)$ for the coherent states respectively incoming on ports A and B of the BS. The complete wave functions of these states are then expressed as

$$\xi_{A,B}(t) = \underset{A,B}{\int} \left(t \mp \frac{\delta}{2} \right) e^{-i\left(\omega \mp \frac{\Delta}{2}\right)t}, \tag{2}$$

where ω stands for the central frequency of the lasers, δ for the time delay and Δ for the frequency offset between the two pulses [10]. The field operator corresponding to the annihilation of a photon in these ports writes

$$E^+_{A,B}(t) = \xi_{A,B}(t)a_{A,B} \tag{3}$$

The BS transformation between the output C and D ports with respect to the input A and B ports gives

$$\begin{cases} E_C^+(t) = \frac{1}{\sqrt{2}}\left[-iE_A^+(t) + E_B^+(t)\right] \\ E_D^+(t) = \frac{1}{\sqrt{2}}\left[E_A^+(t) - iE_B^+(t)\right]. \end{cases} \quad (4)$$

In this experiment, we have worked with a low mean photon number $\mu_A = 0.02$. In this regime, only the events with a total of two photons, either incoming on the same input port or one for each port, are relevant for the coincidence calculation. The most important term for the HOM effect is the interference between a photon arriving at input port A and one at B, producing detections at the output port C at time τ_0 and D at $\tau_0 + \tau$. For an input state $|\psi_{in}\rangle = a_A^\dagger a_B^\dagger |00\rangle$, it is defined as

$$P_{CD}^{11}(\mu_A, \mu_B, \tau, \tau_0, \delta, \Delta) = \langle\psi_{in}|E_C^-(\tau_0)E_D^-(\tau_0 + \tau)E_D^+(\tau_0 + \tau)E_C^+(\tau_0)|\psi_{in}\rangle. \quad (5)$$

The different contributions P_{CD}^{kl} with k photons in C and l in D are then weighted by the probabilities P_{AB}^{mn} that m photons arrive at port A and n photons at port B, given by

$$P_{AB}^{mn}(\mu_A, \mu_B) = \mu_A^m \cdot \mu_B^n \cdot \frac{e^{-(\mu_A+\mu_B)}}{m!n!}. \quad (6)$$

The coincidence probability, called P_c, results from the weighted probabilities of events containing 2 photons in total and integrated over τ_0.

2.3 HOM Dip Measurements

The normalised coincidences on SNSPD[1] C and D as a function of the time delay between the optical pulses is represented in Fig. 5.

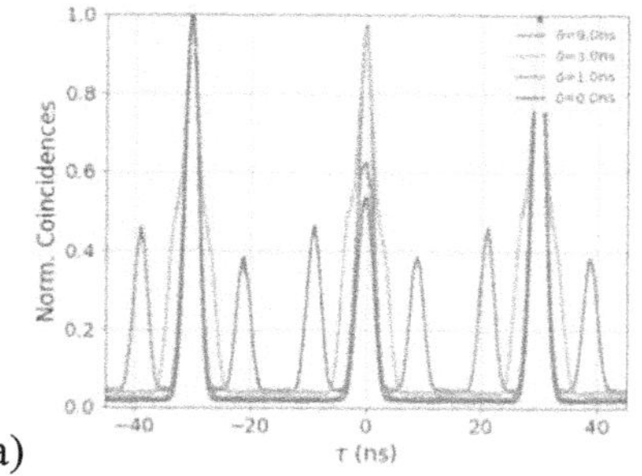

a)

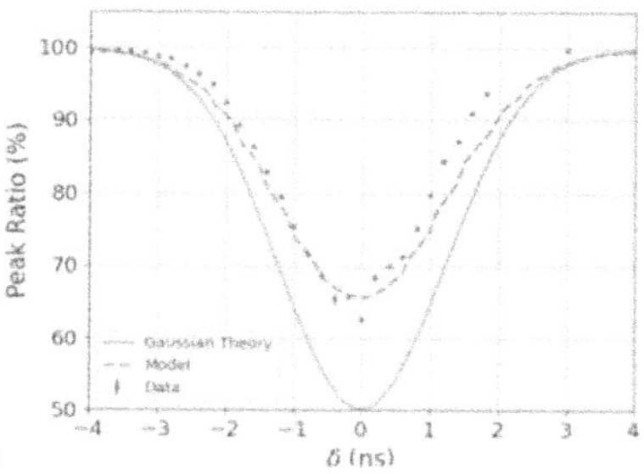

b)

Fig. 5. a) Normalised coincidences as a function of the tagger delay τ for different optical pulses delays δ, with a mean photon number ratio $\frac{\mu_A}{\mu_B} = 1.1$. b) HOM dip defined as the ratio of the central peak over the lateral ones. The orange curve represents an analytical Gaussian approximation, and the green dotted curve is a model based on the experimental wave functions.

---

[1] Quantum efficiency $\geq 80\%$, dark count rate 10Hz, RMS jitter 30ps.

The results of Fig. 5b show that our model, taking into account the actual wave functions and background noises of the experiment (limited by straight light in the laboratory), correctly reproduces the experimental results. The Gaussian theory is computed without any noise on the detectors and shows a well-known maximum HOM dip visibility of 50%. The dip as a function of the frequency difference between the two lasers is represented in Fig. 6.

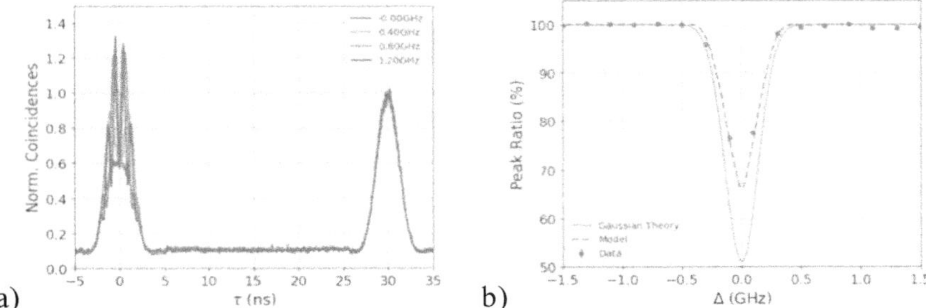

Fig. 6. a) Normalised coincidences as a function of the tagger delay τ for different optical pulses frequency shifts Δ for a mean photon number ratio $\frac{\mu_A}{\mu_B} = 0.8$. b) HOM dip in frequency.

Interestingly, as the time-tagger resolution (200ps) is smaller than the pulses width, we observe a frequency beat note inside the central peak of Fig. 6a. The lateral peaks, produced by photons arriving 30 ns apart, do not interfere and only show the amplitude wave functions overlap. The resulting dip in Fig. 6b is obtained by integrating the central and lateral peaks and computing their ratio. It is only sensitive to the pulses frequency overlap, Fourier limited to 400 MHz FWHM. The beat note frequency stability between the two lasers was typically around 100 MHz over 1 s.

Finally, we have performed a HOM dip visibility analysis as a function of the photon ratio $R = \frac{\mu_A}{\mu_B}$. In the final system, EPS will have a different statistics than the Poissonian photon number distribution of coherent states, but this study is still interesting for use cases such as teleportation between an EPS and a coherent state. The visibility, calculated with the correlation probabilities P_c, is defined as

$$V(\mu_A, \mu_B) = 1 - \frac{\int\limits_{-\infty}^{+\infty} P_c(\mu_A, \mu_B, \tau, 0, 0)d\tau}{\int\limits_{-\infty}^{+\infty} P_c(\mu_A, \mu_B, \tau, +\infty, +\infty)d\tau}. \tag{7}$$

The results for a set of experimental photon number ratios are shown in Fig. 7.

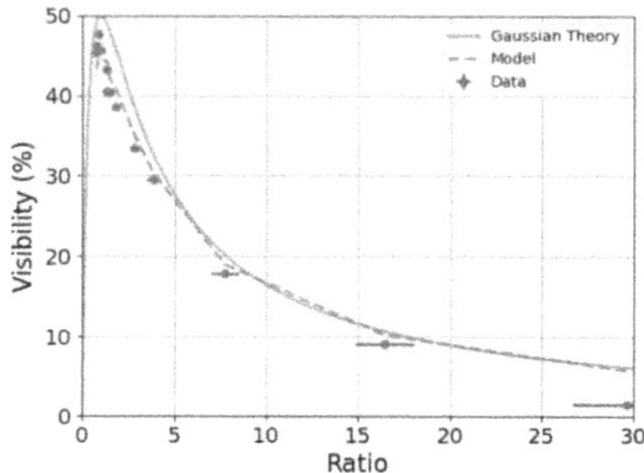

Fig. 7. HOM dip visibility as a function of the mean photon ratio between the two coherent states for $\delta = 0 \pm 0.1\,ns$ and $\Delta = 0 \pm 100\,MHz$. Data points (blue dots), theoretical curve obtained with Gaussian pulses (orange line) and model (green dotted curve) taking into account the experimental wave functions and noises. (Color figure online)

The theoretical visibility obtained with Gaussian coherent states (represented by the orange line in Fig. 7), reaches the maximum of 50% as it is computed without any background noise. The green dotted line, corresponding to our model, is very close to the Gaussian curve shape as the actual pulses wave functions are almost Gaussian, but the visibility is reduced due to experimental noises. We still achieve a maximum of 47.5%. The slight discrepancy between the data points and the models are probably due to the fact that the photon number ratio estimation, based on the experimental wave functions integration, can be biased.

The experimental visibility of the HOM dip will be used to estimate the BSM success probability with EPS in the next phase of the project.

2.4　Conclusion

We have reported on a set of preliminary POC experiments that allows us to assess a simplified architecture for a future satellite-based QIN system. The results on the HOM effect between the equivalent of a satellite payload-emitted photon and a ground source photon reveal the dependency on the arrival time, the frequency offset and the mean photon number ratio of the two pulses. In the final system, we expect a higher visibility when considering entangled photon pairs instead of coherent pulses, as the statistics will be different.

In the next step of the project, this POC testbed will be used to assess more realistic synchronisation scenarios for a satellite system, where the on-board and ground sources have physically distinct timing references.

Acknowledgement. We thank the French space agency *Centre National d'Etude Spatiales* (CNES) for supporting this project.

References

1. Kimble, H.: The quantum internet. Nature **453**, 1023–1030 (2008)

2. Wehner, S., Elkouss, D., Hanson, R.: Quantum internet: a vision for the road ahead. Science **362** (2018)

3. Sangouard, N., et al.: Quantum repeaters based on atomic ensembles and linear optics. Rev. Mod. Phys. **83**(33) (2011)

4. Azuma, K., et al.: Quantum repeaters: from quantum networks to the quantum internet. Rev. Mod. Phys. **95**, 045006 (2023)

5. Klein, B., Degnan, J.: Optical antenna gain. 1: transmitting antennas. Appl. Opt. (1974)

6. Gavin Brennen, G., et al.: Focus on quantum memory. New J. Phys. **17**, 050201 (2015)

7. Hong, C., Ou, Z., Mandel, L.: Measurement of subpicosecond time intervals between two photons by interference. Phys. Rev. Lett. **59**(2044) (1987). https://doi.org/10.1103/PhysRevLett.59.2044

8. Lee, S.-W., et al.: Bell-state measurement and quantum teleportation using linear optics: two-photon pairs, entangled coherent states, and hybrid entanglement, arXiV, p. arXiv:1304.1214v1 (2013)

9. Nölleke, C., et al.: Efficient teleportation between remote single-atom quantum memories. Phys. Rev. Lett. **110**, 140403 (2013)

10. Ferreira da Silva, T., et al.: Spectral characterization of weak coherent state sources based on two-photon interference. J. Opt. Soc. Am. B **32**, 545–549 (2015)

Session 1 Quantum Algorithms, Computing and Simulation – Benchmarking A: Methods and Applications

BenchQC Scalable and Modular Benchmarking of Industrial Quantum Computing Applications

Florian Geissler[1(✉)], Eric Stopfer[2], Christian Ufrecht[2], Nico Meyer[2], Daniel D. Scherer[2], Friedrich Wagner[2], Johannes M. Oberreuter[3], Zao Chen[3], Alessandro Farace[3], Daria Gutina[4], Ulrich Schwenk[4], Kimberly Lange[4], Vanessa Junk[4], Thomas Husslein[4], Marvin Erdmann[5], Florian Kiwit[5], Benjamin Decker[5], Greshma Shaji[5], Etienne Granet[6], Henrik Dreyer[6], Theodora-Augustina Drăgan[1], and Jeanette Miriam Lorenz[1]

[1] Fraunhofer Institute for Cognitive Systems IKS, Munich, Germany
`florian.geissler@iks.fraunhofer.de`
[2] Fraunhofer Institute for Integrated Circuits IIS, Nuremberg, Germany
[3] Reply SE, Machine Learning Reply GmbH, Munich, Germany
[4] OptWare GmbH, Regensburg, Germany
[5] BMW Group, Munich, Germany
[6] Quantinuum, Munich, Germany

Abstract. We introduce BenchQC, a quantum computing benchmarking project that focuses on application-centric benchmarking using diverse industry use cases. Building upon the open-source platform QUARK, we evaluate key metrics across the quantum software stack to identify trends towards quantum utility, and distinguish viable research directions. This initiative contributes to the broader effort of establishing reliable benchmarking standards that drive the transition from experimental demonstrations to practical quantum advantages.

1 Introduction

Quantum computing (QC) offers new avenues to tackle complex data analysis and optimization problems. In light of the ever-growing demand for computational resources, this puts the technology at the center of interest for both scientific researchers and industrial end users. Quantum computing promises to boost industrial applications in particular in the fields of optimization, simulation, and machine learning [7,28]. A 2021 study estimates that, at tech maturity, quantum computing will unlock hundreds of use cases, with potential value creation of $100B and more in *each* of the above fields [15]. Being at the forefront of this emerging technology is therefore a competitive advantage that elicits strategic investments across various industrial domains (see, e.g., [7]).

With limited quantum hardware available in the current noisy intermediate-scale quantum (NISQ) era, quantum utility remains to be demonstrated in real-world applications [20]. To this day, the international quantum communities have each prioritized individual metrics, concepts, and applications for benchmarking (see Sect. 2). Nevertheless, the informative power remains limited for industrial applications, particularly in emerging fields such as quantum machine learning (QML) [10,32]. For end users, it can be challenging to understand whether their problem category is well-suited to benefit from future quantum technology, what problem scales are realistic, or which hardware and algorithms are the best fit. As a lighthouse project of the Munich Quantum Valley (MQV), BenchQC therefore takes a different approach, and makes industrial applications the starting point of the benchmarking workflow. We discuss selected problems from different categories in Sect. 4, after reviewing related work in Sect. 2 and our methodology in Sect. 3.

2   Related Work

Several international quantum computing benchmarking communities have outlined their strategic approaches. Selected examples include the French BACQ consortium [6], focusing on application-specific metrics like the Q-score and scalable aggregation using the Myriad-Q Tool, the Quantum Economic Development Consortium (QED-C) [1] in the USA, centering on hardware-centric metrics like quantum volume [22], or the Quantum Application Score (QuAS) proposed by the Dutch TNO ecosystem [23]. For a more detailed overview, we refer to Refs. [21,28].

Despite these initiatives, the systematic evaluation of quantum utility in industrial problems is still lacking. BenchQC addresses this gap with a structured approach uniquely characterized by i) use cases, which are derived from real industrial problems, ii) utilization of modular open-source frameworks like QUARK, iii) establishing interfaces for full-stack benchmarking, including various quantum hardware types, iv) a holistic range of metrics from hardware to application level, v) the inclusion of benchmarking for quantum machine learning applications, and vi) concepts to simulate problem complexity scaling.

3   Methodology

Problem Decomposition and Metrics: The overall workflow is illustrated in Fig. 1: Starting from a use case (a), we first derive a mathematical formulation (b). A sequence of problem sizes and complexities is formed (c). The problem is encoded as a hybrid algorithm by classical-quantum decomposition (d) in reference to the baseline of a classical solution (e). The hybrid problem is eventually executed on different quantum-computing hardware (f) or simulated on classical hardware (g). Relevant metrics quantify the performance of the various parts of this decomposed workflow to govern the overall benchmarking process (h).

Initiatives towards the standardization of quantum computing benchmarking metrics are ongoing [21]. In BenchQC, we collect application-agnostic metrics along the benchmarking pipeline, which reflect problem and solution architecture, see Fig. 2. Subsequently, we put those in context with the actual application performance to identify trends for potential quantum utility. Our metric categories are, with examples:

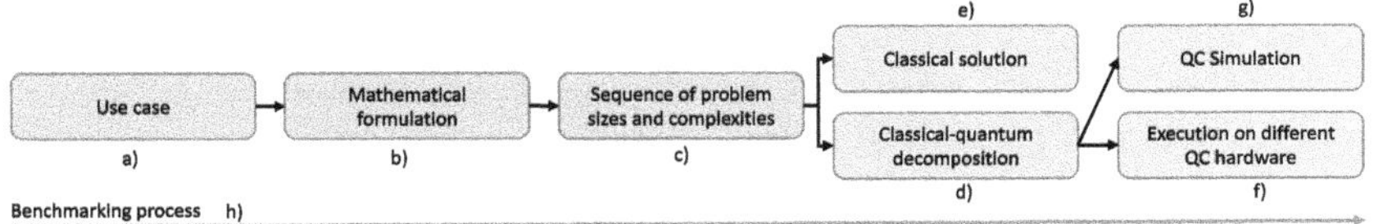

Fig. 1. Decomposition of the benchmarking workflow.

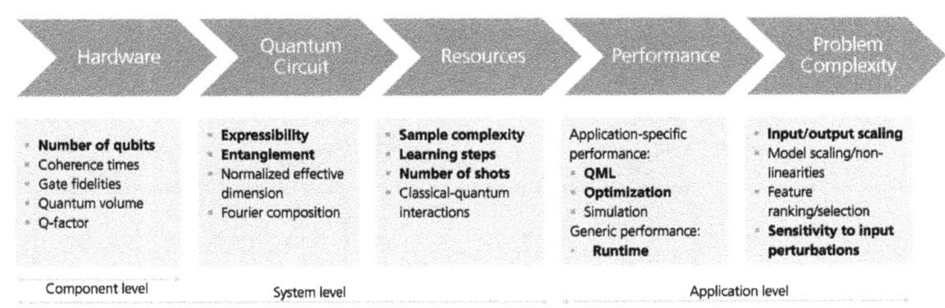

Fig. 2. Metrics across the benchmarking pipeline. Bold letters indicate that a form of this metric is currently implemented in QUARK.

- **Hardware:** Coherence times (qubit lifetime, dephasing time, crosstalk), gate fidelities, the quantum volume (QV), the Q-factor (quality factor) [22,30].
- **Quantum circuit:** Expressibility (here one minus the Jensen-Shannon divergence) and entanglement (Meyer-Wallach measure) [33], normalized effective dimension [3], Fourier coefficients [27].
- **Resources:** Cost budget spent to derive a given solution, e.g., sampling complexity, number of learning steps.
- **Performance:** As a generic measure, we use the runtime or its quantum portion in a classical-quantum hybrid system. For application-specific measures, see the individual use cases in Sect. 4.
- **Problem complexity:** Scaling up the dimensionality of the system's input (e.g., resolution of an image) and output parameters, higher-order characteristics, number of input features, input preprocessing (e.g., adding noise).

The above categories can further be associated with component-level, system-level and application-level metrics, following the taxonomy of [21,30], see Fig. 2. Hardware-related metrics may be associated with either component-level (e.g., gate fidelities) and system-level metrics (e.g., quantum volume). Our integration with the QUARK framework enables a systematic collection and analysis of these metrics across different system configurations, such as different use cases or problem sequences, see Fig. 1). Finally, any metrics aggregation approach (see e.g. Myriad-Q [6] or QuAS [23]) may be employed to arrive at a global utility score.

Tooling: The Quantum Computing Application Benchmarking (QUARK) framework [2] is an open-source platform with a focus on optimization and quantum machine learning applications [17]. Its architecture allows for a simple integration of custom use cases, as well as individual metrics and evaluation pipelines. Users can run benchmarks locally, on quantum simulators, or on quantum computers using interfaces to hardware providers such as IBM, IonQ, or D-Wave. QUARK is organized in plugins (available as pip-packages) and provides standardized messaging interfaces for sequentially run modules. Benchmarking pipelines are constructed in an implementation-agnostic way, such that each module can be connected with another module given that the data type of their interfaces match. An example pipeline is shown in Fig. 3, in which a set cover problem is solved with a quantum annealer.

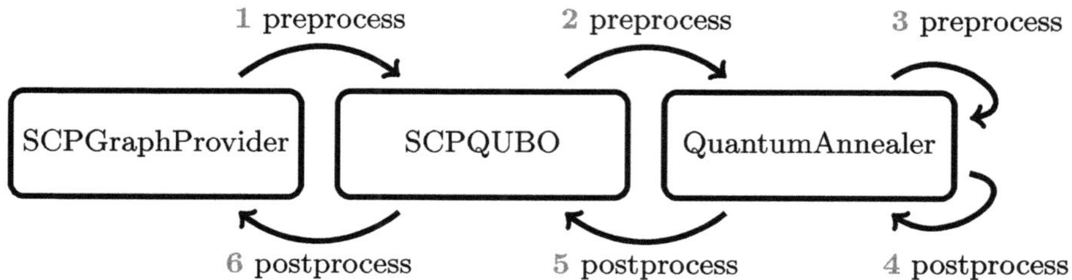

Fig. 3. The QUARK benchmarking pipeline is made up of modules that can pre- and postprocess data. In this case, the set cover problem is solved in two phases: By calling the preprocess functions of each module in order, a graph is created (**1**), mapped to a QUBO formulation (**2**), and solved on a quantum annealer (**3**). Afterwards, calling the postprocess functions of each module in reverse order, the lowest energy sample is obtained (**4**), re-interpreted as a list of nodes (**5**), and compared with the graph (**6**).

In BenchQC, we leverage QUARK as a platform to investigate industrial use cases. The following section gives a review of selected problems from each of the studied categories of optimization, machine learning, and simulation, along with preliminary insights from benchmarking. A full list is given in Table 1.

Table 1. Table of use cases in BenchQC.

Category	Methods	Use case
QML	Classification	Defect detection
	Generative models	Generative design for constructional elements
		Data augmentation for quality inspection
	QRL	6G Beam management
Optimization	Quantum Annealing (QA)	LiDAR sensor configuration
	QAOA and QA	Financial asset optimization
		Assembly line balancing
Simulation	Hamiltonian simulation	Dynamics of electrons in materials
		Low-temperature state preparation of materials
		Nuclear magnetic resonance simulation

4 Selected Use Cases

4.1 Optimization

Mixed-integer linear programs (MIP) represent a class of optimization problems with established solution approaches [11]. Non-linear problems pose additional computational difficulties [18]. For problems where classical solvers reach their limits, quantum methods can possibly offer solution paths with fewer computational steps than their classical counterparts [4,35]. However, the current limitations of quantum hardware restrict their application to relatively small-scale problems. One fundamental metric for benchmarking multi-stage quantum algorithms like QAOA is the *runtime* including both training and sampling [22]. A suitable solution quality metric for the application-specific performance is the approximation ratio, which compares the best found solution relative to the optimal solution found by a classical solver. While it quantifies the gap to the optimal solution, it also remains reliable across different magnitudes of the underlying variables. When the optimal solution is unknown or hard to calculate, we may instead compare to a solution produced by a classical algorithm or even random sampling techniques (a popular metric for the latter is, for example, the *Q-Score* [31]).

The portfolio optimization problem is a fundamental challenge arising in finance. The goal is to select a set of assets that maximizes the return while minimizing the return variance (called *volatility*) [19,29].

Due to the quadratic nature of the volatility, classical solvers scale poorly with an increasing number of assets. We first benchmark the scaling of classical solvers on different versions of the problem to identify its quantum potential. This multi-objective optimization problem can be tackled in various ways [19], like maximizing return with constrained volatility ("Maxret"), minimizing volatility with constrained return ("Minvola"), and maximizing a trade-off between return and volatility ("Multiobj"). Our analysis yields that the Minvola problem formulation requires the longest runtimes for classical optimization methods. For example, for 500 assets, the solving time of SCIP [8] to optimality is $\approx 16\times$ longer for Minvola than for Multiobj, and $\approx 62\times$ longer than for Maxret. This makes it our choice to investigate possible benefits of quantum-assisted solutions, which will be the subject of further investigations in BenchQC. In order to assess the potential of QA and QAOA, we compare the respective results to those obtained by not only exact classical optimizers but also classical heuristics like Simulated Annealing, Tabu Search and Steepest Descent. To maintain consistency across different hardware platforms, we impose a time limit of one minute for each method in order to fairly compare the solution methods on our Minvola test instances.

4.2 Machine Learning

Hybrid classical-quantum neural networks (QNNs) are considered promising candidates for quantum computing [36]. We study two applications from the realms of reinforcement learning (RL) and image classification.

A challenging industrial RL problem is the *beam management* in 6G wireless communication, as next-generation antennas will produce directional beams to serve mobile users [12]. A RL agent learns to select optimal antenna indices and beam directions from a pre-defined codebook based on past interactions, without the need for the exact mobile trajectory [26]. Recent studies highlight potential advantages of quantum reinforcement learning (QRL) over classical RL [25], however, due to various stochastic influences, such conclusions have to be taken with great care. To achieve statistical significance, we hence simulate 100 independent training runs per environment instance [24], by far the largest number performed in this context to date. Using sample complexity [16] - which counts the number of interactions needed to achieve a performance threshold - we compare classical double Q-learning [34] and a quantum version of the algorithm, see left part of Fig. 4. The models of increasing complexity are: i) *small classical* neural network (2 hidden layers, 387 parameters), ii) *small quantum* circuit (4 layers on 14 qubits, i.e., 437 variational parameters), embedded in outer classical layers (with another 101 parameters), iii) *large classical* neural network (2 hidden layers, 4611 parameters). The results indicate that quantum models perform comparably to larger classical models, suggesting further improvement with increasing parameters in future research.

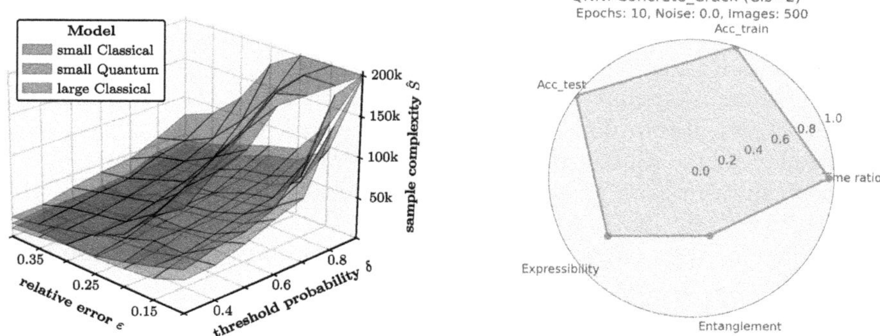

Fig. 4. Performance benchmarks for industrial QML use cases. (left) Empirical sample complexities (lower is better) of classical and quantum RL on a 6G beam management environment. $1 - \epsilon$ is the chosen performance threshold, and δ is the confidence probability. Figure taken from [26]. (right) Defect detection: Selected key metrics for the performance of the classical-quantum hybrid QCNN trained on the road crack dataset.

As an example of image classification, we investigate how hybrid quantum-classical convolutional neural networks (QCNN) architectures can enhance a *defect detection* system, in particular, a classification model on a specialized dataset of road surface images with and without surface cracks [9]. The quantum part of the network is a parametrized quantum circuit with up to 8 qubits, that processes features compressed by preceding dense classical layers, which allow to handle input data of realistic sizes. To evaluate the model performance we use the classification accuracy, while the ratio between the quantum and the overall execution time shows the compute efficiency. A comparison of the hybrid and classical models reveals several key insights, as depicted in the right part of Fig. 4: First, the quantum network achieves a high accuracy ($> 99.5\%$) on the train and test datasets, which is comparable to a purely classical implementation (not shown). Entanglement measurements (≈ 0.4) indicate a limited quantum entanglement in our circuit design, suggesting room for improvement in leveraging quantum correlations. Similarly, the expressibility score (≈ 0.7) indicates that our current circuit architecture may not fully explore the available Hilbert space. These quantum-specific metrics provide valuable guidance for circuit optimization, for example, to investigate the impact of typical quantum properties for final accuracy outcomes.

4.3 Simulation

The simulation of many-body quantum systems is one of the most obvious tasks that quantum computers can perform efficiently [14]. In quantum computing, simulation appears as a sub-routine in many applications, such as in quantum phase estimation to determine the ground state energy of molecules and reaction rates. Its direct industrial applications range from drug discovery to material discovery, enabling efficient simulations of crucial experiments such as nuclear magnetic resonance or neutron-scattering experiments.

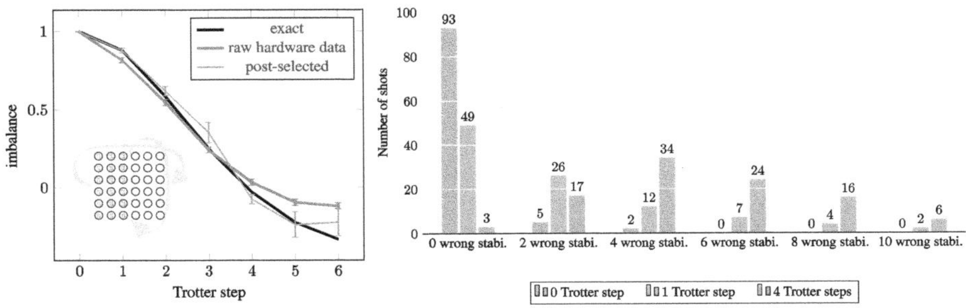

Fig. 5. Performance of Quantinuum's H2-2 ion-trap quantum computer on Bench-QC's simulation use case implemented in QUARK [2], in size 6×6, with Trotter step size dt $= 0.3$ and 100 shots per Trotter step. The circuits involve 54 qubits and around 440 two-qubit gates per Trotter step. (left) Imbalance as a function of number of Trotter steps, comparing exact value, raw hardware data, and data post-selected onto shots with 0 violated stabilizers. The raw data gets a score of 25 056 two-qubit gates, and the post-selected data a score of 59 616 two-qubit gates, which corresponds to the minimal number of gates to run on a perfect hardware to certify that the H2-2 data is incorrect. (right) Histogram of number of wrong stabilizers on hardware data, for different number of Trotter steps.

As a representative industry-relevant use case we consider the simulation of the *dynamics of electrons in materials*. Our analysis starts from the Hubbard model, which is a widely studied model well known for example to describe high-temperature superconductivity in cuprates [5]. The classical computational cost to simulate the time evolution of this system is exponential in either system size for state-vector methods, or in simulation time for tensor networks methods. In order to define a benchmark protocol with an exact, classical reference for comparison, we switch off electron interactions. The evolution can then be classically simulated in polynomial time, while still involving very similar circuits to the Hubbard model, with non-Clifford gates and large entanglement. By time-evolving an initial state and measuring an observable, one can evaluate the accuracy of quantum computing hardware in a scalable and application-oriented way. To implement the fermionic statistics of the electrons, we use a fermionic encoding. This involves stabilizers that can be measured to provide further scalable information on the accuracy of the dynamics implemented on hardware. Varying the simulation time varies multiple features of the hardware panel of Fig. 2, such as the number of qubits and gate fidelity. We give in Ref. [13] a detailed description of the benchmarking protocol.

In Fig. 5, we show results obtained with a Quantinuum's H2-2 ion-trap quantum computing device. The benchmark protocol tests the dynamics of electrons on a 6×6 system, involving 54 qubits and up to $\sim$ 2500 two-qubit gates. It involves computing an imbalance of electron density between two halves of the system, shown in the left panel of Fig. 5, together with the exact, noiseless results. In the right panel, we show how many stabilizers are violated by hardware noise, directly constraining how well the electrons can be simulated. We see, for example, that after one Trotter step, only half of the shots implement the correct fermionic statistics for all the simulated electrons. The performance of the tested hardware can be evaluated with metrics of Fig. 2, e.g., runtime or number of shots to reach a given precision [13]. For example, to reach precision ϵ on all the points of the blue curve of the left panel of Fig. 5, corresponding to raw hardware data, we compute that on the machine tested one requires a runtime t_0/ϵ^2 with $t_0 = 1.1s$. This blue curve displays some difference with the exact black curve due to hardware noise. By post-selecting the shots onto zero violated stabilizers, one can remove a significant part of this noise. From the hardware results we estimate that approximately half of the shots have to be discarded per Trotter step to achieve this post-selection. This way we reach precision ϵ on all the points of the red curve of the left panel of Fig. 5, requiring a runtime t_1/ϵ^2 with $t_1 = 33.5s$. To define a metric that evaluates the precision of the output in a holistic way, we also defined a specific performance metric called the *distinguishability cost*, which helps to guide an optimal choice of quantum hardware. Given the output of a tested hardware, it quantifies the minimal number of gates to be run on an ideal perfect quantum computer to be able to detect an inaccuracy in the output [13]. The hardware data in Fig. 5 scores 25 056 two-qubit gates. This number increases with better hardware quality, as a noisy system with significant bias will require fewer shots to be distinguished from a noiseless quantum computer.

5   Conclusion

This report outlines the importance of an application-centric perspective for quantum computing benchmarking. Classical-quantum hybrid systems, as typical for NISQ, demand a holistic view on the full compute stack for meaningful benchmarking. Our concept is realized with the customizable QUARK framework. Three industry-ready use cases from the fields of optimization, machine learning and simulation are discussed. Preliminary results indicate that quantum solutions can reach competitive performance in such scenarios. Leveraging metrics across the solution pipeline, our work further identifies trends that guide the path towards possible quantum utility.

Acknowledgments. The project BenchQC is funded by grant BenchQC (DIK-2210-0011//DIK0425/02) of the Bavarian Ministry of Economic Affairs, Regional Development and Energy (StMWi). The experimental data reported in this work were produced by the Quantinuum H2-2 quantum computer, Powered by Honeywell, in August 2025.

References

1. The Quantum Consortium - Enabling the Quantum Ecosystem. https://quantumconsortium.org/
2. QUARK: A Framework for Quantum Computing Application Benchmarking (2025). https://github.com/QUARK-framework/QUARK
3. Abbas, A., Sutter, D., Zoufal, C., et al.: Nat. Comput. Sci. **1**(6) (2021)
4. Acuaviva, A., Aguirre, D., Peña, R., et al.: Benchmarking Quantum Computers: Towards a Standard Performance Evaluation Approach (2024). arXiv:2407.10941
5. Arovas, D.P., Berg, E., Kivelson, S.A., et al.: Ann. Rev. Condens. Matter Phys. **13**(1), 239–274 (2022)
6. Barbaresco, F., Rioux, L., Labreuche, C., et al.: BACQ – Application-oriented Benchmarks for Quantum Computing (2024). http://arxiv.org/abs/2403.12205
7. Bayerstadler, A., Becquin, G., Binder, J., et al.: EPJ Quantum Technol. **8** (2021)
8. Bestuzheva, K., Besançon, M., Chen, W.K., et al.: ACM Trans. Math. Softw. **49**(2) (2023)
9. BMW: The road crack dataset (BMW Quantum challenge 2021) (2021). https://github.com/pasqal-io/bmwchallenge/tree/main
10. Bowles, J., Ahmed, S., Schuld, M.: arXiv:2403.07059 (2024)
11. Integer Programming. GTM, vol. 271. Springer, Cham (2014). https://doi.org/10.1007/978-3-319-11008-0
12. Corici, M.I., Franke, N., Heyn, T., et al.: A Fraunhofer 6G white paper: On the road to 6G: Drivers, challenges and enabling technologies. Technical report, Fraunhofer IIS (2021)
13. Granet, E., Dreyer, H.: AppQSim: application-oriented benchmarks for Hamiltonian simulation on a quantum computer (2025). https://arxiv.org/abs/2503.04298
14. Haghshenas, R., Chertkov, E., Mills, M., et al.: Digital quantum magnetism at the frontier of classical simulations (2025). http://arxiv.org/abs/2503.20870
15. Bobier, J.-F., Langione, M., Tao, E., et al.: What Happens When 'If' Turns to 'When' in Quantum Computing? (2021). https://www.bcg.com/publications/2021/building-quantum-advantage

16. Kearns, M., Singh, S.: Finite-sample convergence rates for Q-learning and indirect algorithms. In: Proceedings of the 1998 Conference on Advances in Neural Information Processing Systems II, p. 996. MIT Press, Cambridge, MA, USA (1999)
17. Kiwit, F.J., Wolf, M.A., Marso, M., et al.: KI - Künstliche Intelligenz **38**(4) (2024)
18. Lee, J., Leyffer, S.: Mixed integer nonlinear programming, vol. 154. Springer (2011)
19. Loke, Z.X., Goh, S.L., Kendall, G., et al.: IEEE Access **11**, 33100–33120 (2023)
20. Lorenz, J.M.: Mod. Phys. Lett. A **39**(21n22), 2430006 (2024)
21. Lorenz, J.M., Monz, T., Eisert, J., et al.: Systematic benchmarking of quantum computers: status and recommendations (2025). http://arxiv.org/abs/2503.04905
22. Lubinski, T., Johri, S., Varosy, P., et al.: Application-Oriented Performance Benchmarks for Quantum Computing (2023). http://arxiv.org/abs/2110.03137
23. Mesman, K.J., Schoot, W.V.D., Möller, M., et al.: QuAS: Quantum Application Score for benchmarking the utility of quantum computers (2024). http://arxiv.org/abs/2406.03905
24. Meyer, N., Ufrecht, C., Periyasamy, M., et al.: Qiskit-Torch-Module: Fast Prototyping of Quantum Neural Networks. In: 2024 IEEE International Conference on Quantum Computing and Engineering (QCE), vol. 1, p. 817 (2024). https://doi.org/10.1109/QCE60285.2024.00101
25. Meyer, N., Ufrecht, C., Periyasamy, M., et al.: A Survey on Quantum Reinforcement Learning. arXiv:2211.03464 (2022). https://doi.org/10.48550/arXiv.2211.03464
26. Meyer, N., Ufrecht, C., Yammine, G., et al.: Benchmarking quantum reinforcement learning. In: 2025 Forty-Second International Conference on Machine Learning (ICML). PMLR (2025). https://doi.org/10.48550/arXiv.2501.15893
27. Monnet, M., Chaabani, N., Dragan, T.A., et al.: Understanding the effects of data encoding on quantum-classical convolutional neural networks (2024). http://arxiv.org/abs/2405.03027
28. Riofrío, C.A., Klepsch, J., Finžgar, J.R., et al.: Quantum Computing for Automotive Applications (2025). arXiv:2409.14183
29. Sakuler, W., Oberreuter, J.M., Aiolfi, R., et al.: A real world test of Portfolio Optimization with Quantum Annealing (2023). http://arxiv.org/abs/2303.12601
30. Schoot, W.V.D., Wezeman, R., Eendebak, P.T., et al.: Quantum Inf. Process. **22** (2023)
31. Schoot, W.V.D., Wezeman, R., Neumann, N.M.P., et al.: Extending the Q-score to an Application-level Quantum Metric Framework (2024). http://arxiv.org/abs/2302.00639
32. Schreiber, F.J., Eisert, J., Meyer, J.J.: Phys. Rev. Lett. **131**(10) (2023)
33. Sim, S., Johnson, P.D., Aspuru-Guzik, A.: Adv. Quantum Technol. **2** (2019)
34. Van Hasselt, H., Guez, A., Silver, D.: Deep reinforcement learning with double Q-learning. In: Proceedings of the AAAI Conference on Artificial Intelligence, vol. 30, p. 2094 (2016)
35. Wybo, E., Leib, M.: Missing Puzzle Pieces in the Performance Landscape of the Quantum Approximate Optimization Algorithm (2024). http://arxiv.org/abs/2406.14618
36. Zhao, R., Wang, S.: A review of Quantum Neural Networks: Methods, Models, Dilemma (2021). http://arxiv.org/abs/2109.01840

BACQ - Application-Oriented Benchmarks for Quantum Computing

Frédéric Barbaresco[1](), Félicien Schopfer[2], Emmanuelle Vergnaud[3],
Laurent Rioux[1], Christophe Labreuche[1], Michel Nowak[1], Noé Olivier[1],
Damien Nicolazic[4], Olivier Hess[4], Anne-Lise Guilmin[4], Robert Wang[4],
Tanguy Sassolas[5], Stéphane Louise[5], Kyrylo Snizhko[5], Grégoire Misguich[5],
Alexia Auffèves[6], and Robert Whitney[6]

[1] THALES, Paris, France
`[email protected]`
[2] LNE, Paris, France
`[email protected]`
[3] TERATEC, Paris, France
`[email protected]`
[4] BULL SA, Paris, France
[5] CEA, Paris, France
[6] CNRS, Paris, France

Abstract. Delivering an application-oriented benchmark suite for objective multi-criteria evaluation of quantum computing performance, a key to industrial uses. With the support of the national program on measurements, standards, and evaluation of quantum technologies MetriQs-France, a part of the French national quantum strategy, the BACQ project is dedicated to application-oriented benchmarks for quantum computing. The consortium gathering THALES, BULL SA (previously EVIDEN), CEA, CNRS, TERATEC, and LNE aims at establishing performance evaluation criteria of reference, meaningful for industry users.

Keywords: Quantum Computer · Quantum Algorithm · Quantum Emulator · Quantum Annealer · NISQ · FTQC · Benchmark · Multi-Criteria Decision

1 Introduction

Quantum computing promises to revolutionize multiple technical fields and activity sectors, from optimization in logistics, to simulation for research in physics or chemistry, engineering or industry, passing through cryptography. Measuring the progress towards the quantum advantage and the realization of such promises, with objectivity and reliability, is of high interest for potential end-users and crucial for the future development of the domain, now subject of hype and high competition. The challenges, especially to achieve comparable measurements, comes from the diversity of the hardware platforms, their specificities in terms of physical characteristics and applications, their maturity that can still be low, and the potential rapid evolution of the technologies.

Dedicated to the whole value chain setting up from quantum hardware development to industrial use cases, BACQ is complementary to the benchmarking initiatives only focusing on low-level hardware physical criteria. The envisioned benchmark suite will be based on the resolution of several classes of problems covering important application domains of quantum computing which are meaningful for industrial users (see Fig. 1): simulation of Quantum Physics models, Optimization, Linear System Solving and Prime Factorization. Machine Learning could be included in Optimization application domain.

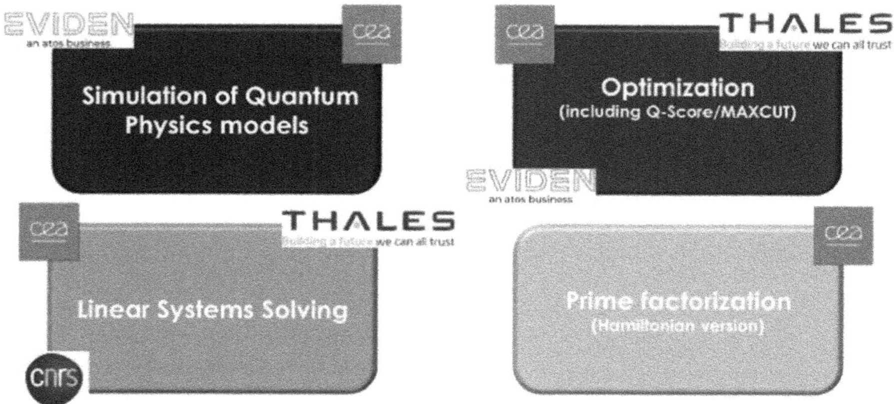

Fig. 1. Families of problems used for the QPUs benchmark

These problems are generic and could be relevant for different branches of industries and services (chemistry, aeronautics, electronics and energy for example). Criteria will be defined for the resolution of each problem, some being hardware-agnostic and others hardware-dependent (low level): computation time, latency, problem size, approximation rate, resolution probability, accuracy, fidelity… Importantly, the project also considers energetic criteria for the evaluation of the energetic performance of the machines.

The proposed methodology consists in the aggregation of low-level technical metrics and a multi-criteria analysis via the tool MYRIAD-Q in order to provide operational performance indicators of the different quantum computing solutions and point out the service qualities of interest to the end-users. The aggregation of the criteria and multi-criteria analysis allows fully explainable and transparent notations, comparisons between different quantum machines and with classical computers, as well as identification of the practical advantages of each quantum machine with respect to specific applications. The project will address both analog machines (quantum simulators and annealers) and gate-based machines, Noisy Intermediate Scale Quantum (NISQ) and Fault Tolerant Quantum Computing (FTQC). The followed practical approach is to have a suite of benchmarks, adaptive to some extent, appropriate to the capabilities of the available machines and able to demonstrate their respective advantages, including, in the longer term, exponential speedup of specific algorithms on FTQC machines.

As part of the project, a first action has already been launched regarding Q-Score, where Eviden developed on the MAXCUT optimization problem, to test and validate the benchmark on different types of quantum machines.

Sharing of the benchmark suite, as broadly as possible, is an important objective to establish common reference measurement methods and ensure no bias with inclusion of all the technologies. Consultations of the technology providers, as well as end-users is key to develop an instrument that meets the needs. Once specified and developed, the benchmark suite will be available to the community for its use.

The BACQ project, with 3-year duration and almost 4 M€ budget, including 2.5 M€ funding by France 2030 via ANR within MetriQs-France. This project precisely aims at developing, exploiting and promoting a reliable, objective and long-lasting measurement instrument for measuring the performance of quantum computing.

2 Objectives

2.1 MetriQs-France Program

MetriQs-France is the French National Program on Measurement, Evaluation, and Standards for Quantum Technologies. MetriQs-France objective is to develop, exploit, and promote measurement capabilities of reference, validated and harmonized, for characterization and performance assessment of quantum technologies with reliability, impartiality and comparability: metrology, test & evaluation, international standardization are at the core of the program. MetriQs-France is coordinated by LNE, the French national metrology and testing laboratory, a public institution under the supervision of the French Ministry for Economy and Finance, in charge of industry.

2.2 BACQ Project

While looking to the future, it is critical to measure the progress towards the realization of the quantum computing promises. Application-oriented benchmarks, enabling to evaluate the real performance of quantum computing from the user perspective appear to be useful within this perspective. The challenge comes from the diversity of the hardware platforms, their specificities in terms of physical characteristics and applications, their maturity and the potential rapid evolution of the technology. Developing an objective, long-lasting, widely shared measurement instrument, to serve as a common reference.

The evaluation of the practical quantum computing performance will be considered through benchmarks close to real applications, meaningful for industrial (as well as academic) end-users. The main objective is to measure the progress towards a practical quantum advantage. In that regard, the consortium plans to perform comparisons between the different quantum computing solutions as well as to compare them with current classical computers. This benchmarking initiative will eventually highlight the assets of each quantum computing solution given specific applications.

The benchmark suite resulting from this project will be maintained by LNE, an independent and trusted third party. Thanks to their interaction with the community of end-users, this tool will be operated in order to analyze the results obtained from the machines tested with the different benchmarks using the explainable aggregation tool. LNE will establish a list of performances, maintain it over time, and update the definition of the tests. Furthermore, adoption of this initiative by French, European and eventually international partners will be encouraged thanks to the development of communication tools regarding the approach taken and visual representations of the aggregation of the results obtained by the machines from the different benchmarks. International dialogues and collaborations on the subject of quantum computers benchmarking will be promoted so that the approach, supported by MetriQs-France, is and remains an international reference. This BACQ project promotes the development of an international standardization in regard of the methods used to evaluate the specifications of quantum machines.

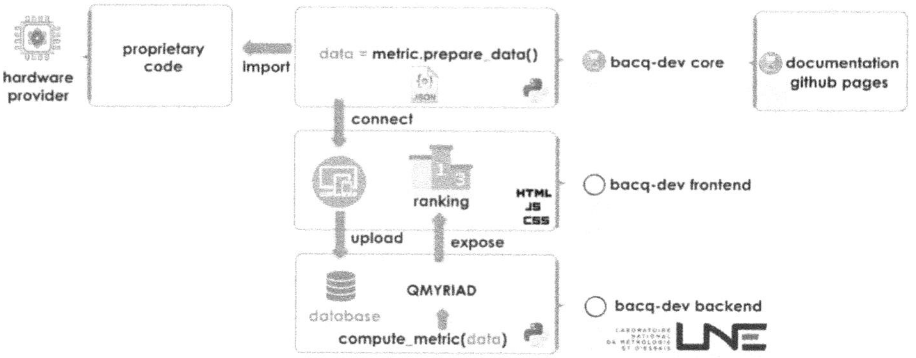

Fig. 2. BACQ Benchmark DevOp

In view of fulfilling these objectives, BACQ has initiated collaborations and dialogue with other international benchmarking initiatives: Fraunhofer IKS & FOKUS (BenchQC Project) in Germany, TNO & TU Delft (QPack Project) in The Netherlands, Qilimanjaro (CUCO Project) in Spain, QuIC (Use cases WG) in Europe, HamLib Project (Intel, LBNL et al.) in US, Unitary Fund (Metriq project) in US and QED-C (QC Benchmarking WG) in US.

3   Benchmark Tests Definition

As seen in the introduction, the set of benchmarks is divided into four categories: Optimization problems (including, but not limited to, EVIDEN's MaxCut QScore), linear system solving, quantum physics simulations and prime factorization. To this end, the set of problems constituting each benchmark family should be viewed as representative enough of the field (although pursuing exhaustivity would be out of the scope of the project at this time).

In order to design useful and relevant benchmark programs, it is necessary to adhere to a set of important principles.

As current quantum machines have a relatively high level of noise, it is of uttermost importance to compare the measurements we produce against a pure random machine.

As these QPUs are good random generators, the usefulness of quantum computing that would be undistinguished from random results must be considered as null.

Every family of benchmark should provide at least two different applications per family whenever it makes sense (i.e. except for Prime Factorization which describe one and only one possible application, even though the implementation can still vary),

As the strong points of a given quantum processor can differ from another, the exact algorithm would better be fitted to match these strong points for a given application. Nonetheless, the design and specification process must be accurately and thoroughly followed and described so that it will be easier to create versions adapted to new hardware platforms. One simple illustration of that can be illustrated by the fact that quantum annealers (e.g. D-Wave QPUs) cannot apply the same algorithm as gate-based QPUs.

Last but not least, each design processes toward the implementation in software of each benchmark program must be thoroughly detailed so that it must be easy for a third party to re-implement the benchmark program from its documentation but also to design easily a new implementation for a new QPU.

The following sections describe each of the four benchmark families.

1) Optimization Benchmarks

Optimization is a field for which quantum computing is believed to be the most apt at for short to mid-term industrial applications. Nonetheless, it is also believed that BQP, the set of problems solvable in polynomial time on a quantum computer is different from NP, therefore one should not have unrealistic expectations from quantum computing approaches. However, it is a domain of application for quantum computing of the highest importance, because it may improve the time to solution, or the quality of approximate solution, or, even more importantly, the energy required to achieve a given level of quality for a solution.

The family of optimization problems and programs will be articulated around 3 to 4 sets of problems:

- MaxCut (as part of Eviden's QScore), as an example of NP-hard quadratic binary unconstrained model,
- Maximum Cardinality Matching, as an example of Polynomial constrained problem which is famously hard for annealing approaches,
- Higher Order Binary Optimization (HOBO) problems, which are not necessarily NP-hard, but will probe the case of non-quadratic problems which can be important in the general case,
- Optionally, some NP-hard constrained problems.

The Maximum Cardinality Matching family of problems is known to be only polynomial but they laid the base to demonstrate the worst-case convergence of Simulated Annealing. CEA-LIST was pioneer in utilizing the G_n series of problem to probe the capabilities of D-Wave Quantum Annealers. This work has since been extended by other research teams in the world. Work is currently underway to extend the study to the latest iterations of D-Wave QPUs and expose the best strategies toward obtaining the best results. These results should be easily extended to other Quantum Analog QPUs (e.g. Pasqal) and QAOA approaches to solve optimization problems.

HOBO problems are important because, while it is always possible to convert them to quadratic problems by introducing ancilla variables (a process called quadratization), doing so introduces many complications for the transformed problems. Therefore, QPUs that allow for avoiding such transformations would take an increased edge against QPUs that would necessitate a quadratization of the problem before solving.

2) Linear Systems Benchmarks

Linear algebra in general provides a set of problem that are as important as optimization problems in term of applications, and is a field where ideal quantum computing has a known exponential speedup. Although it is clear that no short term exponential computing advantage can be expected, it remains an important field to investigate.

To that end, two sets of applications stand out:

- Linear system solving, which is the set for which HHL will be the most relevant algorithm for gate-based QPUs.
- Eigenvalues – eigenvectors of a given matrix, as it is a family of application in which some hybrid algorithms like VQE could be applied.

For the latest, VQE is already known as a family of algorithms which can be a potential short term application of quantum computing.

3) Quantum physics simulations Benchmarks

Quantum physics problems arise in various fields of science. For instance, in solid-state physics the understanding of new quantum phenomena and the design of new materials with potential applications often require to simulate some quantum problems with many electrons. Quantum physics with many electrons also play a central role in chemistry. They are also central problems in other domains of basic research, such as in particle or nuclear physics an even quantum information theory. Among the simple but yet nontrivial many-body Hamiltonians defined on lattices one can mention quantum spin-1/2 models (Heisenberg interactions, Ising model in transverse field, …) spinless fermion models, spin-1/2 fermion models (Hubbard model) or boson models (ex: Bose-Hubbard). Simulating these quantum many-body systems is notoriously difficult on a classical computer, for the same reason as a quantum computer is more powerful than a classical one, namely the Hilbert space dimension of a quantum system grows exponentially with the number of particles. However, with an analogue quantum simulator or with a digital (gate based) quantum computer, one would in principle be able to study many-body systems which are un-tractable on a classical machine. In that context one can mention two distinct tasks: 1) from a given Hamiltonian H, , determine the expectation values of some observables (including H itself) in the ground-state of H. . 2) Given a simple (product) initial state $|\psi_0\rangle$, determine the expectation values of some observables in the time-evolved state $|\psi_t\rangle = \exp(-iHt/\hbar)|\psi_0\rangle$. . The ground-state problem (1) can for instance be tackled using VQE on a gate-based machine or via a (quasi) adiabatic evolution on an analogue simulator. The dynamical (also called quantum quench problem (2) can be studied via a discretization of the time evolution (Trotterization) on gate-based architectures, and somehow more directly with the native Hamiltonian dynamics on an analogue device (provided the interactions between the qubits can be tuned to match the Hamiltonian H of interest).

4) Prime Factorization Benchmarks

The discovery of Shor's algorithm for prime factorization is one of the results which generated immense expectations on the part of quantum computing. While it is clearly unrealistic to expect that this particular algorithm can be put to applications without having first developed fault-tolerant quantum computers, there exists other possible approaches. One is utilizing Hamiltonian based approaches. The basic principle is rather simple because it simply requires the minimization of the following cost function:

$$C(p_1, p_2) = (N - p_1 p_2)^2$$

where p_1 and p_2 are 2 possible prime numbers that ought to be found and N is the integer which is the subject of the prime factorization.

Approaches that are not based on Shor's algorithm are not expected to provide any kind of computational advantage, it can however not be ruled out at the moment that it can provide some kind of advantage on the energy side.

4 Energy Criterion by Resource Estimation for the Resolution of Simple Algorithms on Gate-Based Computers

For any given task, an energy efficiency can be parametrically defined as the metric chosen to quantify the performance of the task, divided by its energy cost. The efficiency is thus tightly dependent on the choice of metric, as well as the list of energy costs.

The objective of the work package is to propose and study the behavior of such energy efficiencies for some of the useful algorithms studied in this project. In principle, all machines can give rise to such figures of merit. Our approach will rely on the MNR methodology introduced, which consists in finding relations between Noise, Metric of performance and Resource cost, then exploit these relations to minimize the resource cost under the constraint of a given performance.

In the spirit of the BACQ project, we will care to remain as hardware-agnostic as possible to propose useful and flexible numerical tools to end-users and industrials. The focus shall thus be put on algorithmic resources, the connection with energy costs being handled via effective models. In close connection with the HQI project of the French quantum strategy, we will build a user-friendly interface accessible to any hardware provider can use to estimate the energy efficiency of the quantum processor at play.

5 Multi-criteria Approach: MYRIAD-Q

The construction of the Multi-Criteria model is organized around three steps with MYRIAD-Q tool.

1) 1st step: structuring phase

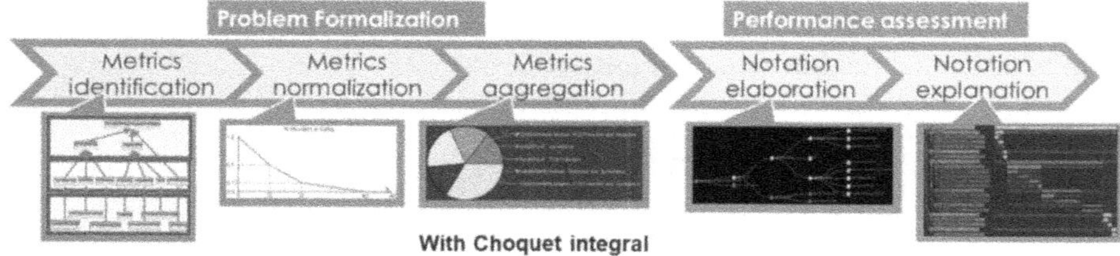

Fig. 3. MYRIAD-Q Multicriteria Agregation Tool

Given the plurality of QPUs, we first need to identify the subsets of QPUs that make sense to be compared. A MYRIAD-Q model is then constructed for each subset. For instance, it might not be relevant to compare Quantum Annealers with NISQ.

Once we have selected the scope of a MYRIAD-Q model, the multi-criteria approach focuses firstly on identifying operational metrics which have a real operational impact (fundamental objectives for experts) and not technical metrics which only have an indirect impact. Then the associated criteria are organized hierarchically.

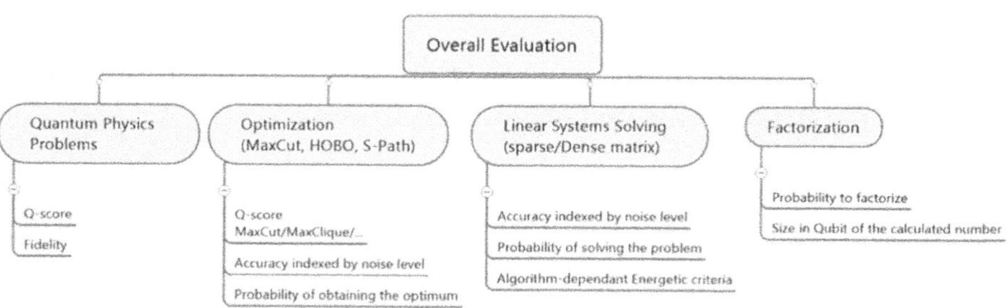

Fig. 4. Sketch of a hierarchy of criteria

2) 2nd phase: normalization

It does not make sense to directly combine the metrics as they are given in different units. The aim of the marginal utility functions is to normalize the metrics and transform them in the same scale. This normalization corresponds to an *interval scale* in the terminology of measurement theory. An interval scale is given up to an affine transformation. One thus needs to define two reference levels to entirely fix it. For scale $S = [0, +\infty)$, we introduce the following two reference elements: B_i (it stands for Bad) is an element of metric X_i that is judged not satisfactory at all, and G_i (Good) is an element of metric X_i that is judged *satisficing*. This latter level is a central ingredient of the Bounded Rationality of Simon. It is a contraction of satisfying and sufficient. The bad element is the saturation level from below (no element has a worst utility); and there exists elements that are preferred to the satisficing level. In order to ensure commensurateness among criteria, we impose that all reference levels of the same nature across the criteria are assigned to the same utility.

3) 3rd phase: aggregation

As already mentioned, the aggregation function F is organized hierarchically. The first phase returns a tree described by a set M of nodes including a root $r \in M$ and by a function $P : M \to 2^M$ returning the set of parents $P(k)$ for every node $k \in M$. The leaves of the tree are the set N of criteria. An aggregation function $F_k : S^{P(k)} \to S$ is defined at each aggregation node $k \in M \setminus N$.

One needs to elicit each aggregation function F_k. The simplest aggregation function is the weighted sum $F_k(a_{P(k)}) = \sum_{l \in P(k)} w_l a_l$, where w_l is the weight of node l. This model cannot capture real-life decision strategies as it assumes that all its inputs are independent.

In order to overcome the limitations of this model, we propose to use the Choquet integral, which has the capacity to represent interactions among criteria. Its so-called 2-additive version is a good compromise between simplicity and expressivity of the model. It takes the form

$$F_k(a_{P(k)}) = \sum_{l \in P(k)} w_l a_l$$

$$+ \sum_{\{l,m\} \subseteq P(k)} \left(w_{l,m}^{\wedge} min(a_l, a_m) + w_{l,m}^{\vee} max(a_l, a_m) \right)$$

where coefficients $w_l, w_{l,m}^{\wedge}, w_{l,m}^{\vee}$ are non-negative coefficients that sum-up to one. Compared to the weighted sum, the min terms represent *complementarity* between the two criteria l, m (necessarily both scores have to be high to produce a high overall score; in other words, if one score is larger than the other one – say $a_l > a_m$ – then the better evaluation on criterion l is penalized by the worst score on criterion m). Likewise, the max terms represent redundancy between the two criteria l, m (it is sufficient that one score is high; in other words, if one score is larger than the other one – say $a_l > a_m$ – then the bad evaluation of criterion m is saved to a degree $w_{l,m}^{\vee}$ by the better evaluation on criterion l.

6 Conclusion

The benchmark tools and methodologies developed in the BACQ project will be used to help select the QPUs for the PROQCIMA program (notified DGA framework agreements to five companies, Alice&Bob, C12, Pasqal, Quandela and Quobly, with a view to identifying solutions enabling the development of these computers). The PROQCIMA program will last for at least 10 years. After four years, only the three most successful projects will continue the program to develop the best logical qubits capable of scaling up. After eight years, the competition will be limited to the two most successful technologies, which will continue the program to move from computer prototypes (target: 128 logical qubits) to industrial products usable by their first customers (target: 2,048 logical qubits).

References

1. BACQ: Application-oriented Benchmarks for Quantum Computing, arXiv:2403.12205v1 [quant-ph]. https://doi.org/10.48550/arXiv.2403.12205

2. Bench-QC: Application-driven benchmarking of quantum computers, Fraunhofer Institute IKS, Germany. https://www.iks.fraunhofer.de/en/projects/bench-qc-application-driven-benchmarking-of-quantum-computers.html

3. QPack: Benchmark for quantum computing. Delft University of Technology, Delft, The Nederlands. https://github.com/koenmesman/QPack?tab=readme-ov-file

4. CUCO project "Computacion Cuantica En Industrias Estrategicas", Spain. https://www.cuco.tech/en/project/

5. Metriq "Community-driven quantum benchmarks", Unitary Fund. https://metriq.info/

6. QED-C benchmark. https://github.com/SRI-International/QC-App-Oriented-Benchmarks

7. DARPA QBI (Quantum Benchmarking Initiative). https://www.darpa.mil/news-events/2024-07-16; https://www.darpa.mil/news-events/2024-08-15

8. HamLib: A library of Hamiltonians for benchmarking quantum algorithms and hardware,in 2023 IEEE International Conference on Quantum Computing and Engineering (QCE), 2023. arXiv:2306.13126v3 [quant-ph]. https://doi.org/10.48550/arXiv.2306.13126

9. HamPerf: A Hamiltonian-Oriented Approach to Quantum Benchmarking, CF '24 Companion: Proceedings of the 21st ACM International Conference on Computing Frontiers: Workshops and Special Sessions, pp. 133–138. https://doi.org/10.1145/3637543.3653431

10. Vergnaud, E., Schopfer, F., Barbaresco, F.: 1st TQCI Seminar "Overview of upcoming application-oriented benchmarks for quantum computing in France and abroad", Thales Research & Technology, Palaiseau, France, 11th May 2023. https://teratec.eu/Seminaires/TQCI/2023/Seminaire_TQCI-230511.html

11. Vergnaud, E., Schopfer, F., Barbaresco, F.: 2nd TQCI Seminar "Overview of upcoming application-oriented benchmarks for quantum computing in France and abroad", 4th & 5th June 2024, Reims, France. https://teratec.eu/activites_quantiques/TQCI_240604_programme.html

12. Vergnaud, E., Schopfer, F., Barbaresco, F.: 3rd TQCI Seminar "benchmarks for quantum computers", 24th & 55th June 2025, Palaiseau, France. https://teratec.eu/Seminaires/TQCI/2025/Seminaire_TQCI-250625_UK.html

13. UK Quantum Computing Metrics Initiative, UK Quantum Metrology Institute of National Physical Laboratory. https://www.npl.co.uk/quantum-metrology-institute

14. Basque Quantum Benchmark, Benchmarking Quantum Computers: Towards a Standard Performance Evaluation Approach. https://arxiv.org/pdf/2407.10941v1

15. GIFAS, Rapport sur les technologies quantiques pour les applications Aéronautique-Spatial-Défense, Groupe de Travail Technologies Quantiques, 2024, Paris

16. Académie des technologies, État de l'art de l'ordinateur quantique tolérant aux fautes – Questions et défis, 2025, Paris. https://www.academie-technologies.fr/wp-content/uploads/2025/06/202505_ordinateur_quantique.pdf

17. EQCBC (European Quantum Computing Benchmarking Committee), Systematic benchmarking of quantum computers: status and recommendations. https://doi.org/10.48550/arXiv.2503.04905

18. Rohe, T., et al.: Quantum Computer Benchmarking: An Explorative Systematic Literature Review. https://doi.org/10.48550/arXiv.2509.03078

19. Koch, T., et al.: Quantum Optimization Benchmarking Library - The Intractable Decathlon. https://doi.org/10.48550/arXiv.2504.03832

Quant^{2}AI – An End-to-End Quantum AI Benchmarking Framework for Both Researchers and Practitioners

Cristobal Corvalan Morbiducci[1]($\boxtimes$), Pascal Halffmann[2], Andrew Barlow[1], Alexander Geng[2], Lautaro Hickmann[1], Ali Moghiseh[2], Sabine Müller[2], Christine Priplata[3], Hans-Martin Rieser[1], Colin Stahlke[3], and Michael Trebing[2]

[1] DLR Institute for AI Safety and Security, Wilhelm-Runge-Straße 10, 89081 Ulm, Germany
`cristobal.corvalanmorbiducci@dlr.de`
[2] Fraunhofer Institute for Industrial Mathematics ITWM, Fraunhofer-Platz 1, 67663 Kaiserslautern, Germany
[3] Conet Solutions GmbH, Bundeskanzlerplatz 2, 53113 Bonn, Germany
`https://www.dlr.de/en/ki`, `https://www.itwm.fraunhofer.de/en.html`, `https://www.conet.de/EN/conet`

Abstract. Quantum AI and Quantum Machine Learning (QML) are among the most promising and dynamic research fields, with a vast variety of QML models. However, standardized benchmarking is lacking, and end users often struggle to determine whether quantum AI—and which specific approach—is suitable for their use cases. Addressing these challenges is essential to evaluate the current state of quantum AI and advance toward quantum utility.

We have developed *Quant^{2}AI*, a holistic benchmarking framework for systematic comparisons of quantum AI pipelines using high performance clusters and both quantum simulators and hardware. Our end-to-end approach evaluates not just QML models but the whole pipeline, including e.g., preprocessing and hyperparameter variations. Its modular design enables easy integration of new components, such as alternative data preparation. We provide standardized and real-world datasets, quantum and classical AI reference pipelines, state-of-the-art evaluation metrics, and intuitive visualizations. Our framework offers benchmarking as a service for both researchers for testing their newly developed quantum AI components as well as end users for an intuitive way to identify promising quantum AI applications.

Keywords: Quantum AI · Benchmarking · Quantum Machine Learning · Use Cases

1 Introduction

Quantum Machine Learning (QML)—harnessing the computational capabilities of quantum systems to improve machine learning (ML) models [5]—has witnessed rapid advancements and substantial attention in recent years [3]. Theoretical propositions of quantum advantage suggest that QML could outperform classical models in training speed or quality [13], however, these prospects remain speculative: Critics point to the limited empirical evidence supporting substantial performance gains over classical baselines, particularly when idealized scenarios for quantum computers are considered [15]. As such, while the theoretical landscape of QML is rich with promise, its actual effectiveness in real-world applications remains largely uncharted. Rigorous benchmarking of QML pipelines [4] is thus essential to determine their practical utility and distinguish genuine quantum utility or even advantage from hype-driven expectations.

1.1 Previous Work

Benchmarking for classical machine learning methods [16] are numerous; an overview is beyond the scope of this paper, as also the major ML packages like Scikit-Learn, PyTorch and TensorFlow provide benchmarking suites. In the quantum realm, benchmarking has mostly been targeted at hardware with e.g., Supermarq [17]. As NISQ hardware becomes increasingly more viable for QML, it has increased the necessity for rigorous and reproducible benchmarking of machine learning methods on quantum devices with first approaches focusing on classification [2] or real-world use cases [10]. Recently, several frameworks have been developed: Besides the QML benchmarks provided by packages such as PennyLane, there exists the QUARK framework [6,11] focusing on application oriented benchmarks for general quantum methods. The MQT Bench framework [12] provides an easy-to-use library for benchmarking a wide variety of quantum tools and algorithms including QML methods like QNN. Automated quantum machine learning (AutoQML) [14] is a framework based on AutoML, which is a unifying interface for deployment of end-to-end QML pipelines. Most recently, BenchQC [7], a general framework to benchmark quantum methods in a modular fashion, also for optimization, has been published.

1.2 Our Contribution

We propose $Quant^2AI$, a framework that encompasses the advantages of the aforementioned benchmarking frameworks for a rigorous, standardized, and reproducible benchmark of QML end-to-end pipelines. Our framework allows for an end-user friendly and fast execution on high performance clusters (hpc) using both simulators and real quantum devices. While we provide a set of quantum and classical reference pipelines comprising of standalone modules, both standard and real-world datasets, and evaluation metrics, users can not only just add their own components, but also fully customize the existing ones, including adding new quantum devices. Therefore, end-users can evaluate their use cases to determine any quantum utility, whereas researchers can seamlessly test their new components or whole pipelines.

Section 2 introduces the Quant^{2}AI framework and its components, while in Sect. 3, we showcase the execution of a benchmark for a QML pipeline.

2   Benchmarking Framework

Robust benchmarking is a step to gain an understanding of the potential advantages, limitations, and applications of quantum machine learning algorithms relative to the classical case. In the following, we present the key ingredients of our benchmarking framework: We discuss the foundations and hpc integration in Sect. 2.1, followed by the presentation of modular end-to-end pipelines in Sect. 2.2. In the next section (Sect. 2.3), we introduce the integration of quantum hardware and conclude with our evaluation module and visualization tool in Sect. 2.4.

2.1   Foundation and HPC Execution

The framework follows the basic idea of allowing non-hpc experts to benchmark complete execution pipelines on cluster systems. Therefore, the actual interaction between user and hpc-backend is reduced to a minimum by different abstraction layers, all of them modular while being straightforward to exchange. The general workflow reflects this regime: The user provides a pipeline configuration, as well as general information about the cluster environment, e.g. the paths where the code and the data resides, to the top most abstraction layer: a python interface that automatically prepares the necessary job scheduling scripts. Here, we currently offer script generation for SLURM. However, tailoring job scheduling e.g. to Portable Batch System (PBS) is reduced to an exchange of this module.

As soon as the job is started, job dependent structures are prepared and all necessary information for a proper job execution are assembled: First, output folders for job-specific information are generated and necessary python environments are locally installed (in case of present local SSD hard drives on node-side). This environment ensures that on the framework side we provide and store the important information of the jobs so that reproducibility of results can be ensured. Once the environment is completely set, the pipeline starts execution. Based on a unique identifier, results as well as general log information about the job run itself combined with pipeline specific log information, are stored in these job-specific output folders. Generally, pipeline modules are executed sequentially, see Sect. 2.2. Once the pipeline is finished, relevant information is collected and stored in the parallel file system in the locations specified in the first step. A database incorporates and processes this information for the visualization of the collected metrics (Sect. 2.4). A schematic view of the framework with further details is given in Fig. 1.

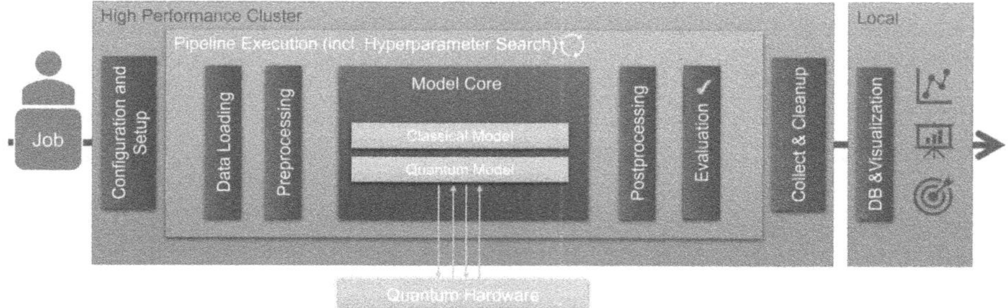

Fig. 1. Abstract visualization of the framework structure. The user submits a job consisting of the pipeline and configuration files (cluster-specific requirements and pipeline information such as hyperparameter values, loss function,...). A script is generated via SLURM that builds output folders and sets up the necessary python environments among others. This is a fresh python installation without any artifacts from previous runs. Once done, the pipeline is executed module by module. The most important module is the model core, where a classical, quantum, or hybrid model is executed. In the latter two cases, the quantum part can be executed on a simulator or on a real quantum device using a connection module. Information from the model core (and the quantum device connection module) is collected to compute suitable metrics. If hyperparameter search is invoked, multiple of these pipeline executions runs are carried out. After the pipeline execution, a cleanup process in invoked and the output is stored, ready to be processed by the visualization tool and its database on a local device.

2.2 Pipelines and Modules

The implementation of pipelines is built around a modular design, enabling flexibility and re-usability across various machine learning and quantum machine learning tasks. The standard pipeline consists of five key modules, see Fig. 1, but users are free to adjust the number and purpose of the modules. The framework already provides implementations:

- Data Loading: Prepares and partitions the dataset for training, validation, and testing. Example: Data loader to load and prepare the iris dataset.
- Preprocessing: Prepares the dataset for the model, e.g., using Principal Component Analysis (PCA).
- Model Core: Implements training of the model, also on a real quantum device, see Sect. 2.3. Example: Classical and quantum SVM, k-means.
- Postprocessing: Counterpart to the preprocessing (e.g., PCA), to reinterpret the results for the original data.
- Evaluation: Evaluates the performance of the pipeline, see Sect. 2.4.

Pipelines are constructed by chaining these modules together, with configuration and data passed between them via a centralized JSON file. This design enables logical interconnection between compatible modules while also enforcing consistency. Each module contains internal validation logic that checks the JSON input for structural and semantic correctness. This ensures that incompatible module combinations—such as applying an image processing pipeline to tabular data—are flagged and prevented early in the workflow making the integration of new modules seamless.

Built-in reference pipelines are provided for a range of use cases including anomaly detection, binary image classification and segmentation, clustering, multi-class classification, and regression – with two datasets for each use case available in the framework. These datasets range from famous toy examples like iris or MNIST to real-world data and application such as identifying cracks on images of concrete or anomaly detection on warranty claims for purchased cars. Suitable data loaders preparing the data are provided. The reference pipelines serve as practical examples of how to structure and connect modules for various machine learning objectives. The user has full flexibility in building or customizing pipelines, for example adding or deleting modules, e.g., add a second pre-processing module or build a new data loader. This architecture ensures a high degree of modularity, robustness, and extensibility, suitable for a wide range of classical and QML applications.

2.3   Quantum Hardware Connection

The framework provides a quantum connection (qc_connect) module which handles communication between the running experiments and the quantum backend. Communication is mainly achieved through the widely used openQASM representation of quantum circuits. Therefore, the qc_connect module supports appropriate creation, interpretation, and storing of these representations for later usage.

The openQASM format allows for a "universal" wide incorporation of quantum computers of different hardware providers. In the same way, incorporation of non-local virtual simulators is also possible. Up to now, connectors for IQM, AQT, and IBM backends have been created. Verification steps such as sending access tokens are integrated in a secure fashion. Further devices can be added to the module with simplified integration steps.

The qc_connect module also handles the backend output, all information from the hardware execution. Different metrics such as the number of qubits and gate depth can also be extracted; further, the integration of new quantum metrics is possible and can then be forwarded to the evaluation module in the next section.

2.4   Evaluation and Visualization

Results of the model cores are processed in the evaluation modules. These modules collect all data from the model execution and compute metrics; besides general ML metrics, the framework already contains task-dependent metrics and evaluation modules for our five use cases. For the quantum pipelines, additional metrics are fetched from the quantum devices, both simulators and hardware, such as number of qubits or gate depth. Similar to the aforementioned modules, the evaluation is fully customizable, allowing the inclusion of newly defined

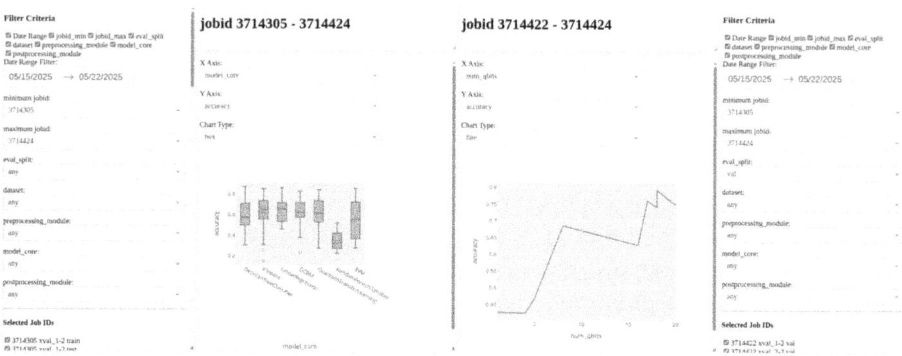

Fig. 2. Dashboard with two linked cohorts. Each filter criteria panel defines the cohort for the results panel next to it. Left: accuracy distributions per model core (boxplots). Right: accuracy vs num_qbits, shown as an aggregate over selected models or, when grouped, as separate curves per model.

metrics and the selection of a hyperparameter search objective. Indeed, a hyperparameter optimization using Optuna can be executed. Further, the aggregation of metrics across different cross validation folds for hyperparameter search is supported, which is crucial for stochastic stability. A prime example is the Inter-Quartile Mean [1] mitigating the impact of outliers to the final result.

In order to support transparent and extensible benchmarking of both classical and quantum-enhanced learning algorithms, we have developed a browser-based dashboard that couples an experiment repository with an interactive analytics layer. For each experiment, all metrics computed by the evaluation module (including hardware metrics stemming from the quantum device) and all metadata – like dataset, model core, or time-stamp – are ingested into a relational database that serves as a single source of truth. With this design, we eliminate hard-coded assumptions allowing the introduction of new metrics, data sets, or modules without refactoring front-end code. Visualization types such as histograms as well as plotted metrics, multi-scale comparisons, and aggregations are user-configurable. Predefined dashboards can be loaded for recurrent benchmarks while still retaining the full flexibility for ad-hoc analysis.

Figure 2 shows two experiment cohorts, arranged side-by-side by the user. Each cohort is defined by its filter criteria panel and rendered in the adjacent results panel. On the left, boxplots summarize accuracy per model core, to reveal variance and outliers. On the right, accuracy is plotted against the number of qubits. The curve aggregates over all model cores as selected in the right filter. The user can restrict the filter to one model core or group by model core to overlay model-specific curves. This pairing enables like-for-like comparisons of models and qubit scaling under identical, reproducible filter settings.

3   Illustrative Example

To demonstrate the implementation and benchmarking of a reference pipeline, we consider a hybrid quantum-classical image classification task—specifically, the real-world example of crack detection in concrete surfaces (HairWidthCracks [8]). The pipeline is based on quantum transfer learning, where a classical neural network is partially replaced by a parameterized quantum circuit [9]. This serves as a practical example of integrating QML within established deep learning workflows. The proposed pipeline is developed similar to Fig. 1. Within the data loading module, the dataset is split into training, validation, and testing subsets. Moreover, a stratified K-Fold procedure and sanity checks are applied.

In the model core, the pipeline employs a pre-trained ResNet18 model as a feature extractor. The final fully connected (FC) layer is replaced with a hybrid structure consisting of a classical linear layer, a quantum circuit layer, and another linear layer. The quantum circuit operates on a number of qubits equal to the output dimension of the preceding linear layer. All training-related parameters are stored either for later use in the evaluation module or for monitoring the training process. The modular structure of the framework supports independent development of individual components. It should be noted, that ResNet18 is not the only possible choice for feature extractor and it could be easily replaced with another backbone (e.g., VGG16).

The evaluation module computes the final performance metrics. For this reference pipeline, these include accuracy, recall, precision, and F1-score. When a quantum circuit is included in the model core, additional quantum-specific metrics are also considered. Pipeline parameters are managed through a JSON configuration file, which defines adjustable hyperparameters such as learning rate, number of qubits, number of epochs, batch size, and Optuna optimization settings.

We have benchmarked both classical and quantum variants of the model on an HPC node using an Nvidia Hopper H100 GPU and dual Intel Xeon Gold 5420+ CPUs. The benchmark uses 5-fold cross-validation on the full Hair-WidthCracks dataset [8].

In addition to the benchmark framework information presented in Fig. 2, the framework also records training-specific parameters, such as the progress shown in Fig. 3. This figure illustrates the training dynamics across different model cores with varying parameters, using a 5-fold cross-validation on the complete HairWidthCracks dataset.

In our example, both the loss and accuracy curves in Fig. 3 and the final evaluation metrics reported in Table 1 illustrate the performance of our pipelines: three neural networks with input and output dimensionality of 512 and 2 respectively. The hidden layer dimensions are set to none layer, 4 and 512. The last pipeline replaces its linear layer by a 4 qubit parametrized quantum circuit. Despite the higher computational overhead introduced by quantum circuit simulation and shot-based measurement, the quantum model achieves performance comparable to classical configurations across all key metrics, underscoring the potential of quantum machine learning (QML) in practical tasks.

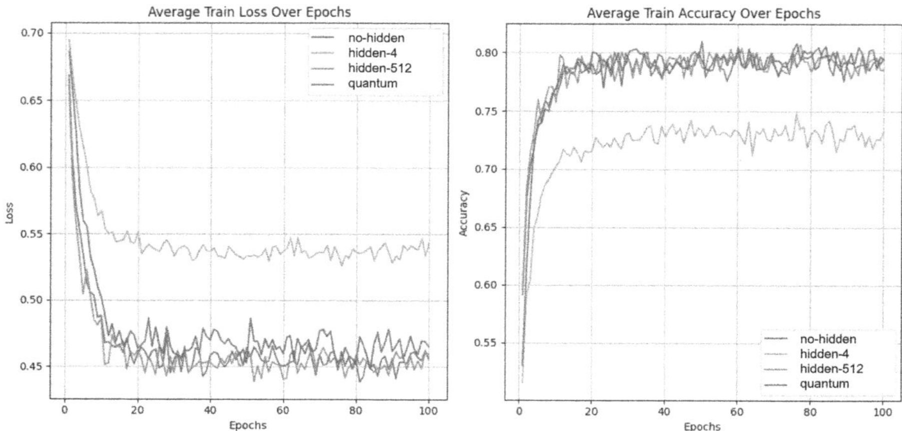

Fig. 3. Training loss and accuracy curves for classical and quantum pipelines over 100 epochs. We present four pipelines. Three classical fully connected linear layers with input and output dimensions of 512 and 2 respectively. The width of the hidden layer is set to none layer ("no-hidden"), 4 ("hidden-4") and 512 ("hidden-512"). The last pipeline replaces the hidden layer with a 4 qubit parameterized circuit ("quantum").

Table 1. Performance metrics on the unseen test dataset for the classic methods (as described in Fig. 3) and the quantum version.

Metric	Classic			Quantum
	no-hidden	hidden-4	hidden-512	
Accuracy	0.8107	0.7640	0.8133	**0.8183**
Precision	**0.8473**	0.7755	0.8357	0.8426
Recall	0.7600	**0.7920**	0.7840	0.7833
F1 Score	0.8000	0.7763	0.8077	**0.8115**

4 Conclusion

This work has introduced a modular benchmarking framework designed to facilitate systematic comparisons between classical and quantum machine learning pipelines. By separating the workflow into well-defined modules, the framework enables reproducibility, transparent evaluation, and extensibility. Reproducibility is addressed by the job generation and environment creation, together with the well-defined available modules. Transparent evaluation is achieved by the synergy between our evaluation modules and its communication with the visualization dashboards, that allows to identify trends in different kind of graphs of aggregated metrics across experiments. Extensibility allows to integrate and test additional functionalities at different levels, such as new evaluation metrics that can be relevant to other use cases. If additional functions or evaluation metrics become relevant in the future, our benchmarking framework can readily accommodate them. Additional datasets, new pipeline steps, and in particular novel (Q)ML models can be integrated with minimal code modifications, allowing researchers to focus on experimentation rather than implementation overhead. Each module is accompanied by sample implementations that can either be used directly or serve as templates for further development. By combining modules in flexible ways, researchers can effectively construct new algorithms or pipelines with minimal additional overhead. This adaptability makes the framework a practical tool for exploring a wide range of scenarios in both classical and quantum machine learning research. The framework comes with a large library of (real-world datasets), QML and ML models, and metrics. Finally, our software is able to integrate its workflow to communicate to different quantum hardware and simulations, highlighting a crucial aspect of the benchmarking paradigm: verification of the functionality of QML needs to consider and compare execution on the devices it is envisioned for. In this way, benchmarking will reveal its utility to do the leap from the research area to the application fields. The presented illustrative example demonstrate not only the validity of our framework but also the feasibility of integrating quantum components into established deep learning architectures and thus highlights the potential of such hybrid approaches for real-world applications. In conclusion, the proposed benchmarking framework is well suited for systematically comparing classical and quantum pipelines. Its modular design enables researchers to gain deeper insights into the impact of different parameter choices and architectural configurations. By recording both training dynamics and final evaluation results, the framework supports transparent and interpretable comparisons across methods.

Acknowledgments. This project was made possible by the DLR Quantum Computing Initiative and the Federal Ministry for Economic Affairs and Climate Action; https://qci.dlr.de/en/quant2ai/.

Disclosure of Interests. DLR, the contractor Fraunhofer ITWM, and the subcontractor Conet Solutions GmbH have a contractual relationship in the context of the project Quant^{2}AI. This relationship is disclosed in the interest of transparency.

References

1. Agarwal, R., Schwarzer, M., Castro, P.S., Courville, A., Bellemare, M.G.: Deep reinforcement learning at the edge of the statistical precipice (2022). https://arxiv.org/abs/2108.13264
2. Aly, M., Fadaaq, S., Warga, O.A., Nasir, Q., Talib, M.A.: Experimental benchmarking of quantum machine learning classifiers. In: 2023 6th International Conference on Signal Processing and Information Security (ICSPIS), pp. 240–245. IEEE (2023). https://doi.org/10.1109/icspis60075.2023.10343811
3. Bauckhage, C., et al.: Quantum machine learning-state of the art and future directions. German Federal Office for Information Security, p. 122 (2022)
4. Bowles, J., Ahmed, S., Schuld, M.: Better than classical? The subtle art of benchmarking quantum machine learning models. arXiv (2024). https://doi.org/10.48550/arXiv.2403.07059

5. Cerezo, M., Verdon, G., Huang, H.Y., Cincio, L., Coles, P.J.: Challenges and opportunities in quantum machine learning. Nat. Comput. Sci. **2**(9), 567–576 (2022). https://doi.org/10.1038/s43588-022-00311-3

6. Finzgar, J.R., Ross, P., Holscher, L., Klepsch, J., Luckow, A.: Quark: a framework for quantum computing application benchmarking. In: 2022 IEEE International Conference on Quantum Computing and Engineering (QCE), pp. 226–237. IEEE (2022). https://doi.org/10.1109/qce53715.2022.00042

7. Geissler, F., et al.: Benchqc – scalable and modular benchmarking of industrial quantum computing applications. arXiv (2025). https://doi.org/10.48550/ARXIV.2504.11204

8. Geng, A., Moghiseh, A.: HairWidthCracks dataset. Fordatis - Forschungsdaten-Repositorium der Fraunhofer-Gesellschaft (2024). https://fordatis.fraunhofer.de/handle/fordatis/429

9. Geng, A., Moghiseh, A., Redenbach, C., Schladitz, K.: Hybrid quantum transfer learning for crack image classification on NISQ hardware. In: The National University of Science and Technology International Conference For Engineering Sciences, p. 130001 (2025). https://doi.org/10.1063/5.0246500. https://pubs.aip.org/aip/acp/article-lookup/doi/10.1063/5.0246500

10. Gujju, Y., Matsuo, A., Raymond, R.: Quantum machine learning on near-term quantum devices: current state of supervised and unsupervised techniques for real-world applications. arXiv (2023). https://doi.org/10.48550/ARXIV.2307.00908

11. Kiwit, F.J., Marso, M., Ross, P., Riofrío, C.A., Klepsch, J., Luckow, A.: Application-oriented benchmarking of quantum generative learning using quark. In: 2023 IEEE International Conference on Quantum Computing and Engineering (QCE), pp. 475–484. IEEE (2023). https://doi.org/10.1109/qce57702.2023.00061

12. Quetschlich, N., Burgholzer, L., Wille, R.: MQT bench: benchmarking software and design automation tools for quantum computing. Quantum **7**, 1062 (2023). https://doi.org/10.22331/q-2023-07-20-1062

13. Ristè, D., et al.: Demonstration of quantum advantage in machine learning. NPJ Quantum Inf. **3**(1) (2017). https://doi.org/10.1038/s41534-017-0017-3

14. Roth, M., et al.: Autoqml: a framework for automated quantum machine learning (2025). https://arxiv.org/abs/2502.21025

15. Schuld, M., Killoran, N.: Is quantum advantage the right goal for quantum machine learning? PRX Quantum **3**(3), 030101 (2022). https://doi.org/10.1103/prxquantum.3.030101

16. Thiyagalingam, J., Shankar, M., Fox, G., Hey, T.: Scientific machine learning benchmarks. Nat. Rev. Phys. **4**(6), 413–420 (2022). https://doi.org/10.1038/s42254-022-00441-7

17. Tomesh, T., et al.: Supermarq: a scalable quantum benchmark suite. arXiv (2022). https://doi.org/10.48550/ARXIV.2202.11045

Application of Multi-criteria Decision Aiding to Quantum Benchmarking: Some Results on Scale Invariance

Christophe Labreuche$^{(\boxtimes)}$

Thales cortAIx-Labs, 1 Avenue Augustin Fresnel, 91767 Palaiseau, France
christophe.labreuche@thalesgroup.com

Abstract. Application-oriented benchmarking of quantum computing requires the evaluation of multiple, often conflicting, Key Performance Indicators (KPIs). To assess the overall quality of a single solution or to compare multiple solutions, these KPIs must be synthesized into a comprehensive metric. Multi-Criteria Decision Aiding (MCDA) provides a suitable methodological framework for this purpose. MCDA involves two main steps to produce an overall score from a vector of KPIs: first, normalizing the KPIs, and second, aggregating the normalized scores. It is important to note that normalization is not uniquely defined and may correspond to different measurement scales. We therefore examine the property of invariance, where switching from one admissible scale to another should not alter the comparative ranking of options. To facilitate aggregation, it is often assumed that some form of commensurability (or comparability) exists across the normalized KPI scores. However, this is a strong assumption that may not always hold in practice. As such, we investigate various aggregation functions that can accommodate different invariance assumptions, providing a more nuanced and robust approach to decision-making in quantum benchmarking.

Keywords: Quantum Computer Benchmarking · Multi-Criteria Decision Aiding

1 Introduction

Quantum computing benchmarking is a critical issue from the user's perspective, as it enables the assessment of performance and the analysis of tangible benefits compared to High-Performance Computing (HPC) or traditional computers [12]. In this context, we concentrate on application-level benchmarking, as opposed to component-level or system-level benchmarking. Application-level benchmarking represents the most comprehensive form of evaluation, subsuming the other levels, and aims to determine the capability of a quantum computer or simulator to solve specific problems of interest to users, such as optimization or fundamental physics problems.

Benchmarking requires the definition of Key Performance Indicators (KPIs) to evaluate different solutions. Application-level KPIs are directly interpretable and relevant to end-users. The Q-Score measures the largest problem size that can be solved with performance significantly above the noise level [13]. Originally defined for the Max-Cut

F. Barbaresco and G. François (Eds.): QUEST-IS 2025, CCIS 2743, pp. 237–247, 2026.
https://doi.org/10.1007/978-3-032-13852-1_24

optimization problem [13], it has since been extended to the Max-Clique problem [19]. The Quantum Application Score (QuAS) is defined as the volume under the curve of a function fitting the Pareto front in the $\langle$accuracy, runtime$\rangle$ space [14]. TAQOS (Tight Analysis of Quantum Optimization Systems) introduces a KPI defined as the percentage of instances for which a solution is found with a cost function within a distance at most ε from the reference value within a time limit t, evaluated for different values of ε and t on Noisy Intermediate-Scale Quantum (NISQ) devices across various optimization problems [7]. The BACQ project aims to develop KPIs for a range of applications (such as optimization, linear system solving, prime factorization, and many-body problems), and also incorporates energy-related considerations [2].

When the user is interested in multiple KPIs, rather than a single one, it becomes necessary to combine these KPIs to enable meaningful comparisons between quantum computing solutions. Multi-Criteria Decision Aiding (MCDA) is a methodology that aims to elicit and represent the preferences of a stakeholder –typically a decision maker – regarding different options based on multiple, and often conflicting, decision criteria represented by KPIs [6].

QPack introduces an overall score that aggregates around a dozen KPIs [5] In this framework, the *pure* score denotes the raw value of the KPI, while the *mapped* score is a normalized value that facilitates the aggregation of different KPIs into a single comprehensive score. A similar methodology is applied in BACQ [2], where KPIs are first normalized and then aggregated. The distinction between BACQ and QPack is twofold: (1) BACQ proposes a MCDA-based elicitation approach that constructs normalization and aggregation schemes from decision maker input, whereas QPack relies on predefined formulas; (2) the aggregation models differ – BACQ employs a Choquet integral [4], while QPack uses a multi-linear model [10].

The aim of this paper is to analyze the applicability of MCDA to quantum benchmarking. The normalization of a KPI corresponds to defining a scale [11]. Scales are always specified up to some degree(s) of freedom, and comparisons among options made by the MCDA model should be invariant under valid transformations of these scales. Moreover, each KPI is normalized independently, and reference levels are introduced to establish relationships among the normalized KPIs. The number of reference levels dictates the degree of interrelation among the normalized KPIs. We derive the form of the MCDA model under varying assumptions about invariance on the reference levels (i.e., by imposing from 0 to 2 reference levels). We demonstrate that there is a tradeoff between the versatility of the model and the ease of defining normalized KPIs. Specifically, the more reference levels are imposed, the more challenging it becomes to normalize the KPIs, but a wider family of MCDA models becomes compatible.

After a brief overview of MCDA in Sect. 2, we discuss the notion of reference levels in Sect. 3. The main results related to invariance are presented in Sect. 4.

2   Reminder on MCDA

The aim of this section is to select a family of MCDA approaches suited for quantum benchmarking. The decision criteria are denoted by $N = \{1, \ldots, n\}$. Each criterion $i \in N$ is characterized by a KPI (also referred to as an *attribute*) that takes values in a set X_i. Thus, each alternative x is described by a vector $(x_1, \ldots, x_n)$, where $x_i \in X_i$ for all $i \in N$, and is therefore considered as an element of $X := X_1 \times \cdots \times X_n$. For $x, y \in X$ and $i \in N$, we denote by (x_i, y_{-i}) the composite alternative in X that takes the value x_i for attribute i and y_j for all $j \in N \setminus \{i\}$.

The central problem in MCDA is to construct a model for comparing the elements of X. This is formalized through a binary relation $\succsim$ defined on X, representing the stakeholder's preferences. For $x, y \in X$, we write $x \succsim y$ to mean that the stakeholder considers x at least as attractive as y, taking all criteria comprehensively into account.

Example 1. For illustration, consider $n = 3$ KPIs corresponding to different optimization problems: (1) Q-Score for the Max-Cut problem; (2) Q-Score for the Max-Clique problem; (3) Maximum problem size for which the exact solution of the Max Cardinality Matching (MCM) problem is obtained. Suppose we have two quantum computers, $x = (50, 30, 40)$ and $y = (30, 20, 50)$, evaluated according to these three KPIs. The user may express a preference for the first option, that is, $x \succsim y$. ∎

2.1 Overview of the Main MCDA Approaches

A taxonomy of the main MCDA methods can be found in Fig. 1.

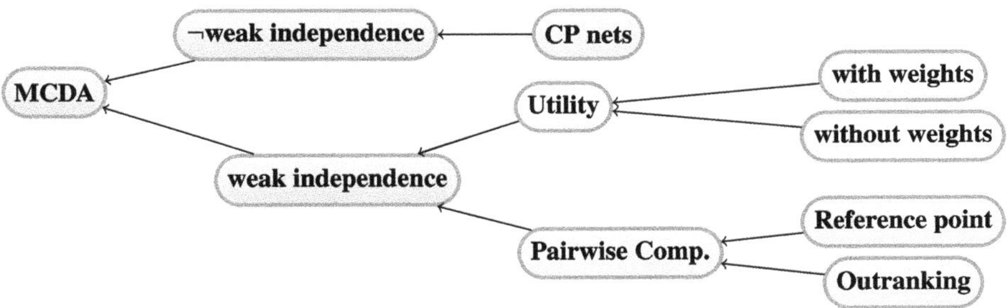

Fig. 1. Overview of the main MCDA methods. (Color figure online)

The first branching in Fig. 1 arises from a property known as *Weak Independence* [11, Sect. 7.2.1]. A binary relation $\succsim$ is said to be *weakly independent* if, for every $i \in N$, and for all $x_i, y_i \in X_i$, $z_{-i}, z'_{-i} \in X_{-i}$, the following equivalence holds:

$$(x_i, z_{-i}) \succsim (y_i, z_{-i}) \quad \Longleftrightarrow \quad (x_i, z'_{-i}) \succsim (y_i, z'_{-i}). \tag{1}$$

This condition states that if value x_i is preferred to y_i for attribute X_i when all other attributes have certain fixed values z_{-i}, then the same preference between x_i and y_i must hold regardless of the values assigned to the other attributes, i.e., for any z'_{-i}. The MCDA method located at the top of Fig. 1 (in red) do not satisfy weak independence. The most prominent example is the CP-net (Conditional Preference network) [3].

Weak independence is satisfied for all KPIs relevant to quantum benchmarking. In Example 1, suppose we prefer a quantum computer with a Q-Score of $x_1 = 50$

over another with $y_1 = 30$, *ceteris paribus* (i.e., holding all other criteria constant). Given Eq. (1), we can define the binary relation $\succsim_i$ on X_i by stating that $x_i \succsim_i y_i$ if $(x_i, z_{-i}) \succsim (y_i, z_{-i})$ for all $z_{-i} \in X_{-i}$. In Example 1, it follows that $50 \succsim_1 30$ for the first KPI.

The main classes of methods that satisfy weak independence are either based on pairwise comparisons between options – see the bottom of Fig. 1 (in green) – or on utility-based approaches (highlighted in yellow in Fig. 1). In the latter case, the preference relation is represented by a utility function $u : X \to \mathbb{R}$ (that is, $x \succsim y$ if and only if $u(x) \geq u(y)$). One of the most widely used models for representing the order relation $\succsim$ is the *transitive decomposable* model [11], in which

$$u(x) = F(u_1(x_1), \ldots, u_n(x_n)), \tag{2}$$

where $F : \mathbb{R} \times \cdots \times \mathbb{R} \to \mathbb{R}$ is an aggregation function, and $u_i : X_i \to \mathbb{R}$ ($i = 1, \ldots, n$) are referred to as *marginal utility functions*. The value function u_i quantifies the preference relation $\succsim_i$ of the stakeholder with respect to the elements of X_i: for $x_i, y_i \in X_i$, $x_i \succsim_i y_i \iff u_i(x_i) \geq u_i(y_i)$. In [5], the *mapped* scores correspond to the scores produced by the marginal utility functions.

2.2 Choice of a Type of MCDA Model

For quantum benchmarking, weak independence is satisfied. Furthermore, utility methods are generally more appropriate than pairwise comparison methods, as they permit the evaluation of a quantum computer independently of others. According to Fig. 1, utility methods can be classified according to whether they incorporate weights or not:

- **Models with weights**: The aggregation function F includes weights. This framework distinguishes between marginal utility functions – which serve to normalize the KPIs – and weights, which reflect the importance of individual criteria or subsets of criteria. Typical aggregation models include the weighted sum $F(a_1, \ldots, a_n) = \sum_{i=1}^{n} w_i a_i$, the Choquet integral [4], and the multi-linear extension model [10].
- **Models without weights**: Here, no explicit weights are used. The marginal utility functions u_i perform both normalization and the impact of each criterion. Common models include the additive utility model $u(x) = \sum_{i=1}^{n} u_i(x_i)$ (i.e., $F(a_1, \ldots, a_n) = \sum_{i=1}^{n} a_i$) and the *Generalized Additive Independence* model, which introduces additional terms $u_S(x_S)$ in the sum for selected subsets $S \subseteq N$ of criteria.

The principal advantage of separating weights from marginal utility functions is enhanced interpretability. This distinction makes it possible to specify, for example, that the value of a metric on a given criterion is *well-satisfied* (or, conversely, *ill-satisfied*), or that another criterion is *very important* (or *not important at all*). Such qualitative interpretation is not feasible with models that do not use weights.

The elicitation of models without weights typically rests on techniques such as UTA [9]. This process is challenging, as it involves the comparison of numerous vectors of values corresponding to quantum computing metrics. In contrast, elicitation of models based on weights usually separates the identification of marginal utility functions from the determination of weights, as in MACBETH [1] or AHP [17]. We believe this approach is better suited and less demanding for quantum computer benchmarking.

To differentiate between the importance weights of criteria and the level of satisfaction with individual metrics, it is necessary to introduce an additional assumption regarding the existence of reference levels. This is the object of the next section.

3   Existence of Reference Levels

In Eq. (2), the marginal utility functions and the aggregation function are not uniquely determined. The marginal utility function u_i defines a *scale* for measuring the preference relation $\succsim_i$ on X_i [11, 16]. Multiple *admissible* scales u_i can be constructed from a given relation $\succsim_i$. The scale type is characterized by the set Φ_i of admissible transformations $\phi_i \in \Phi_i$, mapping one admissible scale u_i to another, $u'_i = \phi_i \circ u_i$. In an *ordinal* scale, admissible transformations are all strictly increasing functions, allowing for only basic operations on the scale. A more expressive form is the *interval* scale, where Φ_i consists of all positive linear transformations,

$$\phi_i(a_i) = \alpha_i a_i + \beta_i \quad \text{with } \alpha_i > 0,\ \beta_i \in \mathbb{R}. \tag{3}$$

An interval scale is commonly constructed using a quaternary relation $\succsim_i^\star$ defined on $X_i \times X_i$ that assesses differences in satisfaction. For $x_i, y_i, z_i, t_i \in X_i$ with $x_i \succsim_i y_i$ and $z_i \succsim_i t_i$, the relation $(x_i, y_i) \succsim_i^\star (z_i, t_i)$ means that the difference in satisfaction between x_i and y_i is at least as great as that between z_i and t_i. This quaternary relation directly corresponds to a difference in utility u_i [11, Chap. 4, Theorem 2]:

$$(x_i, y_i) \succsim_i^\star (z_i, t_i) \quad \Leftrightarrow \quad u_i(x_i) - u_i(y_i) \geq u_i(z_i) - u_i(t_i). \tag{4}$$

Both $\succsim_i$ and $\succsim_i^\star$ are meaningful to stakeholders, enabling u_i to be elicited from these relations, for example using the MACBETH method. Since an interval scale has two degrees of freedom (α_i and β_i in Eq. (3)), the marginal utility function can be fully determined by specifying its value at two *reference elements* in X_i. These two reference levels are denoted $\mathfrak{B}_i, \mathfrak{G}_i \in X_i$ and their associated utility values are $\mathfrak{b}_i, \mathfrak{g}_i \in \mathbb{R}$, with $\mathfrak{g}_i > \mathfrak{b}_i$. Typically, $\mathfrak{B}_i$ (the *bad* element) represents a completely unsatisfactory outcome, set to the utility value $\mathfrak{b}_i$:

$$u_i(\mathfrak{B}_i) = \mathfrak{b}_i. \tag{5}$$

This serves as a saturation point, meaning for all $x_i \precsim_i \mathfrak{B}_i$, $u_i(x_i) = \mathfrak{b}_i$. In quantum benchmarking, $\mathfrak{B}_i$ might correspond to the performance delivered by a randomized algorithm. The upper reference level $\mathfrak{G}_i$ (the *good* element) is defined as the highest satisfactory level attainable given current technological capabilities. Its utility is set to $\mathfrak{g}_i \in \mathbb{R}$:

$$u_i(\mathfrak{G}_i) = \mathfrak{g}_i. \tag{6}$$

4   Various Qualitative Invariances to Ease the Commensurateness Assumption of the Scales

In Sect. 3, we considered the marginal utility scales independently. The concept of commensurateness arises when comparing values from different marginal utility functions. Two scales u_i and u_k are said to be *commensurate* if $u_i(x_i) = u_k(x_k)$ implies that x_i and x_k represent the *same level of satisfaction*. Conversely, if $u_i(x_i) = u_k(x_k)$ does not convey any relationship of satisfaction, then the scales are not commensurate.

A convenient way to construct commensurate marginal utility functions is to assume that all reference levels $\mathfrak{B}_1, \ldots, \mathfrak{B}_n$ and $\mathfrak{G}_1, \ldots, \mathfrak{G}_n$ are commensurate (i.e., have the same meaning) and are therefore assigned identical utilities:

$$u_1(\mathfrak{B}_1) = \cdots = u_n(\mathfrak{B}_n) \quad \text{and} \quad u_1(\mathfrak{G}_1) = \cdots = u_n(\mathfrak{G}_n). \tag{7}$$

Example 2. (Example 1cont.). All three KPIs are completely unsatisfactory for value 0: $\mathfrak{B}_1 = \mathfrak{B}_2 = \mathfrak{B}_3 = 0$. The upper reference level is assigned to the best value achievable with current technology. We thus set $\mathfrak{G}_1 = \mathfrak{G}_2 = 1000$ (achieved by the NEC computer with a simulated annealing [18]), and $\mathfrak{G}_3 = 512$ (attained by the NEC computer). ∎

When the scales are commensurate (as per Eq. (7)), we have $\mathfrak{b}_1 = \cdots = \mathfrak{b}_n$ and $\mathfrak{g}_1 = \cdots = \mathfrak{g}_n$. Figure 2-Left illustrates two commensurate scales, i and k. The blue arrows represent the unit interval between the Bad and Good levels, which are identical in length due to commensurateness. In Fig. 2-Right, different linear transformations ϕ_ℓ are applied to the scales; here, $\mathfrak{b}'_\ell = \phi_\ell(\mathfrak{b}_\ell)$ and $\mathfrak{g}'_\ell = \phi_\ell(\mathfrak{g}_\ell)$. As a result, the unit intervals between the Bad and Good levels now differ between the two KPIs.

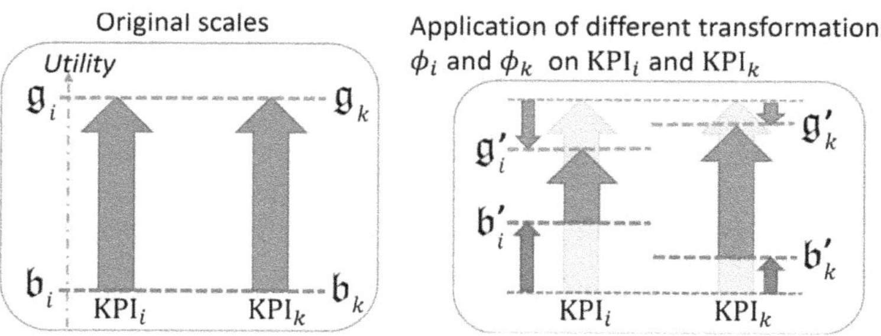

Fig. 2. Effect of affine transformations on the unit distance between the Bad and Good levels for two KPIs.

Let Φ denote the set of admissible tuples $(\phi_1, \ldots, \phi_n)$ of transformations. The property of scale invariance requires that the ranking of any set of options remains unchanged whether we use the original aggregation $F(u_1(x_1), \ldots, u_n(x_n))$ or any normalized version obtained by applying admissible transformations, i.e., $F(\phi_1 \circ u_1(x_1), \ldots, \phi_n \circ u_n(x_n))$ for $(\phi_1, \ldots, \phi_n) \in \Phi$. In other words, for $x, y \in X$, if x is at least as good as y under the aggregation F with marginal utility functions $u_1, \ldots, u_n$ (i.e.,

$F(u_1(x_1), \ldots, u_n(x_n)) \geq F(u_1(y_1), \ldots, u_n(y_n)))$, this ordering must also hold for any admissible transformation of the utilities. Letting $a_i = u_i(x_i)$ and $a_i' = u_i(y_i)$, we express the invariance property for the aggregation function F as follows.

Definition 1. F *is* scale invariant under Φ *if, for all* $(\phi_1, \ldots, \phi_n) \in \Phi$ *and all* $a, a' \in \mathbb{R}^n$,

$$F(a_1, \ldots, a_n) \geq F(a_1', \ldots, a_n') \iff$$
$$F(\phi_1(a_1), \ldots, \phi_n(a_n)) \geq F(\phi_1(a_1'), \ldots, \phi_n(a_n')). \tag{8}$$

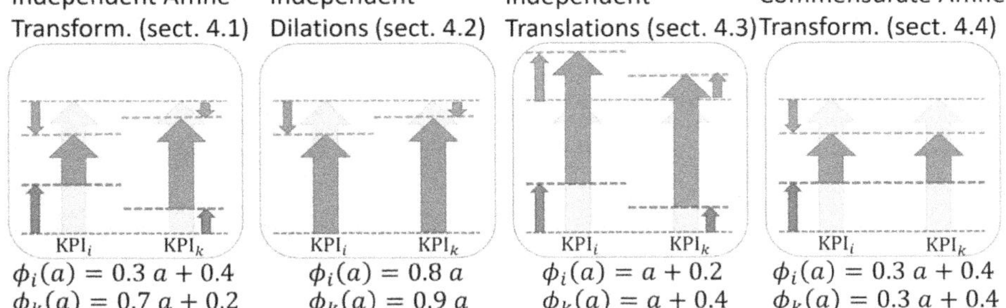

$$\phi_i(a) = 0.3\,a + 0.4 \qquad \phi_i(a) = 0.8\,a \qquad \phi_i(a) = a + 0.2 \qquad \phi_i(a) = 0.3\,a + 0.4$$
$$\phi_k(a) = 0.7\,a + 0.2 \qquad \phi_k(a) = 0.9\,a \qquad \phi_k(a) = a + 0.4 \qquad \phi_k(a) = 0.3\,a + 0.4$$

Fig. 3. Examples of different transformation classes Φ. We illustrate with explicit expressions for ϕ_i and ϕ_k, assuming $\mathfrak{b}_1 = \cdots = \mathfrak{b}_n = 0$ and $\mathfrak{g}_1 = \cdots = \mathfrak{g}_n = 1$.

In some situations, requiring the full commensurateness property of Eq. (7) may be overly restrictive. We now consider weaker assumptions. Let $\mathbb{R}^n_{\mathfrak{b}+} = \{a \in \mathbb{R}^n : \forall i \in N, a_i \geq \mathfrak{b}_i\}$.

4.1 Invariance w.r.t. Independent Affine Transformations on the KPIs

We first examine the extreme case in which the condition of Eq. (7) is simply omitted, meaning that there is no assumption of commensurateness among the reference levels. In this case, the invariance property must allow each marginal utility function $u_1, \ldots, u_n$ to be transformed independently (see Fig. 3):

$$\Phi_{\mathrm{indep}} = \left\{ (\varphi_{\alpha_1, \beta_1}, \ldots, \varphi_{\alpha_n, \beta_n}) \mid (\alpha_1, \ldots, \alpha_n) \in \mathbb{R}^n_+, (\beta_1, \ldots, \beta_n) \in \mathbb{R}^n \right\}$$

where $\varphi_{s,t}(a) = s\,a + t$. In other words, the aggregation function must be able to combine scales that are not commensurate. The following result demonstrates that there is no aggregation function which satisfies scale invariance under Φ_{indep}, together with several mild conditions, such as continuity of F. Proofs of all results are omitted due to space constraints.

Theorem 1. *There does not exist a function F that is non-trivial (not all alternatives are indifferent), continuous, monotone, and scale invariant under Φ_{indep}.*

It is thus necessary to impose some level of commensurateness among the marginal utility functions. Various commensurateness conditions are explored in the next sections.

4.2 Invariance w.r.t. Independent Dilations on the Different Metrics

Here, we assume that all bad levels are commensurate and share the same interpretation, whereas the good levels may differ. This means that we can transform each utility function u_k into another u_k', provided the utility on the bad level remains unchanged:

$$\forall k \in N \qquad u_k(\mathfrak{B}_k) = u_k'(\mathfrak{B}_k) = \mathfrak{b}_k. \tag{9}$$

Example 3. (Example 2cont). Requiring commensurate bottom levels is not restrictive in quantum benchmarking. For example, in most metrics that measure the largest problem size solvable, the bottom reference level is 0: $\mathfrak{B}_1 = \mathfrak{B}_2 = \mathfrak{B}_3 = 0$.

According to Eq. (9), since u_k and u_k' are related by an affine transformation, we have:

$$\forall k \in N \qquad u_k'(x_k) = \alpha_k \left(u_k(x_k) - \mathfrak{b}_k\right) + \mathfrak{b}_k, \tag{10}$$

for some $\alpha_k > 0$. The set of admissible transformations is (see Fig. 3):

$$\Phi_{\text{dilat}} = \left\{ (\varphi_{\alpha_1, \mathfrak{b}_1(1-\alpha_1)}, \ldots, \varphi_{\alpha_n, \mathfrak{b}_n(1-\alpha_n)}) \mid (\alpha_1, \ldots, \alpha_n) \in \mathbb{R}_+^n \right\}.$$

The following result states that, under scale invariance with respect to Φ_{dilat}, there is only one compatible family of aggregation functions: the weighted geometric mean, up to a strictly increasing transformation.

Theorem 2. *If F is non-trivial (not all alternatives are indifferent), continuous, monotone, and scale invariant under Φ_{dilat}, then there exist $w_1, \ldots, w_n \geq 0$ with $w_1 + \cdots + w_n = 1$ and a strictly non-decreasing function $g : \mathbb{R} \to \mathbb{R}$ such that, for all $a \in \mathbb{R}_{\mathfrak{b}+}^n$,*

$$F(a) = g\left(\prod_{i \in N} (a_i - \mathfrak{b}_i)^{w_i} \right).$$

This outcome is intuitive: aggregating heterogeneous variables normalized by their worst value using a geometric mean is meaningful. For instance, in Cooperative Game Theory, the Nash bargaining solution relies on the geometric mean [15], and invariance to input scale is a key axiom. Notably, the geometric mean is also employed as an aggregation function in some optimization benchmarking settings [8, Sect. V.C]. However, the geometric mean has a limitation: a poor score on any KPI (a value close to its bad reference) cannot be compensated by high scores on other KPIs. Thus, it rewards solutions that show consistently strong performance across all KPIs, rather than those that are exceptional in some areas and weak in others.

4.3   Invariance w.r.t. Independent Translations on the Different Metrics

The next type of invariance we consider assumes that the dilation factor is fixed, while the translation factor is variable. This means that the transformation from one admissible utility function to another takes the form $\phi_i(a_i) = a_i + \beta_i$, where the translation factors $\beta_i \in \mathbb{R}$ may differ across criteria. The set of admissible transformations is given by (see Fig. 3):

$$\Phi_{\text{translat}} = \left\{ (\varphi_{1,\beta_1}, \ldots, \varphi_{1,\beta_n}) \mid (\beta_1, \ldots, \beta_n) \in \mathbb{R}^n \right\}.$$

Example 4. (Example 3cont.). Consider measuring the temperature of a quantum computer, assuming this metric is relevant to the user. A unit interval of $100°\text{C}$ between the Bad and Good levels may be meaningful. However, assigning the Bad level to $0°\text{C}$ is arbitrary. Therefore, the Bad level could be shifted, while the unit interval is preserved.

The following result establishes that, under scale invariance with respect to Φ_{translat}, the only compatible family of aggregation functions is the weighted sum, up to a strictly increasing transformation.

Theorem 3. *If F is non-trivial (not all alternatives are indifferent), continuous, monotone, and scale invariant under Φ_{translat}, then there exist $w_1, \ldots, w_n \geq 0$ with $w_1 + \cdots + w_n = 1$, and a strictly increasing function h such that for all $a \in \mathbb{R}^n_{b+}$,*

$$F(a) = h\left(\sum_{i \in N} w_i a_i \right).$$

4.4   Invariance w.r.t. Commensurate Affine Transformations

In this final case, we assume that the scales remain commensurate after applying transformations to the marginal utility functions. Two admissible scale tuples $(u_1, \ldots, u_n)$ and $(u'_1, \ldots, u'_n)$ (satisfying Eq. (7)) are therefore related by the same positive linear transformation, as in Eq. (3), across all criteria, with $\alpha_1 = \cdots = \alpha_n =: \alpha$ and $\beta_1 = \cdots = \beta_n =: \beta$. Thus, the set of admissible transformations is (see Fig. 3):

$$\Phi_{\text{commens}} = \left\{ (\varphi_{\alpha,\beta}, \ldots, \varphi_{\alpha,\beta}) : \alpha \in \mathbb{R}_+, \beta \in \mathbb{R} \right\}.$$

Unlike in previous cases, we do not obtain a uniqueness (representation) theorem, since many aggregation models satisfy scale invariance under Φ_{commens}. For example, the weighted sum, the Choquet integral, and the multilinear model all fulfill this property.

5 Conclusion

MCDA is an essential tool for aggregating various KPIs in quantum benchmarking. Among the different MCDA approaches, we advocate using utility-based methods that distinguish between normalization and aggregation. The utility function associated with each KPI is treated as an interval scale, and positive linear transformations can be applied to each utility without affecting the overall benchmarking result. Traditionally, the reference levels are assumed to be commensurate. While this is a strong assumption, it enables a wide range of aggregation functions. We have demonstrated that no aggregation function remains invariant under independent positive linear transformations of each utility function. Therefore, it is necessary to consider more restrictive forms of invariance. Aggregation functions that are invariant when only the bad reference levels are commensurate (but not the good levels) are homomorphic to a weighted geometric mean. If the distance between the two reference levels is fixed but their positions can be translated freely, the only invariant aggregation function is homomorphic to a weighted sum. With these findings, the decision maker can choose the most suitable MCDA model, balancing the model's expressiveness with the practical challenge of defining appropriate reference levels.

Acknowledgment. As part of the MetriQs-France program, this work is supported by France 2030 under the French National Research Agency grant number ANR-22-QMET-0002 (BACQ project).

References

1. Bana e Costa, C.A., De Corte, J., Vansnick, J.C.: MACBETH. Int. J. Inf. Technol. Decis. Making **11**, 359–387 (2012)
2. Barbaresco, F., et al: BACQ – application-oriented benchmarks for quantum computing. arXiv:2403.12205 (2024)
3. Boutilier, C., Brafman, R., Domshlak, C., Hoos, H., Poole, D.: CP-nets: a tool for representing and reasoning with conditional ceteris paribus preference statements. J. Artif. Intell. Res. (JAIR) **21**, 135–191 (2004)
4. Choquet, G.: Theory of capacities. Annales de l'Institut Fourier **5**, 131–295 (1953)
5. Donkers, H., Mesman, K., Al-Ars, Z., Moller, M.: QPack scores: quantitative performance metrics for application-oriented quantum computer benchmarking. CoRR abs/2205.12142 (2022)
6. Figueira, J., Greco, S., Ehrgott, M. (eds.): Multiple Criteria Decision Analysis: State of the Art Surveys, 2nd edn. Kluwer Acad. Publ. (2016)
7. Gilbert, V., Louise, S., Sirdey, R.: TAQOS: a benchmark protocol for quantum optimization systems. In: International Conference on Computational Science (ICCS) (2023). https://doi.org/10.1007/978-3-031-36030-5_13
8. Hansen, N., Auger, A., Brockhoff, D., Tusar, T.: Anytime performance assessment in black-box optimization benchmarking. IEEE Trans. Evol. Comput. **26**, 1293–1305 (2022)
9. Jacquet-Lagrèze, E., Siskos, Y.: Assessing a set of additive utility functions for multicriteria decision making: the UTA method. Eur. J. Oper. Res. **10**, 151–164 (1982)
10. Keeney, R.L., Raiffa, H.: Decision with Multiple Objectives. Wiley, New York (1976)

11. Krantz, D., Luce, R., Suppes, P., Tversky, A.: Foundations of Measurement, vol. 1: Additive and Polynomial Representations. Academic Press (1971)
12. Lorenz, J.M., et al: Systematic benchmarking of quantum computers: status and recommendations. arXiv:2503.04905v1 (2025)
13. Martiel, S., Ayral, T., Allouche, C.: Benchmarking quantum co-processors in an application-centric, hardware-agnostic and scalable way. IEEE Trans. Quantum Eng. **2** (2021)
14. Mesman, K., van der Schoot, W., Moller, M., Neumann, N.: QuAS: quantum application score for benchmarking the utility of quantum computers. arXiv:2406.03905 (2024)
15. Nash, J.: The bargaining problem. Econometrica **18**, 115–162 (1950)
16. Roberts, F.: Measurement Theory. Addison-Wesley (1979)
17. Saaty, T.L.: A scaling method for priorities in hierarchical structures. J. Math. Psychol. **15**, 234–281 (1977)
18. van der Schoot, W., Leermakers, D., Wezeman, R., Neumann, N., Phillipson, F.: Evaluating the Q-score of quantum annealers. In: IEEE International Conference on Quantum Software (QSW), pp. 9–16. Barcelona, Spain (2022)
19. van der Schoot, W., Wezeman, R., Neumann, N., Phillipson, F., Kooij, R.: Extending the Q-score to an application-level quantum metric framework. In: IEEE International Conference on Quantum Computing and Engineering (QCE), pp. 941–951. Montreal, Canada (2024)

Session 2 Quantum Algorithms, Computing and Simulation – Benchmarking B: Methods and Applications

Benchmark of MHT Algorithm on Quantum Computer

Paul Rousset-Rouard[1(✉)], Bing Hong Teh[2], and Jean-Marc Divanon[2]

[1] CentraleSupélec, University of Paris-Saclay, Gif-sur-Yvette, France
paul.rousset-rouard@student-cs.fr
[2] Thales Research & Technology, Singapore, Singapore
{bing-hong.teh,jean-marc.divanon}@thalesgroup.com

Abstract. This paper explores the application of quantum computing, specifically quantum annealing using D-Wave systems, to address the Multiple Hypothesis Tracking (MHT) problem in tracking systems. This involves the critical task of assigning sensor detections to relevant targets amidst various challenges such as false alarms, noise, and target manoeuvres. The combinatorial nature of MHT presents a significant obstacle, with the number of hypotheses escalating exponentially over time, leading traditional solvers to struggle under substantial computational load. By benchmarking classical solvers against D-Wave's hybrid quantum annealing approach, the study has demonstrated remarkable improvements in computational time, achieving up to 15-fold speed-ups during high false alarm density scenarios. This work highlights the potential of quantum computing over classical methods in efficiently tackling complex combinatorial optimization tasks, thereby offering a promising solution for demanding real-time tracking challenges. The research further suggests avenues for future exploration with different quantum computing technologies, underscoring the evolving landscape of technology in optimizing tracking system efficiency.

Keywords: Multiple Hypothesis Tracking (MHT) · Data Association Problem (DAP) · Maximum Weighted Independent Set (MWIS) · NP-hard · Quantum Computing · D-Wave · Quantum Advantage

1 Introduction

1.1 The Data Association Problem (DAP)

In tracking systems, sensors generate numerous detections that must be accurately attributed to relevant targets. The Data Association Problem (DAP) is the process of assigning these detections (or measurements) to corresponding targets such that each measurement is associated with at most one target, and each target receives at most one measurement per scan. This problem must be addressed following each scan and subsequently for all collected scans of measurements. Several complicating factors introduce significant complexity to the DAP:

F. Barbaresco and G. François (Eds.): QUEST-IS 2025, CCIS 2743, pp. 251–259, 2026.
https://doi.org/10.1007/978-3-032-13852-1_25

- Presence of spurious plots, often referred to as false alarms,
- Noise contaminated measurements,
- Target manoeuvres,
- Missed detections of targets,
- Real-time requirements.

1.2 Multiple Hypothesis Tracking (MHT)

One of the well-known solution methods for DAP is Multiple Hypothesis Tracking (MHT), which was introduced by Reid [1]. At the heart of MHT lies the hypothesis tree structure (see Fig. 1), representing every feasible assignment of plots. For each potential assignment, MHT calculates a likelihood of truth [1, 3, 4]. Indeed, the MHT algorithm consists in evaluating possible tracks at each time step and updates these beliefs thanks to new observations. It basically keeps in memory all detections that happened at timesteps {0, 1,..., k} to make decisions about the most likely track at time step k. This is what makes this tracking algorithm powerful, but also memory and calculation intensive. The most probable hypotheses (meaning whose likelihood of truth are highest) are selected as current solution, though it is not definitive. Less probable hypotheses are nonetheless preserved, as they might become relevant with the receipt of new plots in future scans. This is illustrated in Fig. 1 below, extracted from [3].

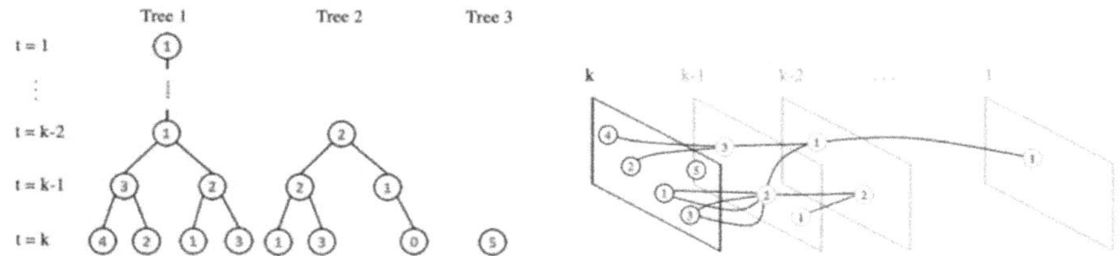

Fig. 1. Hypothesis tree structure built from radar measurements (*left*) and association at different timestep from 1 to k for each plot spatially (*right*). Extracted from Fig. 1 and Fig. 2 of [3].

A notable challenge in MHT is the combinatorial nature of this problem, where the number of associations escalates exponentially over time. This is further illustrated with example in Sect. 3 below. Conventionally, this complexity is managed through various pruning techniques, eliminating less promising hypotheses and thus preventing combinatorial explosion. While this helps reduce the complexity of the DAP, some possible truths may be lost as a trade-off.

A brute force implementation of MHT endeavours to encompass all conceivable scenarios to accurately compute the most probable associations. In this exhaustive approach, every potential assignment of tracks and measurements is considered. For example, when dealing with 10 tracks and 100 measurements, assuming a detection probability of 1, the number of possible associations reaches approximately 17 trillion.

With the growing interest in quantum computing technology, the work presented in this paper provides a benchmark of the MHT algorithm implementation on D-Wave's quantum computer against state-of-the-art classical solvers.

1.3 Quantum Annealing (DWAVE)

Quantum annealing [2, 9] is currently attracting significant interest for its effectiveness as a quantum algorithm in efficiently and quickly solving certain large-scale combinatorial optimization problems, particularly those with unknown structures, where quantum advantage has been indisputably proven [8]. It is commercially implemented almost exclusively by D-Wave.

D-Wave's native sets of problems are more specifically Quadratic Unconstrained Binary Optimization (QUBO) problems, like spin-glass systems, where the energy (in other words the Hamiltonian H) is the quantity to be minimized and can be written as:

$$\min_{\vec{x} \in \{0;1\}^n} H(\vec{x}) = \min_{\vec{x} \in \{0;1\}^n} \vec{x}^T Q \vec{x} = \min_{\vec{x} \in \{0;1\}^n} \sum_{i,j} Q_{ij} x_i x_j \tag{1}$$

where Q is real and symmetric. The fundamental principle quantum annealing is inspired from—adiabatic quantum computation—involves starting with an 'easy' Hamiltonian H_0 (one whose eigenvalues and corresponding system energies are known). The system then gradually evolves (typically over a few milliseconds) to the problem Hamiltonian H_P, whose eigenvalues are the target of optimization.

$$H(t) = A(t)H(0) + B(t)H_P \tag{2}$$

where A (resp. B) is a decreasing (resp. Increasing) function of t. Once the problem Hamiltonian is implemented, the system—under assumptions such as a slow enough evolution—naturally evolves to its lowest energy states, which correspond to the minimum of the optimization problem, according to the adiabatic theorem [10]. The qubits are then measured to determine the solution to the QUBO problem. Because D-Wave's quantum annealing chips are noisy and meant to be fast, they sample low-energy solutions, occasionally reaching the desired optimum. Such quantum annealing solvers are therefore more heuristics based.

Quantum annealing technology is quite mature, with D-Wave's quantum processing units (QPUs) featuring thousands of functional qubits—around five thousand. Additionally, D-Wave has developed a hybrid solver [14, 15] that can tackle optimization problems ranging from 20,000 to 1 million variables by employing quantum annealing for subproblems in parallel with classical optimization techniques.

2 Methodology

2.1 MHT Framework

The framework for the implementation of Multiple Hypothesis Tracking (MHT) involves modelling a 2-D radar tracker with a linear Kalman filter and a constant velocity model to estimate target trajectories. Radar noise is considered normally distributed, and clutter appearance is uniformly distributed across the radar surface, with the number of clutter measurements per timestep following a Poisson distribution. Additionally, the radar's probability of detection is adjustable.

2.2 MHT Implementation

In the context of the MHT algorithm, gating, pruning, and other hypotheses reduction techniques are crucial to manage the computational complexity and optimize performance:

- **Gating** is a process that reduces the number of possible associations between measurements and tracks by setting a threshold based on the distance metric (e.g., Mahalanobis distance) from predicted positions to actual measurements using the Linear Kalman filter (taking measurement noise into account). Only measurements within this threshold are considered for association with existing track hypotheses.

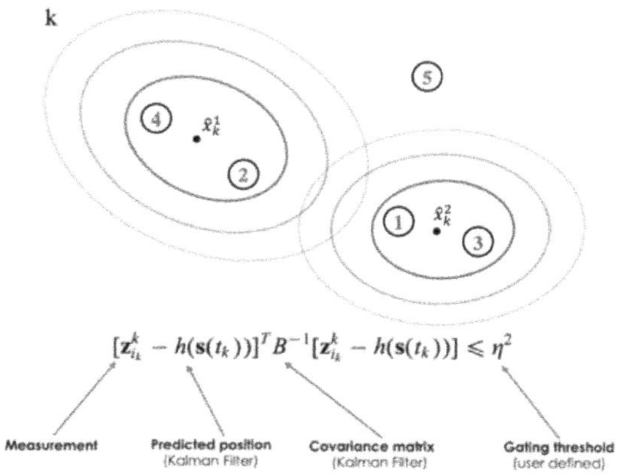

Fig. 2. Gating visualized *(top, extracted from* [3]*)*: Measurements 2 and 4 (resp. 1 and 3) fall close to the prediction $\hat{x}_k^1$ (resp. $\hat{x}_k^2$) and are therefore considered as physically possible continuation of existing track hypotheses. 5 is not. Mathematical formulation *(bottom)* of the Mahalanobis distance between measurements and predictions.

- **Pruning** is a technique that reduces the problem size by eliminating hypotheses with low likelihood of truth from the hypotheses tree. This also involves cutting down branches in the hypothesis tree. Pruning [1] helps to keep the computational load reasonable by discarding hypotheses that have low probability or have become redundant, ensuring that the algorithm focuses resources on the scenarios that are most likely to be.
- Other techniques involve discarding track hypotheses with too many consecutive missed detections, or tracks that involve excessively high velocity [3].

2.3 MWIS (Maximum Weighted Independent Set)

As it turns out, solving the DAP amounts to solving a MWIS problem. The latter can be described as an undirected graph problem, whose aim is to find the subset of non-conflicting hypotheses with the maximum total weight.

In this graph, each node represents a track hypothesis with a weight equal to its likelihood (or score). Edges connect nodes that conflict (i.e. that share a common measurement). The goal is to select a set of non-conflicting nodes (an independent set) with

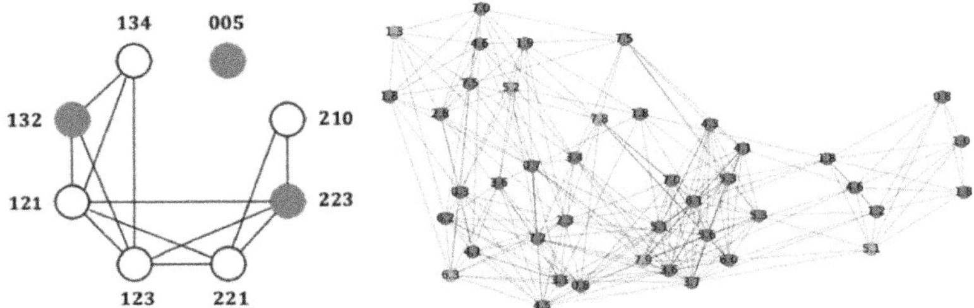

Fig. 3. A small undirected graph with edges (*left, extracted from* [3]): node 132 refers to the track hypothesis "Measurement 1 (M1) at timestep 1" + "M3 at timestep 2" + "M2 at timestep 3". Weights are not represented. The MWIS solution at timestep 3 are the blue nodes. An actual (yet simple) MHT-derived MWIS solution (*right*): on each node is written the score of the track hypothesis. Orange nodes are the MWIS solutions.

the highest total weight. Hence, solving the data association problem involves identifying this weighted independent set. The proper mathematical formulation implies dealing with binary values, where $x_i = 1$ means the node i is taken as a solution of the DAP at a given timestep. The exhaustive problem formulation is:

$$Given\ a\ graph(V, E) : \ \text{find} \operatorname*{argmax}_{\vec{x} \in \{0;1\}^V} \sum_l w_l x_l s.t. x_i + x_j \leq 1, \forall (i,j) \in E \qquad (3)$$

where w_l is the score (likelihood) associated to the node l.

As the graph scales up to thousands of nodes, finding this optimal set becomes increasingly complex, thus necessitating advanced approaches like quantum computing for efficient resolution. It is of interest to note that the MWIS problem is nothing but the Maximum Weight Clique problem (MWC) on the complementary graph.

2.4 Scenario for Benchmarking

The goal of the study was to simulate complex enough scenarios that may arise in real settings and to solve it using the MHT algorithm with different solvers. The scenario in this paper involves a track moving along a straight line for 20 timesteps. Between frames 3 to 10, the expected number of false alarms increases by a factor of 100, introducing significant noise into the track's trajectory and adding complexity to the resolution of the MWIS problem.

The input parameters for the MHT algorithm are detailed in Table 1 above.

Various solvers were tested in this work and are summarized below:

1. Qualex-MS [11], a C-based heuristics solver with time complexity of $O(n^3)$.
2. NetworkX [5], an exact solver using Branch & Bound.
3. SCIP [7], a well-known heuristics solver that uses LP-relaxation.
4. Cliquer [12], an exact solver running on C using Branch & Bound.
5. TSM-WC [6], an exact solver of MWC implemented on C.
6. DWAVE's hybrid heuristics solver [14, 15].

Table 1. Input parameters for the scenario to be evaluated

Parameters	Description	Values taken
dth	Gating area for new detections	**30**
q	Kalman filter process variance on the acceleration ($m^2.s^{-3}$)	10^{-5}
r	Estimate of measurement variance	**1**
n	Number of scan(s) before pruning	**2**
n_{miss}	Maximum consecutive missed detections	**4**
p_d	Probability of detection of the radar	**0.9**
$\lambda_{clutter}$	Expected number of false alarms per area unit per frame	$[10^{-5}, 10^{-4}]$
$\lambda_{newtracks}$	Expected number of new tracks per area unit per frame	$[10^{-8}, 10^{-7}]$
v_{max}	Maximum velocity of the detected objects (m/s)	**15**

3 Results

Firstly, all tested solvers that were exact, other than TSM-WC and DWAVE, become significantly slower once the graphs reached 1000 nodes—typically for a false alarm density greater than $5\,10^{-5}$. Qualex-MS and SCIP, the heuristics solvers, could not return good enough solution quality to the MHT problem and were discarded. For each false alarm parameter setting, classical solvers often exhibited a peak in computation time, around frame 8 or 9 (see Fig. 4).

In response, DWAVE's hybrid solver [14] was implemented to handle these computational peaks, while the classical solver continued to manage the simpler instances. A comparison was then made between the TSM-WC solver alone and the TSM-WC solver supplemented by DWAVE during intensive computations.

As DWAVE's hybrid solver is a black box algorithm, no detailed information on its specific way of solving is disclosed, hence the absence of more sophisticated explanation. The QPU access time is however available after each call, and quantum annealing makes up approximately ten percent of the solver's computing time. It should also be noted that DWAVE's hybrid solver has a minimum runtime of 3 s, explaining the lower bound observed on Fig. 5.

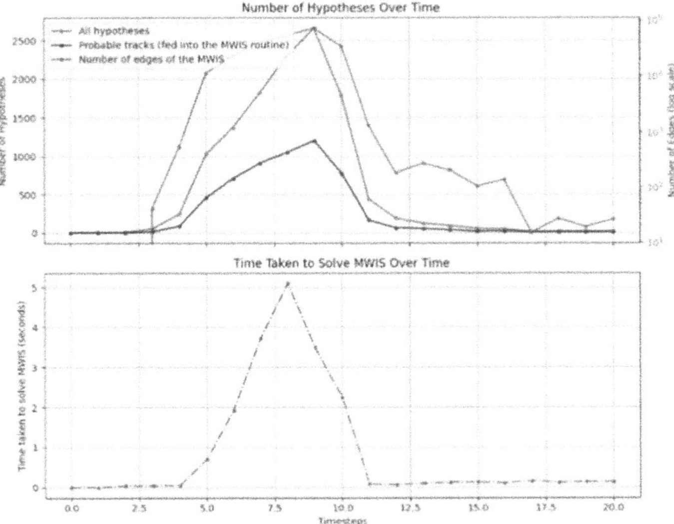

Fig. 4. Number of nodes (hypotheses) and edges (conflicting hypotheses) at each timestep for a given MHT routine (*top*): the probable tracks are those with positive score and are fed to the MWIS routine. (*bottom*): time taken to solve the MWIS routine for TSM-WC.

Table 2. Comparison on TSM-WC and DWAVE time to solution for the hardest instance.

$\lambda_{clutter}$	Time Taken (seconds)	
	Solver	
	TSM-WC	DWAVE
10^{-5}	0.1	3
$7.5 \cdot 10^{-5}$	1.48	3
$9 \cdot 10^{-5}$	6.8	3.6
10^{-4}	11.09	3.5
$1.1 \cdot 10^{-4}$	94	6

Table 2 above illustrates the time taken for TSM-WC and DWAVE for the hardest MWIS problem in each scenario (as depicted in Fig. 4 above).

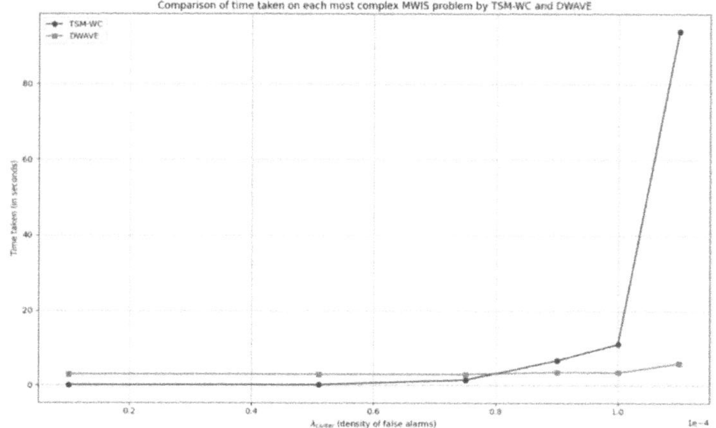

Fig. 5. Comparison on TSM-WC and DWAVE on those high clutter density calculations

Results are displayed in Fig. 5. When comparing TSM-WC and DWAVE during particularly demanding calculations (peaks), the advantage of DWAVE's hybrid solver is clear.

As observed, quantum computing is beginning to offer substantial time advantages when the density of false alarms reaches 10^{-4}. False alarms densities higher than this value yielded MWIS problems too hard to be solved in reasonable times and were not explored. Nevertheless, a measured 15-fold speed-up in computation time was achieved, while maintaining similar results and solution quality. We emphasize this fact, as most classical heuristics solvers did not meet the required solution quality and were therefore not considered for the rest of the study. For this benchmark, if the track hypothesis corresponding to the ground truth (track going in a straight line) was not part of the MWIS solution at the end of the 20 frames—up to negligible missed detections and false alarms—then, the solution quality was deemed mediocre.

As the cost of QPU usage was non neglectable, no other runs were made, and no statistical values have been extracted from this study.

4 Conclusion

The integration of DWAVE with classical solvers shows promise in handling computationally intensive tasks more efficiently, marking a significant step forward in leveraging quantum computing for complex problems.

Current work is underway to evaluate MHT algorithms on C12's novel quantum computer using methods such as developed in [16–18]. C12 carbon nanotubes's spin qubits offer long coherence times [19], low error rates [20], and highly connected qubits [21], enabling both digital and analog modes with operation speeds comparable to superconducting qubits. The native use of spins also allows for a more direct implementation of Ising models.

All computations were performed on a MacBook Air (2022) equipped with an Apple M2 chip featuring 8 cores (4 performance cores and 4 efficiency cores), 8 GB of RAM, and running macOS Sonoma.

Acknowledgements. The authors wish to acknowledge DWAVE for the cloud access to its QPU.

References

1. Reid, D.: An algorithm for tracking multiple targets. IEEE (1979)
2. Sebenik et al. Quantum annealing: a new method for minimizing multidimensional functions. Chem. Phys. Lett. (1994)
3. Ciptadi, A., Rehg, J.M., Kim, C., Li, F.: Multiple hypothesis tracking revisited. IEEE (2015)
4. Stormsa, P.P.A., Spieksma, F.C.R.: An LP-based algorithm for the data association problem in multitarget tracking. IEEE (2000)
5. Hicks, I.V., Warren, J.S.: Combinatorial branch-and-bound for the maximum weight independent set problem. Technical Report, Texas AM University (2016)
6. Liu, Y., Manya, F., Jiang, H., Li, C.-M.: A two-stage maxsat reasoning approach for the maximum weight clique problem. In: The Thirty-Second AAAI Conference on Artificial Intelligence (2018)
7. Tobias Achterberg, S.C.I.P.: Solving constraint integer programs. Math. Program. Comput. **1**(1), 1–41 (2009)
8. Denchev, V.S., et al.: What is the computational value of finite-range tunneling? Phys. Rev. X **6**, 031015 (2016)
9. Farhi, E., et al.: A quantum adiabatic evolution algorithm applied to random instances of an NP-complete problem. Science (2001)
10. Born, M., Fock, V.A.: Beweis des Adiabatensatzes. Zeitschrift für Physik A **51**(3–4), 165–180 (1928)
11. Busygin, S.: A new trust region technique for the maximum weight clique problem (2006)
12. Niskanen, S., Östergård, P.R.J.: Cliquer User's Guide, Version 1.0. Communications Laboratory, Helsinki University of Technology, Espoo, Finland, Technical Report T48 (2003)
13. Raymond, J., et al.: Hybrid quantum annealing for larger-than-QPU lattice-structured problems. ACM Trans. Quantum Comput. **4**(3), Article 17 (2023)
14. D-Wave: Hybrid solvers for quadratic optimization (2022)
15. Osaba, E., Miranda-Rodriguez, P.: D-Wave's nonlinear-program hybrid solver: description and performance analysis (2024)
16. Cazals, P., et al.: Quantum Optimization on Rydberg Atom Arrays with Arbitrary Connectivity: Gadgets Limitations and a Heuristic Approach. arXiv preprint arXiv:2508.06130 (2025)
17. Garrigues, M., et al.: A Scalable Heuristic for Molecular Docking on Neutral-Atom Quantum Processors. arXiv preprint arXiv:2508.18147 (2025)
18. Tomesh, T., et al.: Divide and conquer for combinatorial optimization and distributed quantum computation. In: 2023 IEEE International Conference on Quantum Computing and Engineering (QCE), vol. 1. IEEE (2023)
19. Neukelmance, B., et al.: Microsecond-lived quantum states in a carbon-based circuit driven by cavity photons. Nat. Commun. **16**(1), 5636 (2025)
20. Wu, Y.-H., et al.: Simultaneous High-Fidelity Single-Qubit Gates in a Spin Qubit Array. arXiv preprint arXiv:2507.11918 (2025)
21. Mammola, A., Desjardins, M.M., Schaeverbeke, Q.: Optimal connectivity from idle-qubit residual-coupling crosstalk in a cavity-mediated entangling gate. Phys. Rev. A **111**(4), 042621 (2025)

G-Score, the Maximum Cardinality Matching G_n Series as a Benchmark for Quantum Computers

Stéphane Louise$^{(\boxtimes)}$ 🆔

Université Paris-Saclay, CEA, List, F-91120, Palaiseau, France
`stephane.louise@cea.fr`
`https://list.cea.fr/`

Abstract. The Maximum Cardinality Matching (MCM) G_n series represents a well-established set of problems designed to evaluate the capabilities of optimization heuristics. These problems have been instrumental in probing the optimization potential of various DWave quantum computers, offering insights into their performance and limitations. Complementary to the NP-hard yet unconstrained MaxCut problem used in the Q-Score benchmark, the MCM G_n series provides a constrained series of problems that further challenge quantum optimization techniques. In this paper, we introduce the G-score benchmark specifically tailored to assess the capabilities of quantum optimization heuristics such as Quantum Annealing (QA) and the Quantum Approximate Optimization Algorithm (QAOA) or any other heuristic that can be relevant on quantum computers. This score aims to offer a comprehensive metric for evaluating the effectiveness of these methods in solving complex optimization problems.

Keywords: optimization · quantum computing · maximum cardinality matching · benchmark

1 Introduction and Context

Optimization problems are ubiquitous across various scientific and industrial domains, ranging from logistics and supply chain management and financial modeling. The ability to efficiently solve these problems can lead to significant advancements and economic benefits, making optimization one of the most critical applications in computational science.

Techniques like Quantum Annealing (QA) [3] and the Quantum Approximate Optimization Algorithm (QAOA) [2] have shown potential in solving optimization problems more efficiently than classical methods. However, the practical realization of this potential hinges on the ability to accurately benchmark and evaluate the performance of quantum computers. Benchmarking is crucial for understanding the capabilities and limitations of quantum hardware and algorithms, guiding their development, and ensuring their effective deployment in real-world applications.

F. Barbaresco and G. François (Eds.): QUEST-IS 2025, CCIS 2743, pp. 260–269, 2026.
https://doi.org/10.1007/978-3-032-13852-1_26

One well-established set of problems used to probe the capabilities of optimization heuristics is the Maximum Cardinality Matching (MCM) G_n series. These problems have been instrumental in assessing the performance of Simulated Annealing (SA), one of the major main-stream optimization heuristic. Recently, it was also used to characterize the family of DWave quantum computers, providing valuable insights into their optimization capabilities [8,12,13]. The MCM G_n series complements the NP-hard but unconstrained MaxCut problem utilized in ATOS's Q-Score benchmark [7], by offering a set of constrained problems which are widely spread in real-world applications.

In this paper, we introduce a benchmark score, the G-Score specifically designed to evaluate the capabilities of quantum optimization heuristics such as QA and QAOA on quantum computers. Nonetheless, it is worth noting that any new and relevant quantum heuristic can be tested by the protocol and can provide a value of G-score. This score aims to provide a comprehensive metric for assessing the effectiveness of these methods in solving complex optimization problems.

Applicative benchmarks can be a cornerstone of measuring the gap to quantum utility. Several such initiatives are already into inception e.g. in the U.S.A. with QED-C [6], in Germany with BenchQC [4], or France with BACQ [1]. We place our contribution here presented in the French initiative MetriQs/BACQ as part of the evaluation of Quantum Optimization heuristics, in complement of the aforementioned Q-Score.

2   The Evaluation of Quantum Optimization Heuristics

The set of problems included in the optimization benchmark of BACQ encompasses a wide variety of classes so that to test different aspects of the capabilities of optimization heuristics on Quantum Processing Units (QPUs) (Table 1).

Table 1. Classes of optimization problems addressed by the BACQ benchmarking protocol in its current and planed form

Types	Quadratic Unconstrained	Quadratic Constrained	Higher order
Polynomial	–	MCM G_n	HORN-Sat
NP or NP-hard	QScore MaxCut	–	Max-Sat

While the MCM G-series is not NP-hard, it is specifically designed to be a challenging optimization problem for heuristics. Thus it is important to probe this class of problems, especially because most of real-world optimization problems are indeed constrained.

Higher-order problems that appears in the last column of Table 2 pose additional challenges for quantum optimization heuristics, particularly for Quantum Annealing (QA), due to the quadratization step. This step can easily increase the number of variables by one or even two orders of magnitude, depending on the problem at hand. It is also important to note that constrained optimization problems are already difficult for QA. Addressing higher-order problems will represent a step further the current work in assessing the capabilities of quantum heuristics, a will constitute future work.

3 Maximum Cardinality Matching on the G_n Series

3.1 Definitions and Problem Statement

Let $G = (V, E)$ be an non directed graph. Each edge $e \in E$ can considered as a set of two vertices (i.e.: $e = \{v_0, v_1\}$ with $v_0, v_1 \in V$).

Definition 1 (neighbor). *A vertex $v \in V$ is a neighbor of a vertex $u \in V$ if and only if $\{u, v\} \in E$. An edge $e \in E$ is a neighbor of an edge $e' \in E$ if and only if $e \cap e' \neq \emptyset$ and $e \neq e'$. The set of all neighbors of a vertex or an edge x is denoted $\Gamma(x)$. For undirected graphs, for all $(x, x') \in V^2 \cup E^2$, $x' \in \Gamma(x) \Leftrightarrow x \in \Gamma(x')$.*

Definition 2 (line graph). *The line graph of G is the graph $LG(G) = (E, E')$ where E' is defined as follows:*

$$(e_1, e_2) \in E' \text{ if and only if there exists } v \in V \text{ such that } e_1 \cap e_2 = \{v\} \qquad (1)$$

Definition 3 (matching). *A matching of G is a set of edges $M \subseteq E$ without common vertices, that is, for all $\{e, e'\} \in M$, $e \cap e' = \emptyset$.*

The maximum cardinality matching (MCM) problem is the following: Given an undirected graph $G = (V, E)$, find a matching M with the largest number of edges (i.e.: for all M' matching of G, $|M| \geq |M'|$).

3.2 G_n Series

Definition 4 (Complete bipartite graph $K_{n,m}$). *Let be $n, m \in \mathbb{N}^*$, a complete bipartite graph $G = (V, E)$ is a graph that can be partitioned into two subsets of vertices $V_l, V_r \subseteq V$ such that:*

- *$V_l \sqcup V_r = V$, that is, $V_l \cup V_r = V$ and $V_l \cap V_r = \emptyset$.*
- *for all $u, v \in V$, $\{u, v\} \in E \Leftrightarrow u \in V_l$ and $v \in V_r$.*

Such a graph is denoted $K_{n,m}$ with $n = |V_l|$ and $m = |V_r|$.

Definition 5 (G_n graph). *Let be $n \in \mathbb{N}^*$, the graph $G_n = (V_n, E_n)$ is composed of n bipartite graphs $K_{n+1,n+1}$ linked together with $n + 1$ edges between them. Bipartite graphs on each end are linked with $n+1$ edges to $n+1$ 1-degree vertices (see an example on Fig. 1 for $n = 3$). The total number of vertices is $|V_n| = 2(n + 1)^2$ and the total number of edges is $|E_n| = (n + 1)^3$.*

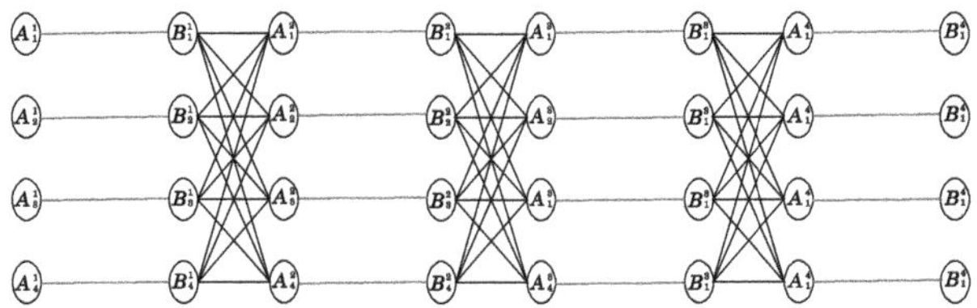

Fig. 1. G_3 graph with its maximum cardinality matching highlighted in red. The matching is done between A nodes and B nodes with a single match per node at most. (Color figure online)

The solution of the maximum cardinality matching problem on each graph of that series is unique, composed of all edges between all the $K_{n+1,n+1}$ subgraphs (the red edges in Fig. 1), for a total of $(n + 1)^2$ edges. That series is interesting because G. Sasaki and B. Hajek proved that simulated annealing cannot find the solution of the maximum matching problem over the G_n graphs in polynomial time [9]. As simulated annealing can be considered as a classical equivalent (but without quantum properties) of the Quantum Annealing (QA) [3] and the Quantum Approximate Optimization Algorithm (QAOA) [2], two of the most commonly used quantum algorithms in the field of optimization, the same tool can be used for probing the optimization heuristics in quantum computing. Solving the MCM problem on a graph G is also equivalent to solve the MIS problem on its line graph $LG(G_n)$.

4 Measuring Utility and Setting the Benchmark Principles

Given the inherent randomness of QPU outcomes, any evaluation of quantum utility must discriminate between cases where a QPU achieves a solution through genuine computational effort versus sheer luck. This distinction is especially important for noisy quantum devices, which struggle to reliably execute complex computations or maintain the fidelity of the quantum states required for meaningful problem-solving. A robust benchmarking framework must therefore ensure that correct results reflect real algorithmic performance, not random sampling or noise-induced errors. This is critical to avoid overstating the capabilities of noisy quantum hardware.

4.1 Number of Variables and Number of Qubits Utilized

A key criterion in addressing this issue is the number $\mathcal{N}$ of binary variables (for computational tasks) or spin states (for physics simulations) relevant to the problem's outcome. Reducing this number $\mathcal{N}$ would be considered unfair, as it would artificially shrink the solution space and falsely enhance the perceived quantum utility measured by the benchmark.

It is, therefore, crucial to preserve this number $\mathcal{N}$ and ensure that the actual implementation on the QPU utilizes $\mathcal{N}$ as the basis for the benchmark application. Notably, for gate-based QPUs, the number of qubits required for the implementation of the problem is typically greater than $\mathcal{N}$, due to overheads such as error correction and auxiliary qubits. For analog QPUs (quantum simulators), the required number may also exceed $\mathcal{N}$, not due to the algorithm itself but because of embedding constraints imposed by the hardware.

In either case, it should be transparent in the benchmark outcome that at least $\mathcal{N}$ qubits are effectively used. Under such conditions, the likelihood of a random outcome from the QPU would be proportional to $\frac{1}{2^{\mathcal{N}}}$. To ensure that the observed results are not merely a consequence of random chance, it is essential to apply a rigorous statistical validation. Specifically, a 5-sigma (5σ) criterion—commonly used in physics to signify a result beyond reasonable doubt—should be employed if possible. Considering the cost of doing heavy computation and sampling on nowadays QPUs, in the short term, a 3-sigma criterion can be regarded as sufficient for a first evaluation.

As a general good practice, this number $\mathcal{N}$ will be pivotal in defining the scale at which a given test is performed and in case of a positive result, it is advised that it is a base of defining one element of the outcome of the benchmark.

4.2   Utility Evaluation

In some scenarios, an approximate result may be seen as an acceptable outcome of the program—for example, obtaining an energy level sufficiently close to the ground state of a Hamiltonian or achieving a near-optimal solution to an optimization problem. However, in other cases, the solution is binary: it is either correct or incorrect. In these instances, the utility of an incorrect result is effectively zero, regardless of how many qubit values in the output were accurate. This distinction highlights the need for a case-by-case approach when evaluating each application of a given benchmark.

4.3   Measure Tools for the Evaluation

Hamming Distance. The Hamming distance is a metric that measures the number of bit positions at which two binary strings of equal length differ. In the context of evaluating the output of a quantum or classical program execution, it can serve as an effective measure of correctness and performance by quantifying how close the computed output is to the expected or optimal solution.

In the case of our specific need in Benchmarking, the Hamming distance is used for

- Correctness Assessment: For outputs where the exact solution is known (e.g., prime factorization or solving a linear system), the Hamming distance from the expected result indicates the effectiveness of the program and hardware. A low or zero Hamming distance reflects higher accuracy.

– Thresholding Success: A predefined Hamming distance threshold can be used to determine whether the execution is acceptable. For example, outputs with a Hamming distance below a certain value might still be considered useful approximations in certain applications.

Thus, the Hamming distance provides a practical, quantitative measure to evaluate and compare the effectiveness of programs and hardware implementations, especially in noisy or approximate computational environments.

Optimality of the Solution. Whenever an exact solution is not necessarily required—such as in combinatorial optimization, where a suboptimal but sufficiently good solution can be acceptable as long as the computational cost is low (i.e., a favorable quality-to-cost or quality-to-energy ratio)—a measure of distance to the optimal solution can be used for evaluation. While this is particularly challenging for NP-hard problems, the Q-Score Maxcut provides a framework to asymptotically assess the optimality of the best-found solution.

For problems involving constraints, which is often the case with real-world optimization tasks, it may be important to account for how well the proposed solutions satisfy those constraints. For instance, solutions that violate constraints could be penalized to reflect their reduced utility. Striking the right balance in weighting these factors is a challenging task, but the goal is to propose initial approaches and, through further development and testing, determine whether these choices are meaningful and effective.

4.4 Defining a Metric for the MCM G_n Series

In the case of the G_n series the problem is heavily combinatorial and allow only one optimal solution comprising $(n + 1)^2$ over $\mathcal{N} = (n + 1)^3$ variables for the problem. What we aim to obtain by the specified metric would be to have an evaluation of the number of variables that the QPU can handle and for which a relevant solution can be obtained. We can define the relevant thresholds and associated parameters:

– The value n_s for which the problem G_{n_s} can be solved exactly with a significant probability.
  - For G_1 which comprises 8 variables, the threshold should be $\frac{1}{32} \simeq 3.2\%$ over a minimum sampling of 1000 (the random case being $\frac{1}{256} \simeq 0.4\%$,
  - From G_2 as the number of variables is already significant, the threshold can be placed at $\frac{1}{1000}$ when a minimum of 10000 runs/samples are obtained. Otherwise, it should be at least $\frac{10}{S}$ if S is the number of runs/samples, with a minimum number of sample of 1000 to be significant.
– For the first "failed" graph of the series (n_s+1), we can fine tune the evaluation by utilizing the two following metrics:
  - The relative Hamming distance to the optimal solution $r_H = \frac{d_H}{(n_s+1)^3}$ where d_H is the Hamming distance (see 4.3)

- An evaluation of the optimality taking into account the invalidation of constrains into the best found solution: $r_d = \left(0, \frac{\mathcal{L}-f}{(n_s+2)^2}\right)$ where $\mathcal{L}$ is the number of links in this best found solution and f being the number of failed constraints in it
- The choice of the "best" failed solution should be a reliably found solution (i.e. whose probability of occurrence should be higher than 1%) but that may take into account some light post-processing to account for glitches in the mapping of the variables (e.g. if we found inconsistencies in the values of the qubits corresponding to a single variable, it would be possible to do a "majority vote" to decide the value of this variable as the outcome of the QPU processing, and it may take into account degenerate solutions, i.e. solution that have the exact same optimality score)
- By these choices, we can define a ratio to optimality that we can choose to be

$$r_o = \frac{1}{2}(1 - r_H + r_d) \tag{2}$$

- Then the base metric in number of variables can be defined as:

$$S_{G_1} = (n_s + 1 + r_o)^3 \tag{3}$$

because $(n_s+1)^3$ is the number of edges in the G_{n_s} graph and $(n_s+2)^3$ is for the first failed G_{n_s+1} graph, so it interpolates smoothly between the two.

Illustration on DWave-2000Q: without pre-processing, according to the results presented in the PhD thesis of D. Vert [11], shows $n_s = 2$, and for G_3 the Hamming distance is 8, so $r_H = \frac{8}{64}$ and that there are 15 links with 3 broken constraints also on the best found G_3 so $r_H = \frac{8}{64} = \frac{1}{16}$ and $r_d = \frac{15-3}{16} = \frac{12}{16}$ therefore $r_o = \frac{1}{2}(1 - \frac{1}{16} + \frac{12}{16}) = \frac{29}{32}$ and the result of the DWave-2000Q for the G-Score is:

$$S_{G_1,dw2000q} = \left(2 + 1 + \frac{29}{32}\right)^3 = 59.6$$

The S_G score is open-ended and gives an approximation of the capability of the QPU utilizing the specified algorithm to operate simple cost functions with hard constraints.

One important caveat of S_{G1} score, is that it does not take into account the probability of finding the solution at n_s and this creates an obvious bias toward $(n_s + 2)^3$ because of the non-linearity of the power function. This is noticeable on the experimental results shown in Sect. 5. As a consequence, we decided to put more emphasis on the n_s value by taking into account the probability of finding the optimal solution for G_{n_s} by defining the G-score as follows:

$$S_{G_1} = (n_n + 1 + r_o \times p_{n_s})^3 \tag{4}$$

where p_{n_s} is the probability of finding the solution for G_{n_s} using the considered heuristic. As both r_o and p_{n_s} are in the interval $[0, 1]$ so is their product. This gives a much stronger bias toward n_s which is a more relevant way of characterizing the capabilities of the tested heuristic.

Example On the same evaluation of DW2000Q, the probability of finding the optimal solution for G_2 is $p_2 = 0.06$ (6%) so the S_G score for the DW-2000Q is:

$$S_{G,\text{dw2000q}} = (2 + 1 + \frac{29}{32} \times 0.06)^3 = 28.5$$

which is a more accurate evaluation of the real solving capability of the DW2000Q without using pre-processing.

5 S_G Score on the DWave QPU Family

In this set of experimental validations, we used the same methods and principles (and a part of the open-source code) from the authors in [13]. We obtained the score over 3 generations of the annealers, thus measuring the improvements in the evolution of this family of QPUs.

Table 2. Results on several generations of DWave QPUs. In the table, n_s represent the highest problem solved with statistical significance, r_o is a ratio of optimality at the first failed rank $n_s + 1$ (see Eq. 2), $p_n s$ is the probability of finding the solution at rank n_s and S_{G_1} and S_G are the values of the two (naive and improved) G-scores. It is worth noting that the experiments are done with enough sampling such that the scores are accurate at the number of significant digits written in the table. Scores from classical heuristics from [5]. For classical heuristics, the time to solution is enforced at 60 s maximum.

QPU architecture	n_s	r_o	S_{G_1}	p_{n_s}	S_G
DWave-2000Q naive embedding	2	0.906	59.6	0.06	28.5
DWave-2000Q (Chimera)	2	0.938	61	0.849	54.7
DWave Advantage (Pegasus)	3	0.904	116	0.146	70.5
DWave Advantage-2-proto (Zephir)	3	0.936	122	0.315	79.2
Classical Heuristics tested					
Simulated Annealing	5	0.92	331	4%	220
NEC's Vector Annealing V.2	7	0.93	712	0.7%	513
NEC's Vector Annealing V.3	7	0.95	717	100	717

These results were sampled over sampling sets between 10^4 and 10^5 on several architectures when it was possible (i.e. at the moment, only on the Advantage QPUs). The difference between our results on the DWave-2000Q and the results derived in Sect. 4.4 can be solely attributed to a pre-processing step for improving the minor-embedding of the problem on the QPU. This procedure is described and available as open-source code in [13] and was reused mostly verbatim in these new tests.

6   Potential Critics

One another fair critic that arose of this series of problems is the reduction of the spectral gap rapidly when n grows. In [8] the authors evaluate that there is a risk of an exponential reduction of the spectral gap. While not rigorously demonstrated, the point is nonetheless valid. But it is also worth noting that such a statement is probably also true for any real-world optimization problem, as it is though that $BQP \neq NP$ and any heuristic would be poised by its ability to solve a NP-hard problem. Therefore, it is highly probable that for real-world problem, the spectral gap will also collapse exponentially as the number of variables increases.

Instead of viewing this risk as a limitation, our intuition leads us to consider it as a probing tool for assessing the utility of quantum heuristics in solving real-world problems. We expect that heuristics capable of accounting for the hard constraints of the G_n problem will perform much better than standard QA and QAOA. As the heuristic to be used is not specified, this benchmark will serve as a valuable basis also for comparing various quantum heuristics.

We noted in Sect. 3.2 that solving the G_n series is equivalent to solving the Maximum Independent Set problem (MIS) on its line graph. MIS is also formally equivalent to the Maximum Clique (MaxClique) problem which is NP-hard in the general case. A well-known study in the literature defines a Q-Score MaxClique [10] by analogy to the Q-Score MaxCut from Eviden, which is already integrated into BACQ. One valid critique, therefore, is that the G-score might be redundant with the Q-Score MaxClique. However, experimental results indicate that the Q-Score MaxClique on DWave machines yields values nearly identical to those of the Q-Score MaxCut. Moreover, both scores on DWave machines appear to be primarily constrained by the hardware's capacity to embed a $(N, \frac{1}{2})$-Erdős-Rényi graph under either definition of the Q-Score. This limitation does not apply to the G-score, which provides a metric more closely aligned with the actual capabilities and utility limits of DWave hardware in solving optimization problems.

7   Future Work

In this paper, we defined the G-Score benchmark that probe the capabilities of quantum optimization heuristic for constrained combinatorial problem, thus evaluating their real-world quantum utility. It will complement nicely with the Q-Score MaxCut aimed at evaluating these heuristics on unconstrained NP-hard problems.

The next steps will aim at generalizing the benchmark results on several other QPU families including potentially Neutral Atom based and Gate-based QPUs. The code and the instances as well as the code for computing the G-Score will be provided as open-source software, and will permit reproducibility tests. As noted previously, some future work will also be done toward a utility score based on the MaxSat family of problems.

This study was funded by the MetriQs/BACQ program under the France 2030/National Strategy for Quantum Technology initiative. MetriQs France 2030, ANR-22-QMET-0002. The authors have no competing interests and operate independently of any QPU provider.

References

1. Barbaresco, F., et al.: Bacq – application-oriented benchmarks for quantum computing (2024)
2. Farhi, E., Goldstone, J., Gutmann, S.: A Quantum Approximate Optimization Algorithm. arXiv preprint arXiv:1411.4028 (2014)
3. Farhi, E., Goldstone, J., Gutmann, S., Sipser, M.: Quantum Computation by Adiabatic Evolution. arXiv preprint quant-ph/0001106 (2000)
4. Geissler, F., et al.: Benchqc – scalable and modular benchmarking of industrial quantum computing applications (2025). https://arxiv.org/abs/2504.11204
5. Louise, S.: Vector annealing, a quantum-inspired technique: benchmarking performance against quantum and simulated annealing within the bacq framework. In: IEEE International Conference on Quantum Computing and Engineering, QCE 2025, Alburqueque, NM, USA, Aug 31-Sept.5, 2025. IEEE (2025)
6. Lubinski, T., et al.: Quantum algorithm exploration using application-oriented performance benchmarks (2024). https://arxiv.org/abs/2402.08985
7. Martiel, S., Ayral, T., Allouche, C.: Benchmarking quantum coprocessors in an application-centric, hardware-agnostic, and scalable way. IEEE Trans. Quantum Eng. **2**, 1–11 (2021). https://doi.org/10.1109/tqe.2021.3090207
8. McLeod, C.R., Sasdelli, M.: Benchmarking D-wave quantum annealers: spectral gap scaling of maximum cardinality matching problems. In: Groen, D., de Mulatier, C., Paszynski, M., Krzhizhanovskaya, V.V., Dongarra, J.J., Sloot, P.M.A. (eds.) Computational Science – ICCS 2022. ICCS 2022. LNCS, vol. 13353, pp. 150–163. Springer, Cham (2022). https://doi.org/10.1007/978-3-031-08760-8_13
9. Sasaki, G.H., Hajek, B.: The time complexity of maximum matching by simulated annealing. J. ACM (JACM) **35**(2), 387–403 (1988)
10. van der Schoot, W., Wezeman, R., Neumann, N.M.P., Phillipson, F., Kooij, R.: Extending the q-score to an application-level quantum metric framework (2024). https://arxiv.org/abs/2302.00639
11. Vert, D.: Étude des performances des machines à recuit quantique pour la résolution de problèmes combinatoires. (Study of the performance of quantum annealing machines for solving combinatorial problems). Ph.D. thesis, University of Paris-Saclay, France (2021). https://tel.archives-ouvertes.fr/tel-03208838
12. Vert, D., Sirdey, R., Louise, S.: Benchmarking quantum annealing against "hard" instances of the bipartite matching problem. SN Comput. Sci. **2**(2), 106 (2021). https://doi.org/10.1007/s42979-021-00483-1
13. Vert, D., Willsch, M., Yenilen, B., Sirdey, R., Louise, S., Michielsen, K.: Benchmarking quantum annealing with maximum cardinality matching problems. Front. Comput. Sci. **6** (2024). https://doi.org/10.3389/FCOMP.2024.1286057, https://doi.org/10.3389/fcomp.2024.1286057

Benchmarking Data Encoding Methods
in Quantum Machine Learning

Orlane Zang[1,2]($\boxtimes$), Grégoire Barrué[2], and Tony Quertier[2]

[1] Université de Picardie Jules Verne, Amiens, France
`orlane.zang@u-picardie.fr`
[2] Orange Innovation, Rennes, France
`{orlane.zang,gregoire.barrue,tony.quertier}@orange.com`

Abstract. Data encoding plays a fundamental and distinctive role in Quantum Machine Learning (QML). While classical approaches process data directly as vectors, QML may require transforming classical data into quantum states through encoding circuits, known as quantum feature maps or quantum embeddings. This step leverages the inherently high-dimensional and non-linear nature of Hilbert space, enabling more efficient data separation in complex feature spaces that may be inaccessible to classical methods. This encoding part significantly affects the performance of the QML model, so it is important to choose the right encoding method for the dataset to be encoded. However, this choice is generally arbitrary, since there is no "universal" rule for knowing which encoding to choose based on a specific set of data. There are currently a variety of encoding methods using different quantum logic gates. We studied the most commonly used types of encoding methods and benchmarked them using different datasets.

Keywords: Quantum Machine Learning · Data embedding · Optimization

1 Introduction

Quantum Machine Learning (QML) is a research area that focuses on the development of Machine Learning (ML) algorithms that can be executed by a quantum computer. For solving any specific problem, QML algorithms can be designed as a quantum circuit with a sequence of different quantum gate operations. However, designing QML algorithms that perform well as or better than conventional ones is still a challenge, due to limited ressources of Noisy Intermediate Scale Quantum (NISQ) computers [1]. Quantum encoding involves the conversion of classical information into quantum states, enabling QML algorithms to operate efficiently. This process is fundamental, as the way in which data is encoded can significantly influence the results obtained by learning models [2]. Various encoding methods have been proposed in the literature [3], and in this work we make a comparative study of encoding methods that are recurrent

F. Barbaresco and G. François (Eds.): QUEST-IS 2025, CCIS 2743, pp. 270–279, 2026.
https://doi.org/10.1007/978-3-032-13852-1_27

in the literature. To this end, we have used a quantum neural network classification model, presented in Fig. 1, and trained it on real datasets, noting the results obtained for each encoding. The rest of the paper is divided in four parts. The encoding methods used for our benchmarking are presented in Sect. 2, and Sect. 3 covers presentation of individual datasets used and preprocessing step. The results for each dataset are presented in Sect. 4, with interpretations. We end with a conclusion and outlook (Sect. 5).

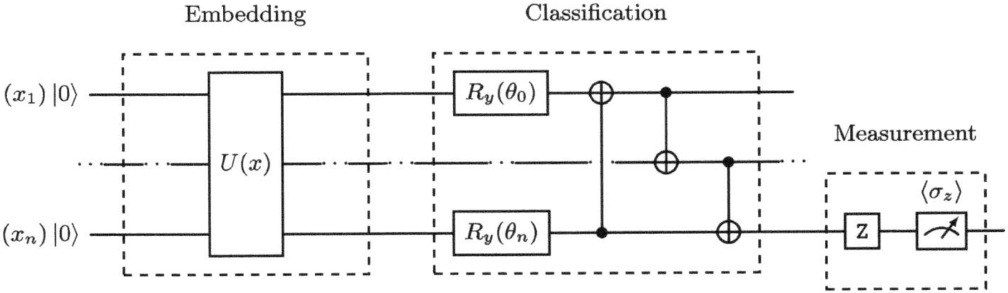

Fig. 1. A layer of the QML model used and the measurement part. Only the embedding part varies. The classification and measurement parts are the same for all the encodings tested.

2 Encoding Methods

2.1 Angle Encoding

The first type of encoding used is angle encoding [4]. For a classical dataset $D = (x, y)$, where $x \in \mathbb{R}^N$ is a N-dimensional input data (the features) and y the class label for the corresponding data, we can encode x into the quantum state $|\psi\rangle = \bigotimes_{j=1}^{N} |\psi_j\rangle$. Depending on the type of angle encoding used, the $|\psi_j\rangle$ state can take one of the following forms:

$$\begin{cases} |\psi_j\rangle = \mathrm{R_X}(x_j)\,|0\rangle = \cos\frac{x_j}{2}\,|0\rangle - i\sin\frac{x_j}{2}\,|1\rangle, & \textbf{Simple Angle,} \\[2ex] |\psi_j\rangle = \mathrm{U}(x_j)\mathrm{H}\,|0\rangle = \cos x_j\,|0\rangle + \sin x_j\,|1\rangle, & \frac{\pi}{4}\textbf{-Angle encoding,} \end{cases} \tag{1}$$

where $\mathrm{R_X}$ is the Pauli rotation gate around the x-axis, H is the Hadamard gate, and $\mathrm{U}(x)$ is an unitary operation [5] with the following square matrix

$$\mathrm{U}(x) = \begin{pmatrix} \cos\left(\frac{\pi}{4} - x\right) & \sin\left(\frac{\pi}{4} - x\right) \\ -\sin\left(\frac{\pi}{4} - x\right) & \cos\left(\frac{\pi}{4} - x\right) \end{pmatrix}. \tag{2}$$

Fig. 2 presents these two types of angle encoding.

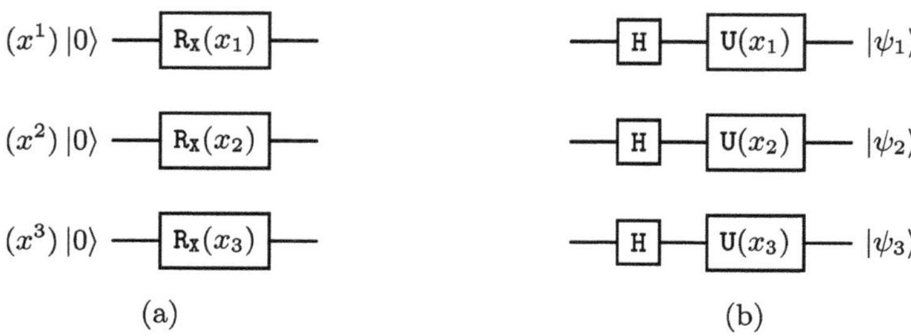

Fig. 2. Simple Angle encoding (a) and $\frac{\pi}{4}$-Angle encoding (b).

Another extension of angle encoding used for our simulations is the so called **Entangled Angle encoding** [6], which is constructed by applying a *Hadamard gate* to each qubit, followed by the typical Simple Angle encoding (with R_Y gate), and a *Controlled-NOT* gate between each pair of adjacent qubits (Fig. 3). In all three cases of angle encoding, the number **n of qubits** required to encode **data of size N** is such that **n = N**.

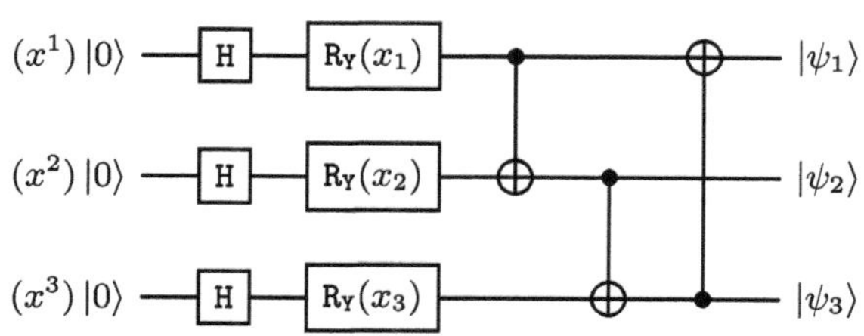

Fig. 3. Entangled Angle encoding.

2.2   Amplitude Encoding

Amplitude encoding embeds classical features of the form $\sum_{i=1}^{N} |x_i|^2 = 1$, in an amplitude vector of a n-qubits quantum state $|\psi\rangle = \sum_{i=1}^{N} x_i |i\rangle$, where $N = 2^n$, $|i\rangle$ is the i-th computational basis state, and x_i is the i-th element of x. Among the methods used to carry out this encoding, we used *Algorithm 1* of the paper [7] called the *divide and conquer algorithm*. Indeed, it is a two-part algorithm, the first part computes the encoding rotation angles as follows: *starting from the input vector x of dimension N , a new vector x_1 of dimension $\frac{N}{2}$ is created such that*

$$x_1 = \left(\sqrt{|x[2k]^2 + x[2k+1]|^2} \right)_{k=0}^{\frac{N}{2}}. \tag{3}$$

As long as the dimension of the new vector x_1 is greater than 1, x is assigned x_1 and the procedure is repeated. Then, for each new vector x_1, the rotation angles *are calculated as follows*:

$$\alpha = \begin{cases} 2\arcsin\left(\frac{x[2k+1]}{x_1[k]}\right), & \text{if } x_1[k] > 0 \\ 2\pi - 2\arcsin\left(\frac{x[2k+1]}{x_1[k]}\right), & \text{else.} \end{cases} \tag{4}$$

The second part of this algorithm generates the encoder circuit as presented in Fig. 4 for an input data of dimension 8.

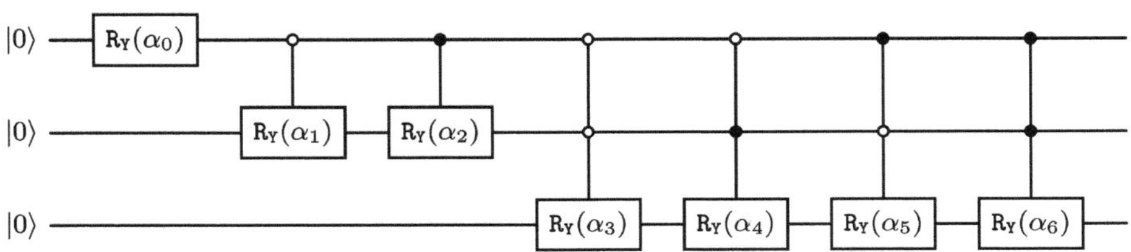

Fig. 4. Architecture of our Amplitude encoding with $N = 8$.

With n qubits, we can encode N data points, such that

$$\begin{cases} 2^{n-1} \leq N \leq 2^n, & \text{if } n = 1, \\ 2^{n-1} < N \leq 2^n, & \text{if } n > 1. \end{cases} \tag{5}$$

The cases $2^{n-1} < N < 2^n$ and $N = 2^n$ require the same number of qubits. However, the first case generate fewer rotation angles than the second. Hence, when $2^{n-1} < N < 2^n$, the remaining gates are set up with 0 as rotation angles.

2.3 Instantaneous Quantum Polynomial-Time (IQP) Encoding

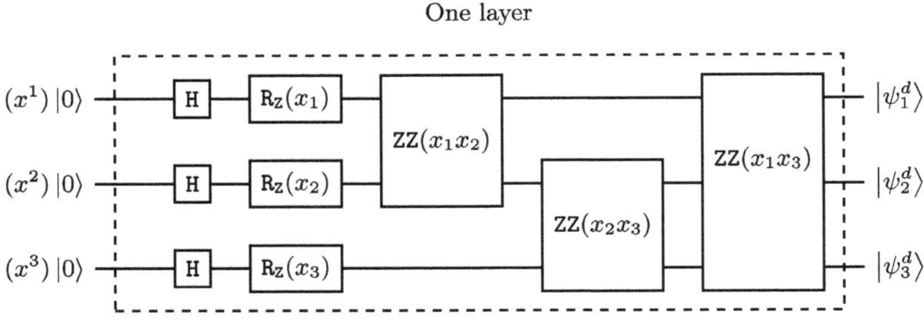

Fig. 5. An IQP circuit on 3-qubits: the IQP embedding is composed of at least two layers of this IQP circuit.

The instantaneous Quantum Polynomial-time (IQP) embedding [8] encodes the features into qubits using Hadamard gates H, R_Z gates and the two-qubits R_{ZZ} for entanglement that encode higher-order data diagonal gates.

IQP circuit was originally introduced by [9] as a response to the difficulty of experimentally demonstrating a quantum advantage without having a complete universal computation. Indeed, in the standard circuit model, a universal set of quantum gates is used to generate the special unitary group $SU(2^n)$, which is non-abelian. Hence, the order in which the gates are applied is crucial and reflects the temporal complexity of the computation. IQP circuits are limited to Hamiltonians that commute with each other, thus generating an abelian subgroup of the unitary group. This commutative property means that all gates can be seen as applied simultaneously, hence the term *instantaneous*. Although mathematically much more limited than universal computation, these circuits generate probability distributions that are believed to be classically inapproximable. This makes it possible to design Alice and Bob type protocols, where Alice encodes a hidden structure in a problem and Bob must, using his IQP device, produce samples that respect a specific correlation that no classical computer can reproduce efficiently.

An IQP Embedding, as presented in [10] and [11], takes as its input the $|0\rangle^{\otimes n}$ state, and is composed of at least two layers of the IQP circuit configuration, as presented in Fig. 5. The IQP embedding is capable of embedding N data points $(x_1, x_2, \cdots, x_j)$ into the state $|\psi\rangle = U_Z\, H^{\otimes n}\, U_Z\, H^{\otimes n}\, |0\rangle^{\otimes n}$ of $n = N$ qubits, with

$$U_Z = \exp\left(\sum_{\substack{j=1 \\ k=1 \\ j \neq k}}^{n} -i\frac{x_j x_k}{2} Z_j Z_k + \sum_{j=1}^{n} -i\frac{x_j}{2} Z_j \right).$$

3 Datasets and Preprocessing

We used four benchmark binary datasets. The first is the popular dataset ***Wisconsin Diagnostic Breast Cancer (WDBC) dataset***, which comprises 569 total instances, including 357 benign cases and 212 malignant ones, and a total of 30 distinct attributes. For our simulations we used 455 samples for train and 114 for test (split ratio of 80 : 20). Next, two of the four datasets are binary datasets formed from the ***Modified National Institute of Standards and Technology (MNIST) dataset***, which consists of 60,000 training images and 10,000 testing images, black and white, of handwritten digits, normalized centered with 28 pixels on each side. The two datasets are **MNIST01** *(classes 0 and 1, 12665 training and 2115 test samples)* and **MNIST08** *(classes 0 and 8, 11774 training and 1954 test samples)*. The choice of these classes is motivated by the fact that while 0 and 1 have distinct characteristics that can make them easier to differentiate, the distinction between 0 and 8 presents a moderate challenge, which can be interesting for testing the robustness of the models without being too complex. The fourth dataset, called ***Malware dataset***, is a binary dataset consisting of malicious and benign Portable Executable (PE) files from datasets Bodmas [12] and PE Malware Machine Learning [13]. Bodmas consists of 134,435 files. The 57,293 malicious files are fairly recent (collected between 2019 and 2020), and labeled by category (Trojan, Worm, Ransomware, $\cdots$). As for the PE Malware Machine Learning dataset, 114,737 files are malicious and come from sources such as VirusShare[1] and TheZoo[2]. The benign files (86,812) come from different versions of the Windows operating system (version $\geq$ Windows 7), for a total number of 200,549 files. We chose this dataset for our benchmarking because it corresponds to one of the key areas of our work: *cybersecurity*. Specifically, because of the fact that the dataset is big and not easy to classify, it is offers a valuable application for quantum classification tasks [14].

For the training, *4000 samples* are used for MNIST and Malware datasets. Regarding tests, we used *2000 samples* for Malware and MNIST01 datasets, while all *1954 test samples* of MNIST08 are used. We extracted from each dataset the 8 features with the highest impact and the lowest correlation with each other. The work involved in ranking the most impacting features of the Malware dataset is developed in *Benjamin Marais*'s thesis [15]. For WDBC dataset, we used the *Permutation feature importance* metric from *Scikit-learn*[3] to extract these features. Data from WDBC and Malware datasets were rescaled in $\left[0, \frac{\pi}{2}\right]$. The image data from MNIST was resized using the *Principal Component Analysis (PCA)* to suitable dimensions, and then also rescaled in $\left[0, \frac{\pi}{2}\right]$.

4 Results

For each of the 4 datasets, we carried out simulations using **Qiskit** simulators, on the 5 encodings for **N features** and **M layers**. First, we chose the number of features and layers as a power of 2. Then we looked at the behavior of Amplitude encoding in cases where N is not a power of 2. This is how we set $N \in \{4, 6, 8\}$, $M \in \{2, 4\}$. Test accuracies and F1-scores are presented in Tables 1, 2, 3 and 4 for all the four datasets. training has been done over 5 epochs, with a learning rate $\eta = 0.1$. In a longer version of this work available in pre-print [16], these train results are presented in the form of tables and figures, giving more precise training accuracies for each epoch.

On average, Amplitude encoding gives the best results on Malware and both MNIST datasets. On Malware dataset, it gives the best train and test accuracies for models where the number of features is a power of 2 (with F1-scores varying between 88% and 90%). On MNIST08, it performs best whatever the model, and on MNIST01, this better performance over the other models is only observed on 3 models: (4F, 4L), (6F, 4L) and (8F, 4F). Conversely, on WDBC dataset, Amplitude encoding gives the worst results on most models, with a max F1-score of 73.78%, and a min of 54.76%. As the WDBC dataset has fewer samples than Malware and MNIST datasets, these results could mean that Amplitude encoding is best suited to complex datasets with a large number of samples.

---

[1] VirusShare.
[2] TheZoo.
[3] Permutation feature importance.

Given that classification on MNIST08 is more difficult than on MNIST01, the fact that Amplitude encoding is better than the others on MNIST08 would support this interpretation.

Table 1. Tests and F1-scores of Malware dataset.

	(4F, 4L)		(4F, 2L)		(6F, 4L)		(6F, 2L)		(8F, 4L)		(8F, 2L)	
Encoding	Test	F1-S	Test	F1-S	Test	F1-S	Test	F1-S	Test	F1-S	Test	F1-S
Simple A.	0.87	0.88	0.815	0.830	0.847	0.856	0.811	0.828	0.848	0.857	0.819	0.835
$\frac{\pi}{4}$-Angle	0.882	0.872	0.80	0.821	0.9	**0.892**	0.833	0.830	0.888	**0.881**	0.830	0.828
Entangled A.	0.83	0.843	0.82	0.836	0.892	**0.891**	0.822	0.837	0.828	0.841	0.896	**0.891**
Amplitude	0.894	**0.891**	0.884	**0.884**	0.883	0.876	0.841	**0.851**	0.901	**0.9**	0.885	**0.886**
IQP	0.898	**0.891**	0.825	0.84	0.893	0.883	0.83	0.838	0.869	0.864	0.827	0.84

Table 2. Tests and F1-scores of WDBC dataset.

	(4F, 4L)		(4F, 2L)		(6F, 4L)		(6F, 2L)		(8F, 4L)		(8F, 2L)	
Encoding	Test	F1-S	Test	F1-S	Test	F1-S	Test	F1-S	Test	F1-S	Test	F1-S
Simple A.	0.949	**0.923**	0.912	0.888	0.938	**0.913**	0.833	0.716	0.85	0.753	0.859	0.804
$\frac{\pi}{4}$-Angle	0.947	**0.92**	0.912	0.88	0.903	0.857	0.894	**0.846**	0.868	0.8	0.798	0.646
Entangled A.	0.938	0.909	0.937	**0.907**	0.938	**0.915**	0.868	0.788	0.929	**0.894**	0.929	**0.89**
Amplitude	0.614	0.62	0.684	0.632	0.666	0.547	0.701	0.653	0.763	0.737	0.719	0.673
IQP	0.938	0.917	0.833	0.707	0.912	0.868	0.877	0.815	0.859	0.771	0.807	0.656

Unlike Amplitude encoding, Angle encodings are not as good on MNIST and Malware datasets. The lowest F1-score (82.11%) on Malware dataset comes from the $\frac{\pi}{4}$-Angle encoding, obtained with 4 features and 2 layers. Further on, we note that this encoding gives the lowest precision on all 2-layers models. On **MNIST08**, this encoding never gives the best results, whatever the model. Simple Angle encoding returns, on average, the lowest accuracies on the MNIST01 dataset. Case (6F, 2L) is particularly poor, with train accuracies ranging from 61%–67%, and a F1-score of 73.65%. These results are the worst of all MNIST datasets. Note that this encoding has the same behavior on both MNIST datasets. However, with this encoding we note that, on all the datasets, the training reaches a maximum value from the first epochs. So, the Simple Angle encoding do not need more epoch for the training. This last observation is also made with Entangled encoding on MNIST datasets. On average, of the three angle encodings, $\frac{\pi}{4}$-Angle and Entangled encoding returns the best precisions.

On WDBC, the best results are obtained with Angle encodings, where Simple Angle encoding returns the highest train and test accuracies (>94%), and the

Table 3. Tests and F1-scores of MNIST01 dataset.

Encoding	(4F, 4L)		(4F, 2L)		(6F, 4L)		(6F, 2L)		(8F, 4L)		(8F, 2L)	
	Test	F1-S	Test	F1-S	Test	F1-S	Test	F1-S	Test	F1-S	Test	F1-S
Simple A.	0.971	0.972	0.785	0.824	0.987	0.987	0.616	0.736	0.979	0.98	0.802	0.844
$\frac{\pi}{4}$-Angle	0.97	0.973	0.984	**0.985**	0.991	**0.992**	0.989	**0.99**	0.972	0.974	0.99	**0.99**
Entangled A.	0.994	**0.994**	0.884	0.902	0.967	0.97	0.865	0.885	0.986	0.987	0.971	0.973
Amplitude	0.997	**0.997**	0.982	**0.983**	0.995	**0.995**	0.977	**0.979**	0.991	**0.992**	0.946	0.952
IQP	0.994	**0.994**	0.873	0.893	0.935	0.943	0.781	0.83	0.972	0.974	0.922	0.928

Table 4. Tests and F1-scores MNIST08 dataset.

Encoding	(4F, 4L)		(4F, 2L)		(6F, 4L)		(6F, 2L)		(8F, 4L)		(8F, 2L)	
	Test	F1-S	Test	F1-S	Test	F1-S	Test	F1-S	Test	F1-S	Test	F1-S
Simple A.	0.95	0.949	0.902	0.896	0.949	0.949	0.94	0.939	0.94	0.941	0.938	0.938
$\frac{\pi}{4}$-Angle	0.932	0.934	0.93	0.933	0.925	0.925	0.948	0.948	0.921	0.921	0.927	0.929
Entangled A.	0.925	0.923	0.919	0.917	0.967	**0.967**	0.852	0.868	0.958	0.957	0.905	0.909
Amplitude	0.935	0.931	0.941	**0.948**	0.957	**0.956**	0.94	0.936	0.964	0.963	0.943	0.941
IQP	0.942	**0.941**	0.944	0.942	0.944	0.945	0.802	0.753	0.936	0.938	0.877	0.877

highest F1-score ($>92\%$) with 4 features and 4 layers. One observed behaviour on the Malware dataset, that recurs with WDBC, is the fact that the best accuracies with Simple Angle encoding are achieved with just 4 features. Comparing Simple Angle and $\frac{\pi}{4}$-Angle encodings, it seems that, of the two angle encodings, it is better to use the Simple Angle in cases with very few data, such as the WDBC dataset. The $\frac{\pi}{4}$-Angle encoding, on the other hand, would be a better choice than the simple one for large amounts of data (Malware dataset). We remark that Entangled Angle encoding seems to behave on WDBC dataset the same way as Amplitude encoding behaves on MNIST and Malware datasets. That is, on several models, it is one of the best performers, unlike the two others Angle encodings, which perform very well on some models and less well on others.

On WDBC, IQP encoding seems to have the same behavior as on the Malware dataset. An important observation with IQP encoding is the fact that, with 8 features, the imbalance in the WDBC dataset (number of benign samples greater than the number of malignant samples) seems to be reflected, whatever the number of layers. Indeed, we have an F1-score of 77.14% with a train accuracy of up to 87% and a test accuracy of 85.96% on 4 layers. On 2 layers, the F1-score is lower (65.62%) with learning and test accuracies above 80%. Our intuition is that, with very few data and layers, the model trained with IQP encoding concentrates on learning the excess class in the dataset (benign class), neglecting the other class.

5   Conclusion

The aim of our work is to contribute to the optimization of QML algorithms in the current era of NISQ, by benchmarking 5 types of data encodings. Starting from a Quantum Neural Networks (QNN) model (Fig. 1), we benchmarked these encodings on 4 datasets. The results suggest that encodings such as $\frac{\pi}{4}$-Angle and IQP need few features to return good accuracies. To improve these accuracies, increasing the number of layers is a more effective solution than increasing the number of features. Simple Angle encoding does not require several epochs for training. However, of the three angle encodings, it is the one that performs the worst on the average of the models. On the other hand, the Amplitude encoding was better on several models on the Malware and MNIST datasets, while on WDBC, it returned the lowest performance. These performances tend to decrease as the number of features decreases.

In summary, for datasets with features ranked in order of importance, if the size of the dataset is large enough (thousands of data points), we can build an efficient model based on Amplitude encoding with very few features, and therefore very few qubits. However, to ensure good accuracy, it would be preferable to choose a number of features as a power of 2. In a case where the dataset does not have enough data (as in the case of the WDBC dataset), Angle encodings will be more suitable for a high-performance model, also using fewer qubits and few epochs. Entangled Angle and $\frac{\pi}{4}$-Angle encodings returned the best results compared with the Simple Angle encoding, while the latter requires little epoch to achieve its best accuracies. However, in the three cases, it is important to choose a large enough number of layers (at least 4 layers) to have good performances.

The primary goal of this work was the presentation of the results and their analysis. A possible direction for future work would be to go beyond this analysis by providing an interpretation related to the nature of each encoding. Moreover, the simulations were carried out only once, on simulators. It would also be interesting to repeat the same tests multiple times on the same simulators, and to consider performing them on real noisy quantum computers.

References

1. Preskill, J.: Quantum Computing in the NISQ era and beyond. Quantum **2**, 79 (2018). http://dx.doi.org/10.22331/q-2018-08-06-79
2. Schuld, M., Sweke, R., Meyer, J.J.: Effect of data encoding on the expressive power of variational quantum-machine-learning models. Phys. Rev. A **103** (2021). https://doi.org/10.1103/physreva.103.032430, http://dx.doi.org/10.1103/PhysRevA.103.032430
3. Ranga, D., Rana, A., Prajapat, S., Kumar, P., Kumar, K., Vasilakos, A.V.: Quantum machine learning: exploring the role of data encoding techniques, challenges, and future directions. Mathematics **12** (2024). https://doi.org/10.3390/math12213318, https://www.mdpi.com/2227-7390/12/21/3318
4. Schuld, M., Killoran, N.: Quantum machine learning in feature hilbert spaces. Phys. Rev. Lett. **122**, 040504 (2019). https://link.aps.org/doi/10.1103/PhysRevLett.122.040504

5. Chalumuri, A., Kune, R., Manoj, B.S.: A hybrid classical-quantum approach for multi-class classification. Quantum Inf. Process. **20**(3), 1–19 (2021). https://doi.org/10.1007/s11128-021-03029-9

6. Luca, G.D., Vlasic, A., Vitz, M., Pham, A.: Empirical Power of Quantum Encoding Methods for Binary Classification, 2024. https://arxiv.org/abs/2408.13109

7. Araujo, I., Park, D., Petruccione, F., da Silva, A.: A divide-and-conquer algorithm for quantum state preparation. https://doi.org/10.48550/arXiv.2008.01511(2020)

8. Havlicek, V., et al.: Supervised learning with quantum-enhanced feature spaces. Nature **567**, 209–212 (2019). https://doi.org/10.1038/s41586-019-0980-2

9. Shepherd, D., Bremner, M.J.: Temporally unstructured quantum computation. Proc. R. Soc. A: Math. Phys. Eng. Sci. **465**, 1413–1439 (2009). https://doi.org/10.1098/rspa.2008.0443

10. Koike-Akino, T., Wang, P., Wang, Y.: Autoqml: Automated Quantum Machine Learning for Wi-Fi Integrated Sensing and Communications, 2022. https://arxiv.org/abs/2205.09115

11. Hong, Y.-Y., Arce, C., Huang, T.-W.: A robust hybrid classical and quantum model for short-term wind speed forecasting. IEEE Access 1 (2023)

12. Yang, L., Ciptadi, A., Laziuk, I., Ahmadzadeh, A., Wang, G.: BODMAS: an open dataset for learning based temporal analysis of PE malware. In: 2021 IEEE Security and Privacy Workshops (spw), pp. 78–84 (2021)

13. Lester, M.: PE Malware Machine Learning Dataset. https://practicalsecurityanalytics.com/pe-malware-machine-learning-dataset/

14. Bermejo, P., Braccia, P., Rudolph, M.S., Holmes, Z., Cincio, L., Cerezo, M.: Quantum convolutional neural networks are (effectively) classically simulable, 2024. https://arxiv.org/abs/2408.12739

15. Marais, B.: "Améliorations des outils de détection de malwares par analyse statique grâce à des mécanismes d'intelligence artificielle", Theses (Normandie Université, Dec. 2023). https://theses.hal.science/tel-04416984

16. Zang, O., Barrué, G., Quertier, T.: Benchmarking data encoding methods in quantum machine learning, 2025. https://arxiv.org/abs/2505.14295

Quantum Emulator Benchmarking Made Easy

Anna Leonteva[✉][iD], Maxime Outteryck, and Guido Masella[iD]

QPerfect, 23 Rue du Loess, 67000 Strasbourg, France
`anna.leonteva@qperfect.io`

Abstract. Classical quantum emulators are key tools for algorithm design, error correction, and noise mitigation. They rely on a variety of simulation methods such as statevector, tensor networks, matrix product states, decision diagrams and others. Their performance depends not only on simulation method and circuit properties, but also on emulator-specific parameters, pre-processing and implementation. Systematic benchmarking of emulators is essential for advancing simulation techniques and selecting the most suitable emulator for a given task. To address this need, we present an open-source, user-centric, problem-agnostic framework with a unified API for twelve leading emulators, delivering detailed performance insights for quantum computing research and development.

Keywords: Benchmarking · Quantum Emulator · Optimization

1 Introduction

Classical simulation is a core technology for quantum computing research and development, driving progress on both hardware and software innovation. On the hardware side, emulators allow researchers to pinpoint performance bottlenecks, optimize quantum architectures, and validate error-correction protocols before physical implementation [1–3]. In software, emulators provide an environment for the design, testing, and refinement of quantum algorithms [4,5].

Since classical emulation methods are continuously evolving, they demand systematic evaluation not as a mere leaderboard competition, but as a scientific methodology that bridges quantum computing and its potential applications by revealing each emulator's strengths and limitations. Moreover, robust benchmarking can systematically map which classes of quantum circuits today's emulators can tractably simulate and which remain beyond their reach, providing a clear roadmap for future research in quantum algorithms [6,7].

However, benchmarking quantum emulators is a challenging task, because their performance depends on many factors. A key factor is the core simulation method. For instance, statevector methods typically face exponential demands in computational time and memory with respect to the number of simulated qubits, restricting their effective use to systems of a few tens of qubits [7].

F. Barbaresco and G. François (Eds.): QUEST-IS 2025, CCIS 2743, pp. 280–288, 2026.
https://doi.org/10.1007/978-3-032-13852-1_28

Whereas, Matrix Product States (MPS) methods can achieve polynomial scaling in time and memory, allowing for simulating hundreds or even thousands of qubits for circuits with limited entanglement. Secondly, emulator-specific parameters (e.g., bond dimension and accuracy cutoffs in MPS) can strongly influence performance, depending on the circuit [6]. Furthermore, the programming language used to implement an emulator can significantly affect its performance. For instance, Python implementations are typically slower than those in C++, while Julia often achieves competitive run times thanks to just-in-time (JIT) compilation and efficient memory management. Finally, circuits themselves can affect results: e.g., the same emulator may scale poorly on one circuit, but achieve up to thousands of qubits on another [6,7]. All of these factors introduce biases and lead to unreliable benchmarking results [8]. This implies that quantum emulator performance should not be characterized as a singular "empirical" property, but rather as a set of metrics (including run time, fidelity, and scalability) that are dependent on all mentioned factors.

Several open-source projects have addressed quantum emulator benchmarking from different angles. QUARK [9] is a modular framework automating application-level benchmarking, from problem definition to solver execution. However, its tight coupling to domain-specific encodings and limited backend support restricts its adaptability for broader, scalable comparisons. Jamadagni et al. [7] presented a toolchain for statevector simulators, profiling single- vs double-precision and single- vs multi-threaded regimes across leading libraries. BenchPress [10] measures circuit construction and transpilation performance across seven SDKs, but omits simulator execution metrics. Bowles et al. [11] introduced a PennyLane-based benchmark, reporting only machine learning quality metrics. While each tool excels in its niche—application workflows, simulator profiling, compiler performance, or quantum machine learning evaluation—none, to our knowledge, offers unified access to multiple quantum emulators with execution benchmarking across diverse circuits. Our framework addresses this gap.

The goal of this paper is to present an open-source framework, FENIQS[1], for automated benchmarking of quantum emulators across simulation methods, circuit types, and user-defined scenarios. It aims to streamline the benchmarking process, reducing time and manual effort involved.

2   Framework

FENIQS is an open-source Python framework designed to streamline the quantum emulator benchmarking process. Its architecture is user-centric, problem-agnostic, and language-independent, featuring a modular and extensible design, where all supported emulators are implemented as plugins. FENIQS benchmarking process is shown in Fig. 1 and contains several principal blocks.

Workflow: Users initiate this workflow by supplying quantum circuits, in Open-QASM 2.0 format, to a designated input directory, which is monitored by a

---

[1] https://github.com/qperfect-io/feniqs_lite.

FENIQS daemon (Linux service). Each new circuit is processed by systematically invoking the configured emulator plugins via the `feniqs_lib` core library. Throughout the execution of each simulation, the framework rigorously enforces user-defined timeouts. Upon completion of a simulation run, all pertinent performance metrics, such as run time and fidelity (discussed in Sect. 3), along with associated metadata, are captured and stored. FENIQS provides flexibility in data archiving, offering options for storage in a PostgreSQL database, facilitating robust and queryable long-term storage, or as individual JSON files. These JSON files can then feed into the framework's post-processing tools to generate a variety of insightful visualizations, including comparative plots of run time, fidelity, and scalability across different emulators and configurations.

Core Library: FENIQS currently supports twelve actively maintained quantum emulator frameworks, covering a wide range of simulation methods and hardware targets (summarized in Table 1). Internally, the core library `feniqs_lib` uses plugin and environment managers to load each emulator in an isolated virtual environment, avoiding conflicts and preserving reproducibility. All plugins conform to a unified interface for circuit execution, enabling fair comparisons across different emulators. This modular architecture also allows users to easily switch between simulation methods, versions or hardware platforms.

Configuration and Pre-processing: Before benchmarking, users can modify three distinct configuration files, in YAML format, to define FENIQS setup: `venv_deps.yaml` defines the environment, selecting emulator plugins and ensuring backend isolation; `config_backend.yaml` specifies emulator types and parameters; `optimizator.yaml` (optional) sets parameter ranges for MPS performance tuning. To address the strong impact of emulator-specific parameters (such as bond dimension for MPS, accuracy thresholds, or circuit pre-optimizations) on both run time and fidelity, FENIQS includes a noise-aware hyperparameter optimizer. It supports CMA-ES for single-objective optimization, minimizing run time while satisfying a fixed fidelity constraint, ideal when high accuracy is required. For broader trade-off exploration, NSGA-II and MOEA-D enable multi-objective optimization, generating a Pareto front that balances fidelity and run time. All candidate configurations are evaluated under run time noise, with optimized solution exportable to the backend YAML file for direct use in benchmarking.

Post-processing: FENIQS includes a set of Python scripts for parsing JSON result files and generating comparative plots. These tools facilitate the analysis of benchmark metrics such as run time, fidelity, and scalability across different emulators.

3   Metrics

FENIQS evaluates performance based on two direct core metrics: *run time* and *mirror circuit fidelity*

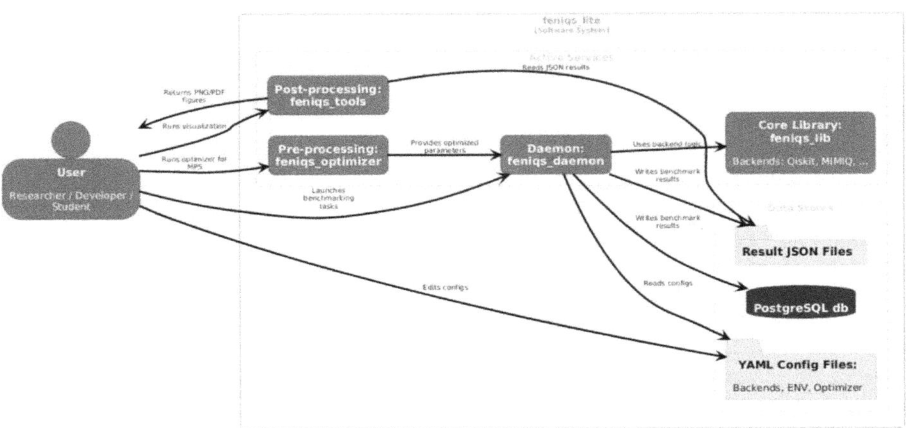

Fig. 1. Architectural diagram of the FENIQS benchmarking platform, illustrating key components, data stores, and user interactions. The large dashed rectangle delineates the system boundary of FENIQS. Boxes within this boundary represent primary software containers and data stores, while boxes positioned externally indicate persons interacting with the system. Arrows indicate the direction of control signals or data flows.

The *run time* metric in FENIQS refers to the total wall-clock duration required to execute a quantum circuit. This includes all necessary operational steps, including circuit parsing from OpenQASM 2.0 format, quantum state initialization, circuit execution, and sampling of measurement outcomes. To account for system warm-up effects and inherent shot-to-shot variability in execution times, each benchmark circuit is executed multiple times, as specified by the users, and FENIQS reports the minimum run time (in seconds) achieved.

For assessing the accuracy of emulators, particularly those not based on statevector methods where direct fidelity calculation with a reference state can be intractable, FENIQS implements the *mirror-circuit fidelity*. This method evaluates an emulator's accuracy without relying on its internal fidelity estimates. It involves executing a circuit followed immediately by its inverse (the mirror circuit), and then measuring the probability of the system returning to its initial computational basis state. To prevent emulators from trivially optimizing the combined circuit to an identity operation, FENIQS supports the insertion of barrier operations after the forward circuit (where the emulator supports them).

Beyond presented primary metrics, FENIQS introduces an indirect metric – *scalability*, which quantifies an emulator's ability to efficiently handle increasing problem size without loss of accuracy or excessive run time. This metric is assessed by analyzing trends in the primary metrics – run time and fidelity – collected by FENIQS across multiple circuit instances of varying qubit numbers for a given algorithm or circuit class. Typically, this involves evaluating how the measured run time $T_{\mathcal{E}}(n)$ for the emulator $\mathcal{E}$ depends on the number of qubits n, subject to maintaining the mirror-circuit fidelity $\mathcal{F}_{\mathcal{E}} \geq 0.99$, and respecting an execution timeout of 300 s. Emulators exhibiting polynomial scaling of run time with n are generally considered more scalable for that class of circuits.

4 Results and Main Features

Table 1. Supported frameworks within FENIQS, detailing their primary simulation methods, API languages, open-source status, and CPU/GPU hardware targets.

Framework	Method	API	Open	CPU	GPU
Cirq [12]	StateVector	Python	✓	✓	✓
Qiskit [13]	StateVector	Python	✓	✓	✓
	MPS	Python	✓	✓	✓
MIMIQ	StateVector	Julia		✓	
	MPS	Julia		✓	
Pennylane [14]	StateVector	Python	✓	✓	✓
Qulacs [15]	StateVector	Python	✓	✓	✓
Yao [16]	StateVector	Julia	✓	✓	✓
ProjectQ [17]	StateVector	Python	✓	✓	
QMatchTea [18]	MPS	Python	✓	✓	✓
MQT-DDS [19]	Decision Diagrams	Python	✓	✓	
Pyqrack [20]	Factorized Ket	Python	✓	✓	
Qrisp [21]	Sparse Matrix	Python	✓	✓	
Quimb [22]	MPS	Python	✓	✓	
	NT	Python	✓	✓	

FENIQS robustness and versatility were demonstrated in a large-scale benchmarking work presented in [6], which assessed seven cutting-edge quantum emulators across diverse simulation techniques. This study included four MPS-based tools (Qiskit-MPS, Quimb-MPS, QMatchaTea-MPS, MIMIQ-MPS), as well as Quimb-TN, Pyqrack, and MQT-DDS. The benchmark suite featured 13 scalable circuits from the MQTBench library[2], ranging from 4 to 1,024 qubits on CPU hardware. To present these results, Fig. 2 reproduces a radar chart from [6], showing the maximum qubit counts achieved by each emulator across the 13 benchmark circuits. The radial axis uses a logarithmic scale, from 12 (innermost circle) to 1,024 qubits (outermost), providing a clear visual summary of scalability. Larger polygon areas indicate emulators capable of solving a greater fraction of circuits at higher qubit counts, highlighting their performance under utility-scale conditions.

Beyond its demonstrated use case, FENIQS offers several key features that make it a powerful tool for diverse benchmarking needs.

---

[2] https://github.com/cda-tum/MQTBench.

Flexibility of the Benchmarking Process: FENIQS enables highly customizable benchmarking by allowing users to group emulators based on simulation method (e.g., statevector, tensor network), hardware platform (CPU/GPU), or criteria such as numerical precision or run time limits. YAML-based configurations let users compare backends, versions, and execution modes without writing code. For example, the left panel of Fig. 3 shows run time scaling of the "Amplitude Estimation" circuit across seven emulators (mentioned above) with increasing qubit counts on CPU; the right panel compares GPU-implemented statevector simulators, Qiskit and PennyLane, executing the `hdyn` task (simulating XYZ-Heisenberg dynamics).

Extensibility: FENIQS supports the addition of new emulators and optimization strategies with minimal core changes. Benchmark suites can be extended by adding OpenQASM 2.0 circuits, and JSON-formatted results integrate easily with external tools for advanced post-processing and visualization, without altering existing workflows.

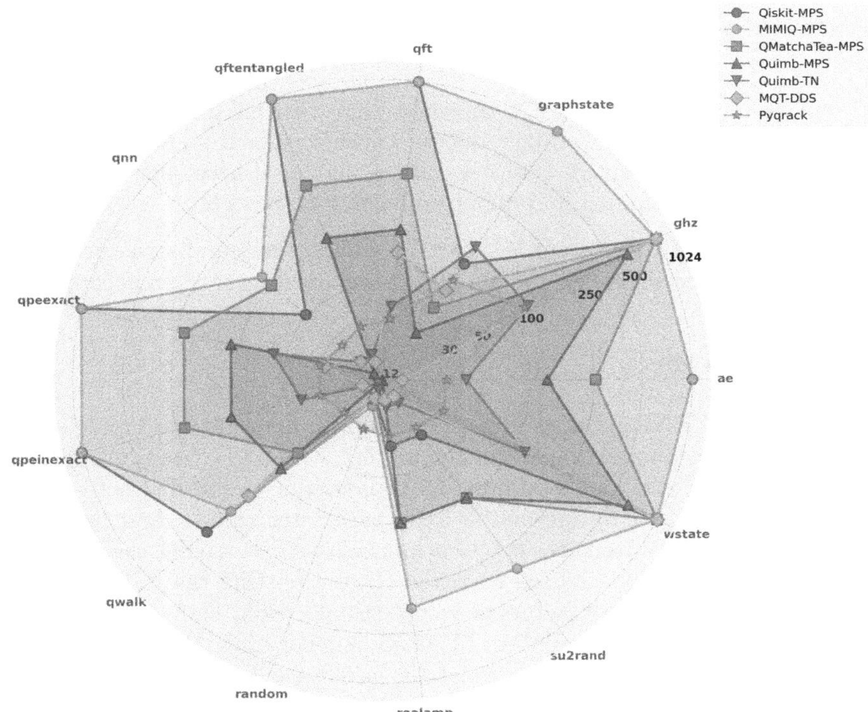

Fig. 2. FENIQS output visualizing comparative emulator scalability [6]. The radar chart displays the maximum achievable qubit counts for seven emulators across thirteen benchmark circuits. Each radial axis represents a specific benchmark algorithm, with concentric circles indicating the qubit scale from 12 to 1,024 (logarithmic). Larger polygon areas generally indicate superior overall scalability across the evaluated benchmark suite.

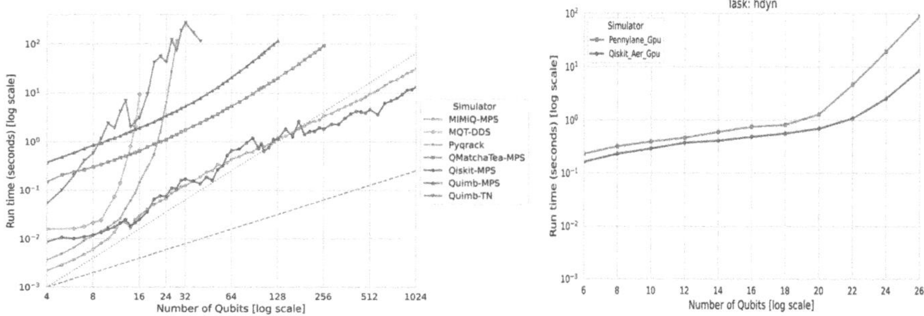

Fig. 3. Run time plots generated by FENIQS, comparing emulator performance as a function of qubit count for two distinct quantum tasks and emulator sets. Both panels present run time in seconds versus the number of qubits. Left panel illustrates run time (in log scales) scaling for the amplitude-estimation benchmark circuit [6] for seven different emulators, as detailed in the legend. Dashed and dotted lines provide visual references for linear and quadratic scaling behaviors, respectively. Right panel shows a performance comparison of two statevector GPU-based simulators (PennyLane and Qiskit) on the **hdyn** benchmarks.

Efficiency and Accessibility: FENIQS streamlines benchmarking by automating setup, execution, and monitoring, reducing workflow time from hours to minutes. YAML configurations replace complex coding, making it accessible to users with less extensive programming and quantum emulators expertise.

Reproducibility and Clarity: YAML-based configurations and structured JSON results ensure reproducibility and simplify comparison across emulators.

5   Conclusions

We introduced FENIQS, a user-centric, problem-agnostic, and plugin-based framework for benchmarking quantum emulators. FENIQS streamlines comparative emulator studies through a unified API supporting twelve distinct backends, a modular plugin system, automated hyperparameter tuning, and robust performance metrics. It automates key benchmarking steps, saving users time and reducing the need for specialized expertise, by supporting the full benchmarking life-cycle, from scenario setup to result visualization, and enabling custom experiments. The practical utility of FENIQS in generating reliable performance insights, such as detailed emulator scalability profiles and assessments of classical simulation boundaries—as demonstrated in a comprehensive utility-scale benchmarking study [6]. FENIQS thus serves as a valuable tool for the quantum computing community, aiding in emulator selection and fostering the advancement of simulation techniques. Future work includes extending support to quantum hardware to address execution variability and cost, and integrating machine-learning-based accelerators for parameter tuning. The full source code is openly available on GitHub[3].

Disclosure of Interests. Guido Masella is a shareholder of QPerfect.

References

1. Arute, F., et al.: Quantum supremacy using a programmable superconducting processor. Nature **574**(7779), 505–510 (2019)
2. Kim, Y., et al.: Evidence for the utility of quantum computing before fault tolerance. Nature **618**(7965), 500–505 (2023)
3. Cross, A.W., Bishop, L.S., Sheldon, S., Nation, P.D., Gambetta, J.M.: Validating quantum computers using randomized model circuits. Phys. Rev. A **100**(3), 032328 (2019)
4. Morvan, A., et al.: Phase transitions in random circuit sampling. Nature **634**(8033), 328–333 (2024)
5. King, A.D., Nocera, A., Rams, M.M., Dziarmaga, J., Wiersema, R., et al.: Beyond-classical computation in quantum simulation. Science **388**(6743), 199–204 (2025)
6. Leonteva, A., Masella, G., Outteryck, M., Piñeiro Orioli, A., Whitlock, S.: Comparative benchmarking of utility-scale quantum emulators, 2025. https://arxiv.org/abs/2504.14027
7. Jamadagni, A., Läuchli, A.M., Hempel, C.: Benchmarking quantum computer simulation software packages: state vector simulators. preprint arXiv:2401.09076, 2024
8. Lubinski, T., et al.: Quantum algorithm exploration using application-oriented performance benchmarks. preprint arXiv:2402.08985, 2024
9. Finžgar, J.R., Ross, P., Hölscher, L., Klepsch, J., Luckow, A.: Quark: a framework for quantum computing application benchmarking. In: 2022 IEEE International Conference on Quantum Computing and Engineering (QCE), pp. 226–237. IEEE, 2022
10. Nation, P.D., et al.: Benchmarking the performance of quantum computing software for quantum circuit creation, manipulation and compilation. Nat. Comput. Sci. 1–9 (2025)
11. Bowles, J., Ahmed, S., Schuld, M.: Better than classical? The subtle art of benchmarking quantum machine learning models. arXiv preprint arXiv:2403.07059, 2024
12. Cirq Developers. Cirq (v1.5.0), 2025. https://doi.org/10.5281/zenodo.15191735
13. Treinish, M., et al.: Qiskit/qiskit: Qiskit 2.0.0 (2.0.0), 2025. qiskit bot. https://doi.org/10.5281/zenodo.15116124
14. Bergholm, V., et al.: Pennylane: automatic differentiation of hybrid quantum-classical computations. arXiv preprint arXiv:1811.04968, 2018
15. Suzuki, Y., et al.: Qulacs: a fast and versatile quantum circuit simulator for research purpose. Quantum **5**, 559 (2021)
16. Luo, X.Z., Liu, J.G., Zhang, P., Wang, L.: Extensible, efficient framework for quantum algorithm design. Quantum **4**, 341 (2020)
17. Steiger, D.S., Häner, T., Troyer, M.: Projectq: an open source software framework for quantum computing. Quantum **2**, 49 (2018)

---

[3] https://github.com/qperfect-io/feniqs_lite

18. Montangero, S.: Introduction to Tensor Network Methods. Springer, Cham (2018). https://doi.org/10.1007/978-3-030-01409-4
19. Wille, R., Hillmich, S., Burgholzer, L.: Decision Diagrams for Quantum Computing. In: Topaloglu, R.O. (eds.) Design Automation of Quantum Computers, pp. 1–23. Springer, Cham (2023). https://doi.org/10.1007/978-3-031-15699-1_1
20. Strano, D., Bollay, B., Blaauw, A., Shammah, N., Zeng, W.J., Mari, A.: Exact and approximate simulation of large quantum circuits on a single GPU. preprint arXiv:2304.14969, 2023
21. Seidel, R., et al.: Qrisp: a framework for compilable high-level programming of gate-based quantum computers, 2024. https://arxiv.org/abs/2406.14792
22. Gray, J.: Quimb: a python library for quantum information and many-body calculations. J. Open Source Softw. **3**(29), 819 (2018). https://doi.org/10.21105/joss.00819

Evaluating Quantum Wire Cutting
for QAOA: Performance Benchmarks
in Ideal and Noisy Environments

Michel R. M. J. Meulen[1,2], Niels M. P. Neumann[1(✉)], and Jasper Verbree[1]

[1] Department of Applied Cryptography and Quantum Algorithms, The Netherlands
Organisation for Applied Scientific Research, The Hague, The Netherlands
{michel.meulen,niels.neumann,jasper.verbree}@tno.nl
[2] Fontys University of Applied Sciences, Eindhoven, The Netherlands

Abstract. Current quantum computers suffer from a limited number of qubits and high error rates, limiting practical applicability. Different techniques exist to mitigate these effects and run larger algorithms. In this work, we analyze one of these techniques called quantum circuit cutting. With circuit cutting, a quantum circuit is decomposed into smaller sub-circuits, each of which can be run on smaller quantum hardware. We compare the performance of quantum circuit cutting with different cutting strategies, and then apply circuit cutting to a QAOA algorithm. Using simulations, we first show that Randomized Clifford measurements outperform both Pauli and random unitary measurements. Second, we show that circuit cutting has trouble providing correct answers in noisy settings, especially as the number of circuits increases.

Keywords: Quantum Computing · Circuit Cutting · Wire Cutting · QAOA · MaxCut

1 Introduction

Quantum computers have the potential to solve problems infeasible for any classical supercomputer [15,19]. Investments in quantum computers and the corresponding algorithms are continuously rising. However, current quantum computers, often called Noisy Intermediate-Scale Quantum (NISQ) devices, are still far away from their full potential. They have a limited number of qubits and high error rates [18]. Improving usability requires increasing the number of qubits and decreasing the error rates. Recently, techniques like error mitigation have been proposed to mitigate, rather than solve, these limitations [12,23].

In 2020, Peng et al. showed how to use an n-qubit quantum computer to run an $n + k$-qubit quantum circuit [17], introducing the subject of *circuit cutting*. Their method decomposes a quantum circuit in different blocks and combines the outcomes of the individual blocks to approximate expectation values produced by the original circuit. Follow-up work improved their method and provided explicit

F. Barbaresco and G. François (Eds.): QUEST-IS 2025, CCIS 2743, pp. 289–298, 2026.
https://doi.org/10.1007/978-3-032-13852-1_29

decompositions for certain quantum gates [1,14,16]. Typically, we distinguish between two different circuit cutting techniques: wire cutting, where a qubit is measured in a certain basis and initialized again, and gate cutting, where a multi-qubit quantum gate is replaced by single-qubit gates and measurements with post-selection [1]. This work focuses on wire cutting.

First, we compare three variants of wire cutting, using exact and noisy simulations, and determine which of these variants performs best. Second, we apply the best wire cutting variant found to the quantum approximate optimization algorithm (QAOA) [5] applied to the MaxCut problem [25]. In this way, we learn the potential of quantum circuit cutting and its overhead for a practical problem showing how the exact theory translates to the noisy reality.

The remainder of this paper is organized as follows: Sect. 2 discusses in more detail wire cutting and the variants considered in this work. Section 3 presents the experimental setup: a benchmark comparison between Pauli-based and randomized measurement-based wire cutting; and the application of the randomized measurement-based wire cutting technique to QAOA in both ideal and noisy simulation environments. Section 4 gives the results for these experiments, and Sect. 5 discusses the results and concludes the work.

2   Wire Cutting

Tang et al. introduced a new method for wire cutting based on decomposing quantum operations into linear combinations of Pauli matrices [22]. By measuring qubits in the X, Y, and Z bases and reinitializing them in the 0, 1, +, or i state, a larger circuit can be approximated by smaller ones.

Once all measurement outcomes are obtained, the probability of a basis state being produced by the uncut circuit can be reconstructed. For instance, consider the state 010 where the second wire is cut. We define two auxiliary states $u_1 = 00$ and $u_2 = 01$ for the first and second qubit, and an output state $d = 10$ for the second and third qubit. The probability of measuring 010 is:

$$\Pr[010] = \frac{1}{2} \sum_{i=1}^{4} U_i D_i, \tag{1}$$

where

$$U_1 = 2u_{1,Z}, \quad U_2 = 2u_{2,Z}, \quad U_3 = u_{1,X} - u_{2,X}, \quad U_4 = u_{1,Y} - u_{2,Y},$$
$$D_1 = d_0, \quad D_2 = d_1, \quad D_3 = 2d_+ - d_0 - d_1, \quad D_4 = 2d_i - d_0 - d_1,$$

and $u_{i,P}$ corresponds to the probability of measuring state u_i in Pauli basis P and d_ϕ corresponds to the probability of measuring state d on input state ϕ. A schematic presentation of wire cutting is given in Fig. 1. Note that cutting the middle qubit wire of a three-qubit circuit results in subcircuits of two qubits.

Lowe et al. introduced an alternative approach using two distinct quantum channels to reduce the sampling overhead of exact methods [14]. Their method

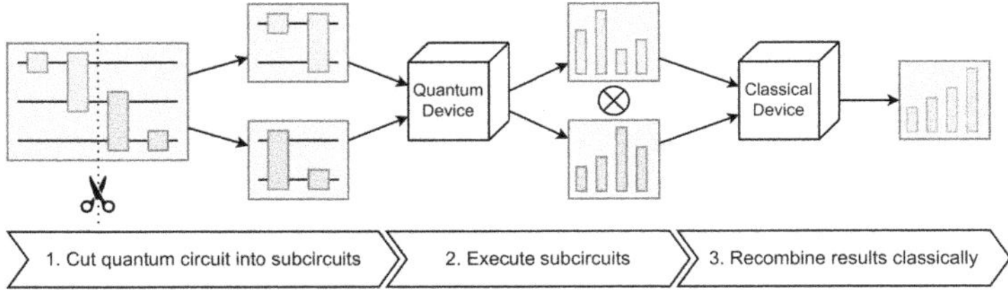

Fig. 1. Visual representation of wire cutting using Pauli measurements. Each subcircuit is run on a quantum device and the results are classically recombined. The tensor product symbol ($\otimes$) refers to the recombination of the different probability distributions as in Equation (1).

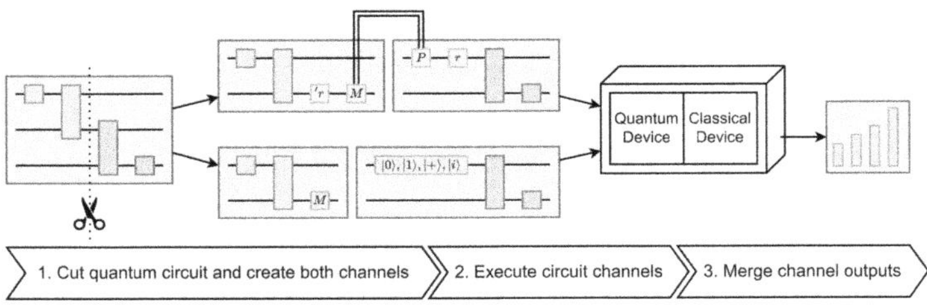

Fig. 2. Visual representation of randomized-measurement-based wire cutting. For every sample, the method randomly chooses between two quantum channels: The random Clifford channel (top) and the depolarization channel (bottom). The chosen channel is then executed on a hybrid-quantum-classical-device. The recombination of the probabilities follows Equation (2).

is based on already existing methods to classically simulate the entanglement between matrix-product states [3,9]. One channel applies a random Clifford gate before the measurement and applies the conjugate of the Clifford gate as qubit reinitialization. The other channel acts as a depolarizing channel, where the qubit is measured and then reinitialized in a random basis state. Their method randomly selects between these two channels with probabilities $\frac{d+1}{2d+1}$ and $\frac{d}{2d+1}$, where $d = 2^k$ for k cut qubits. The probability corresponding to a state b is then given by

$$\Pr(b) = (d + 1) \cdot \Pr_{Clifford}(b) - d \cdot \Pr_{depolarizing}(b), \tag{2}$$

where $\Pr_X(b)$ is the probability found for state b for channel X. Figure 2 gives an abstract overview of this randomized circuit cutting.

The Clifford gates used in the random measurement circuit cutting approach are examples of unitary 2-designs [14]. One might wonder whether other gates improve the results even further. A simple generalization of Pauli gates is rotational Pauli gates. Even though these gates no longer form unitary 2-designs, it is worth investigating whether they still help reduce the sampling overhead. We therefore also consider a modified random measurement circuit cutting approach where we apply random Pauli rotation gates before the measurement and initialize the qubits using its conjugated form.

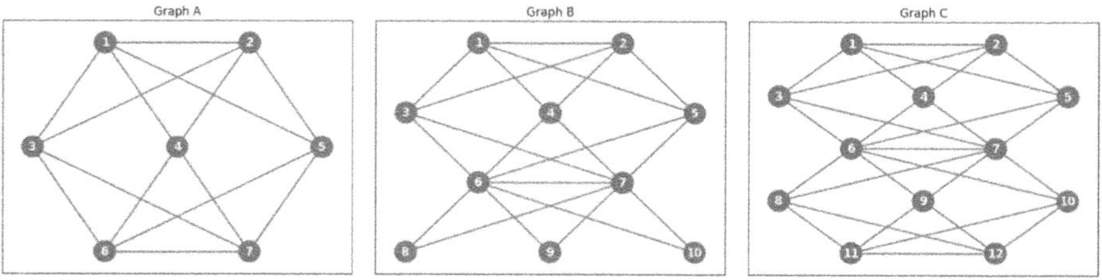

Fig. 3. A visual representation of the three layered graphs used in the experiments. The colors indicate the optimal division of the nodes.

3 Experimental Setup

We now detail the experiments we have run and their implementation framework.

3.1 Comparing Circuit Cutting Methods

In the first experiment we compared the exact wire cutting method based on Pauli measurements [22] with the randomized measurement approach [14] and a version where we apply arbitrary Pauli rotation gates. We consider a simple circuit that prepares a five-qubit GHZ state [8], which we chose due to its wide applicability in many applications such as distributed quantum computing. Additionally, the cascade-like nature of the quantum circuit makes it suitable for wire cutting. We added rotation gates before and after the CNOT gates to complicate the state and assure all measurement bases and initial states contribute.

3.2 Evaluating the Performance for Circuit Cutting in QAOA

In the second experiment, we use the best performing method to test the performance for a QAOA algorithm applied to the MaxCut problem. Given a graph, the MaxCut problem is to partition the vertex set of the graph in two such that the number of edges with vertices in distinct partitions is maximized [6]. For the QAOA algorithm, we assigned every vertex in the graph a qubit in the circuit. The cost operator is implemented by applying parametrized ZZ-gates between two qubits if and only if they are neighbors in the original graph. The mixing operator is implemented using parametrized R_X-gates. We implemented the parametrized ZZ-gates using R_Z-gates conjugated by CNOT-gates (where the R_Z-gate is applied to the target qubits of the CNOT-gates).

Figure 3 shows the three layered graphs used in our experiments. Due to the layered structure of the graph, the cost circuit shows a cascading architecture. This means that few gates have to be cut in our experiment. In addition, determining the optimum for comparison is simple for these graphs, allowing easy verification of the obtained results.

The parameters of the QAOA algorithm are optimized classically using Simultaneous Perturbation Stochastic Approximation (SPSA) [20], known for its potentially faster convergence with fewer quantum circuit executions [21].

3.3 Code Framework

We used a dual-platform approach for our simulation workflow. PennyLane [2] performs the ideal simulations, while Qiskit's Aer simulator [11] facilitates both ideal and noisy simulations. The workflow first constructs circuits in Pennylane and then converts them to the OPENQASM 2.0 format using built-in tools [4]. As the required mid-circuit measurements are not supported in this format, we developed a custom tool to upgrade the code to OPENQASM 3.0 [4]. The framework then sends this OPENQASM 3.0 code to construct a Qiskit quantum circuit object.

All experiments use a fixed shot budget evenly distributed over the subcircuits. We measure the difference between two probability distributions $P = (p_i)_{i=1}^n$ and $Q = (q_i)_{i=1}^n$ with the Hellinger distance [7]:

$$H(P,Q) = \frac{1}{\sqrt{2}} \sqrt{\sum_{i=1}^{n} (\sqrt{p_i} - \sqrt{q_i})^2}. \tag{3}$$

Results are averaged of 120 and 60 independent runs for noiseless uncut and cut experiments, respectively. Due to significant simulation cost in noisy versions, we averaged over 20 and 10 samples for the noisy experiments, respectively.

For the cut circuits, we assume that we have access to a quantum device of five qubits at most, requiring one, two, and three cuts for the graphs A, B and C of Fig. 3. Our simulations use a noise model derived from the IBM Quantum Brisbane device [10].

Table 1 shows the number of qubits used for the different graphs in both the uncut and cut quantum circuits.

Table 1. Qubit requirements and the number of wirecuts for Graphs A, B, and C, demonstrating the reduction in qubits achieved through wirecutting.

Graph Name	Wirecut Status	Qubits	Wirecut(s)
Graph A	Uncut	7	0
	Wirecut	5	1
Graph B	Uncut	10	0
	Wirecut	5	2
Graph C	Uncut	12	0
	Wirecut	5	3

4 Results

This section presents the results of two distinct experiments: In Sect. 4.1, we investigate the efficiency and accuracy per shot of various wire-cutting tech-

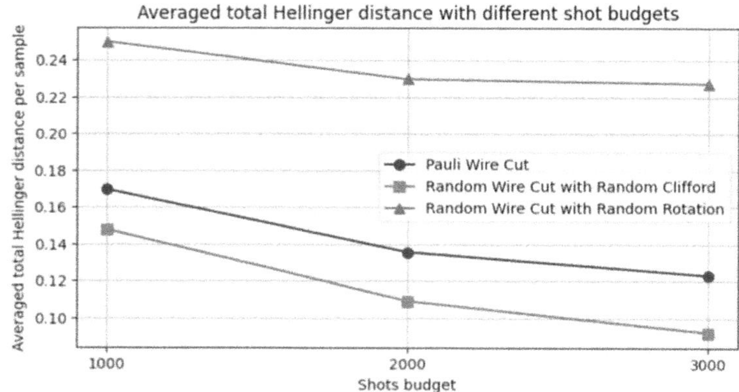

Fig. 4. Total Hellinger distance averaged over 60 independent runs for three wire cutting techniques (Pauli Wire Cut, Random Wire Cut with Random Clifford, and Random Wire Cut with Random Rotation) across varying shot budgets.

niques. Subsequently, in Sect. 4.2, we apply the randomized measurement wire-cutting technique [14] to QAOA for solving the MaxCut problem, evaluating its performance under both ideal and noisy simulation conditions.

4.1 Wire Cutting Techniques

We tested three circuit cutting methods: an exact method based on Pauli-measurements, a randomized method that randomly chooses between a random Clifford channel and a depolarizing channel, and a method that uses random Pauli rotation gates. Figure 4 shows the results for three different shot budgets. The randomized circuit cutting consistently outperforms the exact circuit cutting. Interestingly, applying random rotation gates before measurements and as state preparation (the third method) gives detrimental results.

All three methods show a decreased Hellinger distance with an increased shot budget. As more shots reduce the statistical variance, this is as expected. Lower Hellinger distance implies a better performance. The randomized approach using Clifford gates achieves the best performance with the same shot budget, implying a lower sampling overhead. From the same figure, we also see that, contrary to our hopes, the random Pauli rotations have a worse performance and higher sampling overhead even than standard Pauli cutting. We thus conclude that the randomized approach using Clifford gates works best and therefore use that method in our second experiment.

4.2 Wire Cutting QAOA for MaxCut

This section uses randomized circuit cutting using Clifford gates to test how well a QAOA algorithm can solve MaxCut in both a cut and uncut version and in ideal and noisy settings. We varied the shot budget from 2000 to 4000.

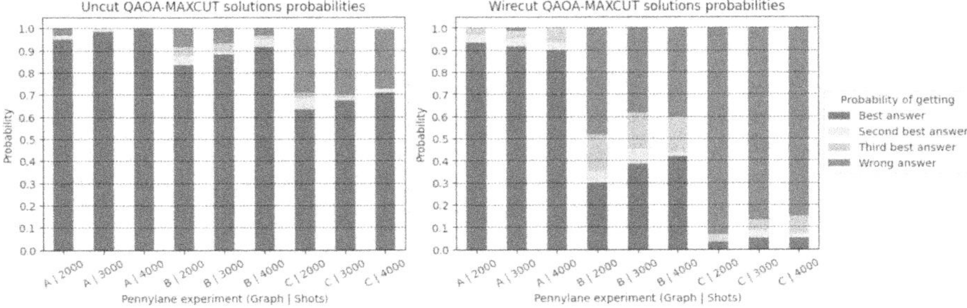

(a) Results for the QAOA-MaxCut experiment with ideal simulation.

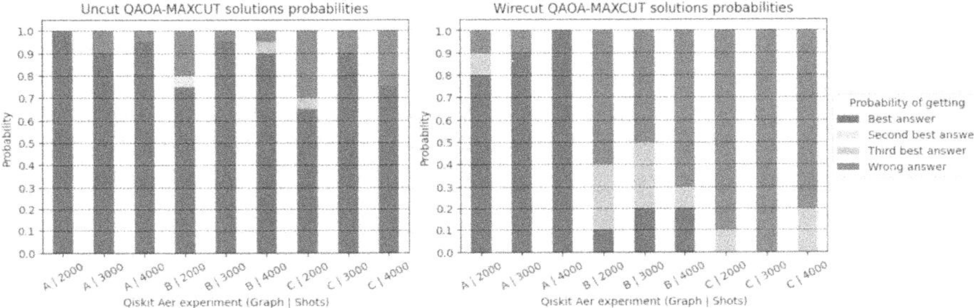

(b) Results for the QAOA-MaxCut experiment with noisy simulation. For the noise model of the IBM Brisbane quantum device was used.

Fig. 5. Numerical results on the performance of circuit cutting for QAOA. The left figures show results using uncut circuits, and the right figures show results using circuit cutting. Each bar shows the fraction of best, second best, third best, and wrong answer measured three graph instances and different shot budgets.

Ideal Simulations. Figure 5a shows how often the optimal cut is found in both an uncut and a cut run of the QAOA algorithm in an ideal simulation. We considered the three graphs shown in Fig. 3.

Again, as shot budget increases, the optimal answer is found more often. As the complexity of the graphs increases, the success probability does however decrease. For the wire cut circuits, we see that especially for the larger graphs, suboptimal answers are found most of the time. This is quickly explained by the number of wire cuts applied (two for graph B and three for graph C), while fixing the shot budget. As a result, statistical variances have a greater effect on the overall performance. For improved performance, the shot budget should be significantly increased for the larger graphs: Every subcircuit requires sufficient samples to obtain the correct outcomes with high probability.

Noisy Simulations. This section details the results of our noisy simulations for QAOA-MaxCut, comparing the performance of uncut and wirecut quantum circuits for the three graphs shown in Fig. 3.

Figure 5b shows the results of the experiments with noisy simulations. We see that noisy simulations for the wirecut circuits are significantly worse than the uncut version. Contrary to what we expect, increasing the shot budget did not always result in higher accuracy. We believe this unexpected behavior stems from the fixed shot budget. With an increasing number of circuit cuts, the number of simulated subcircuits also increases. Each of these subcircuits gets assigned a smaller shot budget, making them more susceptible for statistical inaccuracies. The fact that for Graph C, the correct outcome was never found supports this feeling.

5   Conclusion

In this work we considered circuit cutting, a technique to break down a quantum circuit in smaller blocks. In all of our experiments, we fixed the shot budget, regardless of whether a circuit cut was used or not. In theory, an exponential number of shots is needed in the number of cuts performed. As a result, we expect statistical variance in our results, especially when multiple shots are being used.

We considered three different circuit cutting techniques and found that circuit cutting based on randomized Clifford gates achieves the best performance. We tested this using a circuit to prepare a GHZ state, where additionally random Pauli rotation gates are applied in between the CNOT gates.

In the second experiment we considered the effect of QAOA for solving a MaxCut problem. Here, we considered three different graphs, which required a different number of circuit cuts. Especially for the larger graphs, those that require two or even three circuit cuts, the performance lacks significantly compared to the uncut version. This effect is magnified in the noisy simulations. As mentioned before, we expect that this results in part from the need of an exponential number of shots, whereas a fixed number of shots was used. Overall, it can be concluded that applying wire cutting to QAOA reduces its accuracy significantly, though it enables execution on quantum hardware with fewer qubits. Furthermore, this loss in accuracy can be mitigated by increasing the shot budget for larger graphs with multiple wire cuts in ideal simulations but remains unknown in noisy circumstances.

One can imagine that the impact of individual terms in Equation (1) differs. Some terms have a large impact, while others have a smaller impact. In practice, it might therefore be advantageous to use a skewed shot budget, with more shots being allocated to large impact terms. This is an interesting topic for follow-up research.

For future research, we believe that larger sample sizes can help reduce statistical variances, this includes both larger shot budgets and averaging results over multiple independent runs. Especially the noisy simulations were computationally heavy, particularly due to the need for mid-circuit measurements. Next, it is vital to test the methods on actual hardware to see the performance in practice. Future research can also incorporate error-mitigation techniques, such as Pauli-twirling [24] or zero-noise extrapolation [13,23], to reduce the effect of noise.

Acknowledgments. This work is part of the project *Divide and Quantum* 'D&Q' NWA.1389.20.241 of the program 'NWA-ORC', which is partly funded by the Dutch Research Council (NWO).

References

1. Bechtold, M., Barzen, J., Beisel, M., Leymann, F., Weder, B.: Patterns for quantum circuit cutting. In: Proceedings of the 30th Conference on Pattern Languages of Programs (PLoP '23). The Hillside Group (10 2023). In press
2. Bergholm, V., et al: Pennylane: automatic differentiation of hybrid quantum-classical computations (2022). https://arxiv.org/abs/1811.04968
3. Chen, S., Yu, W., Zeng, P., Flammia, S.T.: Robust shadow estimation. PRX Quantum **2**, 030348 (2021). https://doi.org/10.1103/PRXQuantum.2.030348, https://link.aps.org/doi/10.1103/PRXQuantum.2.030348
4. Cross, A.W., Bishop, L.S., Smolin, J.A., Gambetta, J.M.: Open quantum assembly language (2017). https://arxiv.org/abs/1707.03429
5. Farhi, E., Goldstone, J., Gutmann, S.: A quantum approximate optimization algorithm. arXiv preprint arXiv:1411.4028 (2014). https://arxiv.org/abs/1411.4028
6. Filar, J.A., Haythorpe, M., Taylor, R.: Linearly-growing reductions of Karp's 21 NP-complete problems. Numer. Algebra Control Optim. **8**(1), 1–16 (2018). https://doi.org/10.3934/naco.2018001
7. Gibbs, A.L., Su, F.E.: On choosing and bounding probability metrics. Int. Stat. Rev. **70**(3), 419–435 (2002)
8. Greenberger, D.M., Horne, M.A., Zeilinger, A.: Going beyond Bell's theorem. In: Bell's theorem, quantum theory and conceptions of the universe, pp. 69–72. Springer (1989)
9. Huang, H.Y., Kueng, R., Preskill, J.: Predicting many properties of a quantum system from very few measurements. Nat. Phys. **16**(10), 1050–1057 (2020). https://doi.org/10.1038/s41567-020-0932-7
10. IBM Quantum Computing: IBM Quantum (2025). https://quantum.ibm.com/
11. Javadi-Abhari, A., et al.: Quantum computing with Qiskit (2024). https://doi.org/10.48550/arXiv.2405.08810
12. Kandala, A.: Error mitigation extends the computational reach of a noisy quantum processor. Nature **567**, 491–495 (2019). https://doi.org/10.1038/s41586-019-1040-7
13. Li, Y., Benjamin, S.C.: Efficient variational quantum simulator incorporating active error minimization. Phys. Rev. X **7**, 021050 (2017). https://doi.org/10.1103/PhysRevX.7.021050, https://link.aps.org/doi/10.1103/PhysRevX.7.021050
14. Lowe, A., et al.: Fast quantum circuit cutting with randomized measurements. Quantum **7**, 934 (2023). https://doi.org/10.22331/q-2023-03-02-934
15. McArdle, S., Endo, S., Aspuru-Guzik, A., Benjamin, S.C., Yuan, X.: Quantum computational chemistry. Rev. Mod. Phys. **92**, 015003 (2020). https://doi.org/10.1103/RevModPhys.92.015003
16. Mitarai, K., Fujii, K.: Constructing a virtual two-qubit gate by sampling single-qubit operations. New J. Phys. **23**(2), 023021 (2021). https://doi.org/10.1088/1367-2630/abd7bc
17. Peng, T., Harrow, A.W., Ozols, M., Wu, X.: Simulating large quantum circuits on a small quantum computer. Phys. Rev. Lett. **125**, 150504 (2020). https://doi.org/10.1103/PhysRevLett.125.150504

18. Preskill, J.: Quantum computing in the NISQ era and beyond. Quantum **2**, 79 (2018)
19. Shor, P.W.: Scheme for reducing decoherence in quantum computer memory. Phys. Rev. A **52**, R2493–R2496 (1995). https://doi.org/10.1103/PhysRevA.52.R2493
20. Spall, J.: Multivariate stochastic approximation using a simultaneous perturbation gradient approximation. IEEE Trans. Autom. Control **37**(3), 332–341 (1992). https://doi.org/10.1109/9.119632
21. Szava, A., Wierichs, D.: Optimization using SPSA. https://pennylane.ai/qml/demos/tutorial_spsa (2023), Accessed 04 June 2025
22. Tang, W., Tomesh, T., Suchara, M., Larson, J., Martonosi, M.: CutQC: using small quantum computers for large quantum circuit evaluations. In: Proceedings of the 26th ACM International Conference on Architectural Support for Programming Languages and Operating Systems, pp. 473–486. ASPLOS '21, Association for Computing Machinery, New York, NY, USA (2021). https://doi.org/10.1145/3445814.3446758
23. Temme, K., Bravyi, S., Gambetta, J.M.: Error mitigation for short-depth quantum circuits. Phys. Rev. Lett. **119**, 180509 (2017). https://doi.org/10.1103/PhysRevLett.119.180509
24. Wallman, J.J., Emerson, J.: Noise tailoring for scalable quantum computation via randomized compiling. Phys. Rev. A **94**, 052325 (2016). https://doi.org/10.1103/PhysRevA.94.052325, https://link.aps.org/doi/10.1103/PhysRevA.94.052325
25. Zhou, L., Wang, S.T., Choi, S., Pichler, H., Lukin, M.D.: Quantum approximate optimization algorithm: performance, mechanism, and implementation on near-term devices. Phys. Rev. X **10**, 021067 (2020). https://doi.org/10.1103/PhysRevX.10.021067

Session 3 Quantum Algorithms, Computing and Simulation – Error Mitigation and Fidelity

Quantum Minimal Learning Machine: A Fidelity-Based Approach to Error Mitigation

Clemens Lindner[1(✉)] , Joonas Hämäläinen[1] , and Matti Raasakka[2]

[1] Faculty of Information Technology, University of Jyväskylä,
40100 Jyväskylä, Finland
`clindner@jyu.fi`
[2] Department of Electronics and Nanoengineering, Aalto University,
02150 Espoo, Finland

Abstract. We introduce the concept of quantum minimal learning machine (QMLM), a supervised similarity-based learning algorithm. The algorithm is conceptually based on a classical machine learning model and adopted to work with quantum data. We will motivate the theory and run the model as an error mitigation method for various parameters.

Keywords: Quantum machine learning · Quantum computing · Machine learning · Error mitigation · Quantum information

1 Introduction

Quantum machine learning [2] spans a wide range of possibilities: Leveraging fault tolerant quantum algorithms to accelerate core linear algebra routines [6], adopting quantum counterparts of classical kernel methods [11], and employing parametrized variational quantum algorithms (VQAs) optimized via a hybrid quantum-classical workflow [7]. When run on actual quantum computing hardware these models suffer from hardware noise, therefore methods of error mitigation are becoming another important avenue for the mingling of machine learning with quantum computing [9]. In this paper we will introduce a novel similarity based quantum learning model and explore its use for error mitigation. We will take inspiration from a classical model known as minimal learning machine (MLM) [4] where a linear mapping between distances in the input and output spaces is used to learn the labels of some given data-set. Several quantum k-nearest neighbor approaches have been implemented [1,13], yet most deal with the calculation of distances between classical data instead of quantum data [8]. Since one of the hopes of quantum computers is that they will be able to simulate quantum systems and then provide us with synthetic data we aim to find a way of directly learning from that data. In Sect. 2 we are first going to introduce the MLM model and then establish one possible way to use the same

F. Barbaresco and G. François (Eds.): QUEST-IS 2025, CCIS 2743, pp. 301–309, 2026.
https://doi.org/10.1007/978-3-032-13852-1_30

idea for quantum data. After we reformulated the model to work with quantum data we will use it in Sect. 3 to demonstrate error mitigation of a given noisy quantum state. The main aim in this specific use-case will be to assume we are given a noisy quantum state from a real quantum computer where we don't know the precise noise model and then learn its ideal noise-free version. In our implementation we will simulate this process with depolarizing noise and a dataset of variational quantum circuits using Qiskit and QiskitAer. We will analyze the models performance with different circuit parameter settings (number of qubits, rotational parameter ranges, Ansatz layers, noise levels) and provide the corresponding code.

2 Conceptual Framework and Preliminaries

2.1 MLM

Minimal learning machine (MLM) is a supervised learning method useful for learning the relationship between a set of input and output points. An exemplary use-case is multi label classification (MLC) [5] where the output points correspond to labels of the input points. In the following we will work with that example.

We are given a dataset $\mathcal{D} = \{(\mathbf{x}_1, \mathbf{y}_1), \ldots, (\mathbf{x}_N, \mathbf{y}_N)\}$, where $\mathbf{x}_i \in \mathbb{R}^M$ corresponds to an input vector and $\mathbf{y}_i \in \{0,1\}^L$ to the multi-hot encoded output label vector for $i = 1, \ldots, N$. In $\mathbf{y}_i$, value 1 at index j represents that class-label j is associated with instance i. As an example, $\mathbf{y}_1 = (1,1,0,0)$ could correspond to the vectorized picture data $\mathbf{x}_1$ including class-label 1 and 2 but not 3 and 4. For a MLC dataset, $\sum_{j=1}^{L} \mathbf{y}_i(j) > 1$ at least for one instance.

The approach goes as follows: We calculate the distances between the input space points and some set of reference points, this is done accordingly in the output space. We then store them in distance matrices $\mathbf{D_x}$ and $\mathbf{D_y}$ which are correlated via a parametrized matrix $\mathbf{B}$, i.e. $\mathbf{D_x B} = \mathbf{D_y}$. After having learned $\mathbf{B}$ with a set of training points we can use it to label new data points.

In MLM, the core subroutine is to solve the model's coefficient matrix $\mathbf{B}$ via ordinary least squares $\mathbf{B} = (\mathbf{D_x}^T \mathbf{D_x})^{-1} \mathbf{D_x}^T \mathbf{D_y}$ for feature space and output space distance matrices, $\mathbf{D_x} \in \mathbb{R}^{N \times K}$ and $\mathbf{D_y} \in \mathbb{R}^{N \times N}$, where $\mathbf{D}_{\mathbf{x}(i,j)} = \|\mathbf{x}_i - \mathbf{r}_j\|_2$, $\mathbf{r}_j \in \mathcal{R} = \{\mathbf{r}_k\}_{k=1}^{K} \subseteq \mathcal{X}$ and $\mathbf{D}_{\mathbf{y}(i,j)} = \|\mathbf{y}_i - \mathbf{y}_j\|_2$. The quality of the linear mapping $\mathbf{B}$ is affected by how feature space reference points $\mathcal{R}$ are selected. If $\mathbf{D_x}$ is a full rank matrix for $K = N$, we can use $\mathbf{B} = \mathbf{D_x}^{-1} \mathbf{D_y}$. In the prediction phase of MLM, for a given instance $\mathbf{x}$ distances are computed to $\mathcal{R}$ and then $\mathbf{B}$ is used to estimate output space distances to all training labels. An actual prediction for a given task is then extracted by post-processing of the distance estimates. The simplest approach for this is to identify the smallest predicted distance and its corresponding label vector $\hat{y} \in \mathcal{Y}$ [5]. Note that by selecting $\mathcal{R} = \mathcal{X}$ ($K = N$), MLM is deterministic.

2.2 QMLM

In a 'quantized' version of MLM we can use the same model to work with quantum data. For illustrative purposes we keep with the multi label approach and introduce the space $\mathcal{H}_\mathcal{X}$, a Hilbert space in which our input data lives. More specifically the input data we wish to prescribe multiple labels to is a set of pure quantum states $|\Psi\rangle_i \in \mathcal{H}_\mathcal{X}$ with $i \in \{1, 2, \ldots, N\}$ or their corresponding density matrices $\rho_i = |\Psi_i\rangle\langle\Psi_i|$. As soon as we deal with quantum data we have to use a different distance measure to quantify data similarity. An intuitive way is to introduce the fidelity between two quantum input states as our distance measure, i.e.

$$F(|\Psi\rangle_i, |\Psi\rangle_j) = |\langle\Psi_i|\Psi_j\rangle|^2, \quad \text{where } |\Psi\rangle_i, |\Psi\rangle_j \in \mathcal{H}_\mathcal{X}. \tag{1}$$

We have $F(|\Psi\rangle_i, |\Psi\rangle_j) = F(|\Psi\rangle_j, |\Psi\rangle_i)$ and $0 \leq F(|\Psi\rangle_i, |\Psi\rangle_j) \leq 1$ where a value closer to one means the states are more similar. We choose $\mathcal{R} = \mathcal{X}$ and therefore define the matrix $\mathcal{D}_{\mathcal{H}_\mathcal{X}}$ to hold the fidelity values between all pairs of quantum states in the input space:

$$D_{\mathcal{H}_\mathcal{X}} = \begin{bmatrix} F_{11} & F_{12} & \cdots & F_{1N} \\ F_{21} & F_{22} & \cdots & F_{2N} \\ \vdots & \vdots & \ddots & \vdots \\ F_{N1} & F_{N2} & \cdots & F_{NN} \end{bmatrix}, \tag{2}$$

where $F_{ij} = F(|\Psi\rangle_i, |\Psi\rangle_j)$ and $i, j \in \{1, 2, \ldots, N\}$. In practice, these fidelity values can be determined by implementing a swap test.

For the multi label case we could naively assume an encoding of the labels through the computational basis states, i.e. $|\Phi_i\rangle = |01101\rangle \in \mathcal{H}_\mathcal{Y}$. This approach however would lead to fidelity values of either 0 or 1 and therefore our distance matrix in the label space wouldn't hold much information. We also have to find a way where a large Hamming distance between two label vectors, i.e. the number of bit positions where they differ $d_H(\mathbf{y}_i, \mathbf{y}_j)$, corresponds to a low fidelity between the two corresponding quantum states. This can be achieved by using two overlapping states in our representation. We encode the label vector $\mathbf{y}_i$ in the quantum state $|\Phi_i\rangle$ where $i \in \{1, 2, \ldots, N\}$. If we denote the k-th value in a bit string of length L corresponding to the label $\mathbf{y}_i$ as b_{ik} and it's corresponding quantum state as $|\phi_{ik}\rangle$ we can write

$$|\Phi_i\rangle = \bigotimes_{k=1}^{L} |\phi_{ik}\rangle = \bigotimes_{k=1}^{L} \begin{cases} |+\rangle, & \text{if } b_{ik} = 1 \\ |0\rangle, & \text{if } b_{ik} = 0, \end{cases} \tag{3}$$

where $|+\rangle = \frac{1}{\sqrt{2}}(|0\rangle + |1\rangle)$. The bit string 01101 would hence be represented as $|\Phi_i\rangle = |0 + +0+\rangle$. Since $\langle 0|0\rangle = \langle +|+\rangle = 1$ and $\langle 0|+\rangle = \langle +|0\rangle = \frac{1}{\sqrt{2}}$ we have $\langle\Phi_i|\Phi_j\rangle = (\frac{1}{\sqrt{2}})^{d_H(\mathbf{y}_i, \mathbf{y}_j)}$ and therefore

$$F(|\Phi_i\rangle, |\Phi_j\rangle) = |\langle\Phi_i|\Phi_j\rangle|^2 = \left(\frac{1}{2}\right)^{d_H(\mathbf{y}_i, \mathbf{y}_j)}. \tag{4}$$

This way a large Hamming distance leads to a low fidelity value, as intended. We denote the distance matrix in the output space as $D_{\mathcal{H}_y}$ with entries $F'_{ij} = F(|\Phi_i\rangle, |\Phi_j\rangle)$ where $i, j \in \{1, 2, \ldots, N\}$. The rest of the algorithm can follow analogously to the classical MLM. We assume a linear mapping between the distance spaces, i.e. $D_{\mathcal{H}_x} \mathcal{B} = D_{\mathcal{H}_y}$ and obtain $\mathcal{B}$ by taking the pseudoinverse of $D_{\mathcal{H}_x}$. We then calculate the fidelities for a new test data state $|\Psi_t\rangle$ to all other input states, apply the mapping $\mathcal{B}$ and gain a vector encoding all mapped similarities in the label space. Since this vector encodes fidelities we now have to take the maximum value which corresponds to the highest similarity. The index of that value will give us the label corresponding to $|\Psi_t\rangle$, i.e. $|\Phi_t\rangle$, which after a quick remapping ($|+\rangle \mapsto 1, |0\rangle \mapsto 0$) will yield the labels bit string.

While this specific QMLM example could be used for multi label classification of quantum states it is primarily useful if we want to have the labels stored on a quantum computer and could still be improved upon by e.g. choosing a different label encoding. In the next section we are going to discuss a more near term usecase for QMLM, that being the use for error mitigation of noisy quantum states. Looking further ahead to fault-tolerant devices, QMLM may also prove valuable in any context where a systematic relationship can be established between the similarity structure of quantum states and that of a corresponding output space.

3 QMLM for Error Mitigation

In this section the aim of our QMLM model will be to learn an ideal quantum state when provided with its noisy version. The assumption is that we are given a set of noisy quantum states which could have been obtained from running quantum circuits that suffer from some hardware-specific noise. Our dataset consists of noisy states as inputs, i.e. $\{\rho_{i,\text{noisy}}\}$ and the corresponding ideal pure states $\{|\Psi_i\rangle\}$ as outputs. The fidelity in the input space would therefore take the general form

$$F(\rho_i, \rho_j) = \left(\text{Tr}\sqrt{\sqrt{\rho_i}\rho_j\sqrt{\rho_i}} \right)^2, \tag{5}$$

while the fidelity in the output space reduces to the form we have in Eq. 1. The assumption is now that since there always exists a completely positive trace preserving map $\mathcal{E}$, s.t. $\mathcal{E}(|\Psi_{\text{ideal}}\rangle\langle\Psi_{\text{ideal}}|) = \rho_{\text{noisy}}$, we can learn a useful relationship between the similarity matrices in the input and output space that could provide us with information about the specific hardware noise we encounter. In our implementation of the model we focused on depolarizing noise, i.e.

$$\mathcal{E}_{\text{depol}}(\rho_i) = (1 - p)\rho_i + \frac{p}{d}I \tag{6}$$

where ρ_i is the initial quantum state, p is the depolarizing probability with $0 \leq p \leq 1$, d is the dimension of the Hilbert space (for an n-qubit system, $d = 2^n$) and I is the $d \times d$ identity matrix. Let ρ'_1 and ρ'_2 be the density matrices obtained after applying the depolarizing noise channel $\mathcal{E}_{\text{depol}}$ to the initial pure states ρ_1 and ρ_2 with depolarizing parameters λ_1 and λ_2, respectively:

$$\rho_i' = (1 - \lambda_i)\,\rho_i + \frac{\lambda_i}{d}\,I, \quad i = 1, 2\,. \tag{7}$$

Reordering for ρ_1 and ρ_2 and taking the fidelity formula for pure states $F(\rho_1, \rho_2) = \mathrm{Tr}(\rho_1 \rho_2)$, we can relate the fidelity of our pure states with the trace over the resulting mixed ones:

$$\mathrm{Tr}(\rho_1' \rho_2') = \alpha F(\rho_1, \rho_2) + \frac{1 - \alpha}{d}, \tag{8}$$

where $\alpha = (1 - \lambda_1)(1 - \lambda_2)$. In theory, this would therefore motivate a mapping between fidelities for ideal states and the traces of their noisy counterparts which shows similarities to the approach found in [3], where a linear model is motivated for ideal and noisy expectation values. The benefit of our model (which corresponds to each ideal fidelity being a linear combination of noisy fidelities) is that it could model the underlying noise more generally. This could be useful since we are in reality never limited to just one specific noise model. In our setting, once the ideal version of a quantum state is obtained, the noise-free expectation value of an operator O can be evaluated via $\langle O \rangle_{\text{ideal}} = \mathrm{Tr}(O \rho_{\text{ideal}})$.

Although our QMLM model has the benefit of broad applications, it does suffer from the same problem as quantum kernel methods, that being the problem of exponential concentration [12]. This is obvious since the fidelity matrix in Eq. 2 represents a Gram matrix whose off diagonal values go to zero if the dimension of the Hilbert space grows too large. In our implementation, we try to negate this effect by limiting the region of the accessible Hilbert space through small parameter changes in a specified variational quantum circuit. For real applications this could correspond to the knowledge that a set of given quantum data is limited to a certain region of Hilbert space or exhibits some inherent symmetry akin to approaches in geometrical quantum machine learning [10].

3.1 Qiskit Implementation

We implement the depolarizing noise model with QiskitAer and use the same noise parameters for each training data set. Our output space dataset consists of N variational quantum circuits with an Ansatz corresponding to the following form:

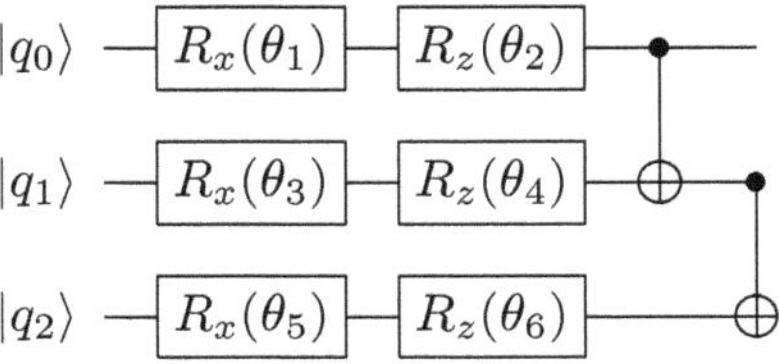

Because of the already mentioned vanishing fidelity problems, we will adjust the range of the randomly drawn θ from $\theta \in [-\pi, +\pi]$ to smaller intervals given by some δ, e.g. $\delta = \frac{\pi}{2}$ s.t. $\theta \in [-\delta, +\delta]$. The amount of repeated Ansatz layers can be specified by a parameter p, the depicted circuit here corresponds to $p = 1$. More specifically our output space dataset is then given by the statevectors we gain from such circuits using an ideal simulator. For each VQA circuit of the output space dataset we apply the depolarizing noise model which applies some error to each 1 and 2 qubit gate in the circuit depending on the chosen parameters p_1 and p_2. For each circuit we then gain the mixed density matrix as an end result of that noisy simulation and set the input space to be the collection of those.

After creating our dataset of ideal statevectors (the set $\mathcal{Y}$) and their noisy counterparts (the set $\mathcal{X}$) we calculate their corresponding fidelity matrices $D_{\mathcal{H}_\mathcal{Y}}$ and $D_{\mathcal{H}_\mathcal{X}}$. Through a standard pseudoinverse of $D_{\mathcal{H}_\mathcal{X}}$ we gain $\mathcal{B}$. Now a new data point is created by randomly drawing a new set of θ_i angles and applying the noise model to the corresponding circuit, meaning the new data point is some noisy state but by the nature of our process we have access to the ideal one. In reality such a new noisy state would be obtained from an actual quantum computation. After the QMLM procedure, we gain the quantum state of our set $\mathcal{Y}$ which exhibits the highest similarity to the ideal version of the new data point. This suggests that using more training samples improves performance, where performance is quantified by the fidelity between the predicted and the true ideal state. Because our data is generated randomly, we evaluate performance across many new test circuits and report it as the "average predicted fidelity".

In Fig. 1a we provide a result for that fidelity in comparison with the number of training dataset points (i.e. the number of VQA circuits) for $Q \in [1, 2, 3, 4, 5]$ qubits. For each visible point in all graphs we have averaged over 400 predictions. We limited the rotation range to $\theta \in [-\frac{\pi}{8}, +\frac{\pi}{8}]$ and have increased the noise to a level above known real hardware noise (approximately $p_1 = 0.001$ and $p_2 = 0.01$) to make its effect more prevalent. We can see that a higher dimensional Hilbert space leads to worse performance since we would need more points in order to approximately cover the whole space. Also while there is an initial increase in performance we can also see a leveling effect as soon as the reachable space is sufficiently covered by the training set data.

In Fig. 1b we compare different $\delta \in [\pi, \frac{\pi}{2}, \frac{\pi}{4}, \frac{\pi}{8}, \frac{\pi}{16}]$ values and focus on three qubits. Despite the averaging we can see some statistical fluctuations which should be smoothed if we increase the number of predictions we average over while increasing model runtime. We observe that when the accessible Hilbert-space region grows, performance degrades, reflecting that both space dimension and VQA circuit expressibility impact results. Since more Ansatz layers also correspond to higher expressibility, the same effect should be a cause of worse performance as seen in Fig. 1c where we compare different Ansatz layer parameters p. The worse performance for higher p values can also in general be traced to the fact that any gate errors now have more chances to accumulate.

At last we will also compare different noise levels in Fig. 1d. Regarding the noise levels, our model runs into two singular cases: one where the noise is zero and $\mathcal{B}$ therefore becomes the identity matrix and the other one where the noise is maximal, leading to all input states being the maximally mixed one. In the

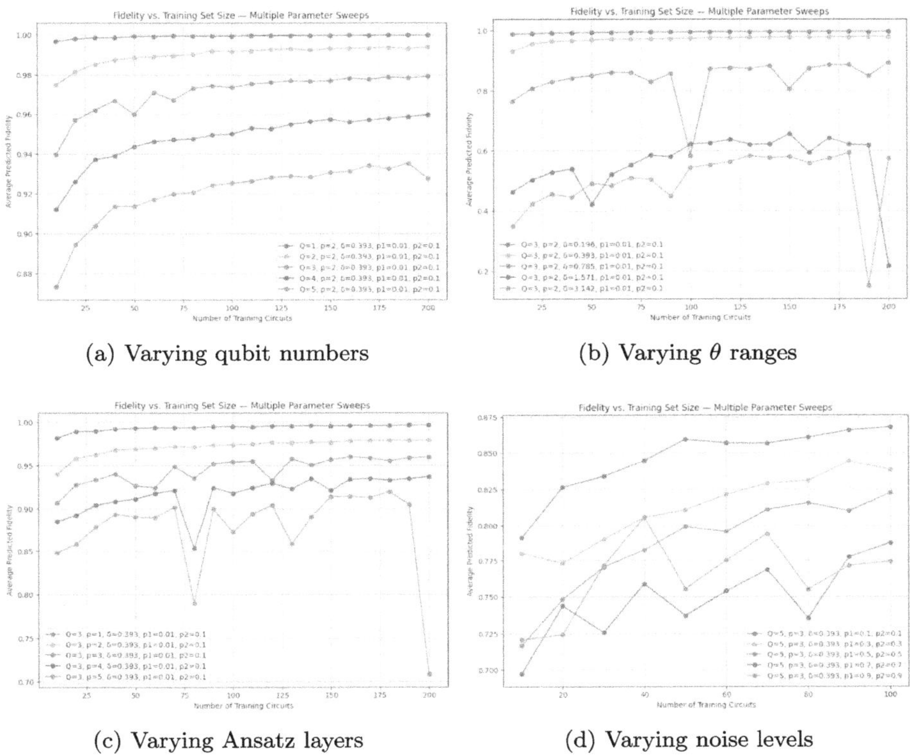

(a) Varying qubit numbers

(b) Varying θ ranges

(c) Varying Ansatz layers

(d) Varying noise levels

Fig. 1. Average predicted fidelity vs. dataset size under different experimental parameters.

first case the model basically acts as a lookup table, i.e. it checks if the new state is in the training data whereas the second case corresponds to $D_{\mathcal{H}_x}$ consisting of only 1's in its entries which leads to the output similarity vector consisting of the same value in each entry hence providing us some trivial average performance depending on $D_{\mathcal{H}_y}$. We plot the results for 5 qubits and note that the noise levels we chose are unrealistic, however increasing them still shows a general decrease in the models performance. This probably corresponds to the fact that the more noise we introduce the more similar the states become, leaving us with less information to learn from. We also observed in general larger fluctuations when increasing noise levels.

While we laid the groundwork for the QMLM model, further improvements can definitely still be done. Examples would be to change the post processing from a simple nearest neighbor approach to some weighted average over the closest points or to test if a different mapping between the distance matrices could lead to a better performance. Furthermore it would be interesting to run the model on actual quantum hardware and see if it sufficiently captures the more general noise models of real case scenarios. This however would lead to other problems, e.g. we can't just run a swap test on real hardware to measure the fidelity of noisy states since the swap test quantum gates themselves would be affected by the noise. We therefore either have to mitigate such effects or access some other methods like quantum state tomography.

4   Conclusion

In this paper, we presented an approach towards a 'quantized' MLM model. This was done by assuming the availability of a given set of quantum states and using the fidelity between them as a similarity measure. We specified use cases and ran the model for error mitigation. Our results show a good performance as long as we limit the accessible range of our created quantum states, more work is however still needed to test if the model is usable for higher dimensional data. QMLM could be a versatile approach to learning features of quantum states and a possible future avenue for learning immediately from the quantum data created on a fault-tolerant quantum computer.

5   Code Availability

Our code was run in Jupyter notebooks using Qiskit and QiskitAer for the quantum simulations. The full implementation is provided in the following GitHub: https://github.com/clemensLin/QMLM.

Acknowledgments. We acknowledge the financial support of the Finnish Ministry of Education and Culture through the Quantum Doctoral Education Pilot Program (QDOC VN/3137/2024-OKM-4) and the Research Council of Finland through the Finnish Quantum Flagship project (JYU 359240, University of Jyväskylä).

Disclosure of Interests. The authors have no competing interests to declare that are relevant to the content of this article.

References

1. Berti, A., Bernasconi, A., Del Corso, G.M., Guidotti, R.: The role of encodings and distance metrics for the quantum nearest neighbor. Quantum Mach. Intell. **6**(2), 62 (2024)
2. Biamonte, J., Wittek, P., Pancotti, N., Rebentrost, P., Wiebe, N., Lloyd, S.: Quantum machine learning. Nature **549**(7671), 195–202 (2017)
3. Czarnik, P., Arrasmith, A., Coles, P.J., Cincio, L.: Error mitigation with clifford quantum-circuit data. Quantum **5**, 592 (2021)
4. De Souza Junior, A.H., Corona, F., Barreto, G.A., Miche, Y., Lendasse, A.: Minimal learning machine: a novel supervised distance-based approach for regression and classification. Neurocomputing **164**, 34–44 (2015)
5. Hämäläinen, J., et al.: Minimal learning machine for multi-label learning (2024). arXiv:2305.05518

6. Harrow, A.W., Hassidim, A., Lloyd, S.: Quantum algorithm for linear systems of equations. Phys. Rev. Lett. **103**(15), 150502 (2009)
7. Jerbi, S.: Quantum machine learning beyond kernel methods. Nat. Commun. **14**(1), 1–8 (2023)
8. Kiani, B.T., De Palma, G., Marvian, M., Liu, Z.W., Lloyd, S.: Learning quantum data with the quantum earth mover's distance. Quantum Sci. Technol. **7**(4), 045002 (2022)
9. Liao, H., Wang, D.S., Sitdikov, I., Salcedo, C., Seif, A., Minev, Z.K.: Machine learning for practical quantum error mitigation. Nat. Mach. Intell. 1–9 (2024)
10. Ragone, M., et al.: Representation theory for geometric quantum machine learning. arXiv preprint arXiv:2210.07980 (2022)
11. Schuld, M.: Supervised quantum machine learning models are kernel methods (2021). https://arxiv.org/abs/2101.11020
12. Thanasilp, S., Wang, S., Cerezo, M., Holmes, Z.: Exponential concentration in quantum kernel methods. Nat. Commun. **15**(1), 5200 (2024)
13. Zardini, E., Blanzieri, E., Pastorello, D.: A quantum k-nearest neighbors algorithm based on the Euclidean distance estimation. Quantum Mach. Intell. **6**(1), 1–22 (2024)

No Scratch Quantum Computing by Reducing Qubit Overhead for Efficient Arithmetics

Omid Faizy[1,2,3,4(✉)], Norbert Wehn[3], Paul Lukowicz[4,5],
and Maximilian Kiefer-Emmanouilidis[4,5]

[1] Laboratoire de Chimie de la Matière Condensée de Paris, UMR CNRS 7574,
Sorbonne Université, 4, place Jussieu, 75252 Paris, France
[2] Laboratoire de Mathématiques Pures et Appliquées Joseph Liouville, Université du
Littoral Côte d'Opale, 50 rue Ferdinand Buisson, CS 80699, 62228 Calais, France
[3] Microelectronic Systems Design (EMS), RPTU Kaiserslautern-Landau,
Kaiserslautern, Germany
[4] Department of Computer Science and Research Initiative QC-AI, RPTU
Kaiserslautern-Landau, Kaiserslautern, Germany
`omid.faizy@rptu.de`
[5] German Research Center for Artificial Intelligence (DFKI), Kaiserslautern,
Germany

Abstract. Quantum arithmetic computation requires a substantial number of scratch qubits to stay reversible. These operations necessitate qubit and gate resources equivalent to those needed for the larger of the input or output registers due to state encoding. Quantum Hamiltonian Computing (QHC) introduces a novel approach by encoding input for logic operations within a single rotating quantum gate. This innovation reduces the required qubit register N to the size of the output states O, where $N = \log_2 O$. Leveraging QHC principles, we present reversible half-adder and full-adder circuits that compress the standard Toffoli + CNOT layout [Vedral et al., PRA, 54, 11, (1996)] from three-qubit and four-qubit formats for the Quantum half-adder circuit and five sequential Fredkin gates using five qubits [Moutinho et al., PRX Energy 2, 033002 (2023)] into a two-qubit, 4 × 4 Hilbert space. This scheme, presented here, is optimized for classical logic evaluated on quantum hardware, which due to unitary evolution can bypass classical CMOS energy limitations to a certain degree.

While the present circuits target classical Boolean inputs (i.e., classically controlled operation) and prepare the outputs from the initialized state $|00\rangle$, we make this scope explicit: the blocks here are not intended as coherent adders acting on superpositions. Instead, they are resource-minimal *classical-to-quantum* primitives that realize full truth tables on two qubits with no scratch registers. Because the input parameters are binary, the resulting unitaries reduce to a finite set of two-qubit permutations that can be implemented exactly without T gates or approximation overhead. We further position these blocks as building units for *FPGA-like quantum-configurable logic*, e.g., programmable interferometer meshes or analog control blocks, where minimizing qubits and circuit depth is advantageous.

Keywords: Quantum Arithmetics · Quantum Hamiltonian
Computing · Adder

1   Introduction

The question of minimal and most efficient arithmetic operators is approaching
the atomic limits. Below a few hundred atoms per transistor, leakage and energy
dissipation at least $kT \ln 2$ per bit become prohibitive [4,7,17]. Reversible quantum logic can bypass these bounds to a certain degree [4,7,32]. Following this
principle, early quantum adders [8,32] implemented binary addition using Toffoli and CNOT gates, but at the expense of extra ancilla qubits for reversibility
and state embedding. In this manuscript, we consider any qubit not needed to
encode the output as an ancillary qubit. This distinguishes our approach from
other ancilla-free methods which do not need any extra qubits when transpiled
into universal gates, such as the three-qubit half adder, the four-qubit full adder.
Here, the qubits necessary for input embedding are not treated as ancillas. To
emphasize this distinction, we refer to scratch qubits or scratch registers instead,
which identifies qubits needed for the embedding, as a form of ancilla.

For instance, standard reversible half-adder and full-adder circuits [21]
require three and four qubits, respectively, in order to produce sum and carry
outputs without erasing the inputs. This holds true even if the inputs are classical so that input and output are product states, and thus separable states. This
"scratch" qubit overhead not only increases the Hilbert space (dimension 8 for
three and dimension 16 for four qubits) but also adds numerous gates to uncom-
pute intermediate results, complicating practical implementation. More recent
designs reduce either the T-count or the ancilla count—sometimes to one or even
zero [13,28,29], where additional ancillas refer to qubits necessary beyond the
aforementioned input embedding for half-adders and four qubits for full-adders.

Quantum Hamiltonian Computing (QHC) encodes Boolean inputs in Hamiltonian matrix elements rather than basis populations, minimizing the state
count for logic gates [1,10,11,22]. Previous QHC studies implemented on single-
molecule platforms realized 4-state half-adders and 8-state full-adders that
required no qubits and relied on intrinsically irreversible dynamics; later work
introduced multi-energy read-out schemes to compress the effective Hamiltonian
even further [11,22].

Theoretical studies show that classical data can be encoded as rotation angles
in a QHC circuit by using a nano-graphene molecule doubly functionalized with
nitro groups and bridged by graphene electrodes, whose nitro-group rotations
serve as binary inputs that collectively implement a half-adder Boolean operation
[31].

Adapting these ideas to quantum computing is to find a unitary $U(\vec{\theta}) =$
$\exp(-iH(\vec{\theta})t)$ that depending on the binary inputs formulates the wanted logical output of any given truth table for the given input $\vec{\theta}$. From this point of
view we argue that we can formulate a quantum representation (in the form of

a quantum gate) which yields a given output (following the corresponding truth table) with only $N = \log_2(O)$ qubits where O is the number of possible outputs.

Using the QHC paradigm, we present constructed reversible half- and full-adder circuits using only two qubits. This represents a significant reduction in the required resources compared to conventional designs. Both our half-adder and full-adder operate within a 4×4 Hilbert-space (two qubits), rather than the 8×8 (three qubit), or 16×16 (four qubits) or larger spaces required by previous methods.

We have to emphasise that our design *does not* implement coherent addition on superpositions of inputs, nor is it meant to be a drop-in replacement for multi-bit quantum adders used inside phase-estimation or modular arithmetic. Rather, it is a resource-minimal *classical-to-quantum* primitive that can be useful in hybrid pipelines [3,19], IO [6,15], quantum routing [26], and programmable analog or photonic fabrics [27].

2   Circuit Design

Throughout, we denote the two input bits by $a, b \in \{0, 1\}$ and the input carry by $c_{\text{in}} \in \{0, 1\}$. The outputs are read as SUM on the second qubit as least significant bits (LSB) and CARRY on the first qubit as most significant bit(MSB). Furthermore, we adopt units where the reduced Planck constant $\hbar = 1$.

Starting with the half-adder as depicted in Fig. 1 (a), the device employs two qubits prepared in $|\psi(t = 0)\rangle = |00\rangle$. A 4×4 unitary $U(a, b)$ confines dynamics to a protected sub-manifold and encodes the angles $(a, b) \in \{0, 1\}^2$ as logical inputs. The $U(a, b)$ filters and maps resulting amplitudes onto the final qubits, which are then measured in the computational basis for SUM (XOR) and CARRY (AND) outputs.

$$U(a, b) = \begin{pmatrix} A & B - F & B + F & 0 \\ B + F & A & B - F & 0 \\ B - F & B + F & A & 0 \\ 0 & 0 & 0 & 1 \end{pmatrix}, \tag{1}$$

where

$$A = \tfrac{1}{3}\left(2\cos\left(\tfrac{2}{3}\pi(a+b)\right) + 1\right),$$

$$B = \tfrac{1}{3}\left(1 - \cos\left(\tfrac{2}{3}\pi(a+b)\right)\right), \tag{2}$$

$$F = \frac{\sin\left(\tfrac{2}{3}\pi(a+b)\right)}{\sqrt{3}}.$$

The corresponding half-adder Hamiltonian is

$$H_{\text{ha}}(a, b) = \frac{2\pi(a+b)}{3\sqrt{3}} h_{\text{ha}} = \frac{2\pi(a+b)}{3\sqrt{3}} \begin{pmatrix} 0 & -i & i & 0 \\ i & 0 & -i & 0 \\ -i & i & 0 & 0 \\ 0 & 0 & 0 & 0 \end{pmatrix}. \tag{3}$$

The time-evolution of this Hamiltonian $U(a,b) = \exp(-ih_{ha}t_{\mathrm{eff}}(a,b))$ shows that the solution of the half-adder can be extracted on the timings given by $t_{\mathrm{eff}}(a,b) = \frac{2\pi(a+b)}{3\sqrt{3}}$, see Fig. 2(b).

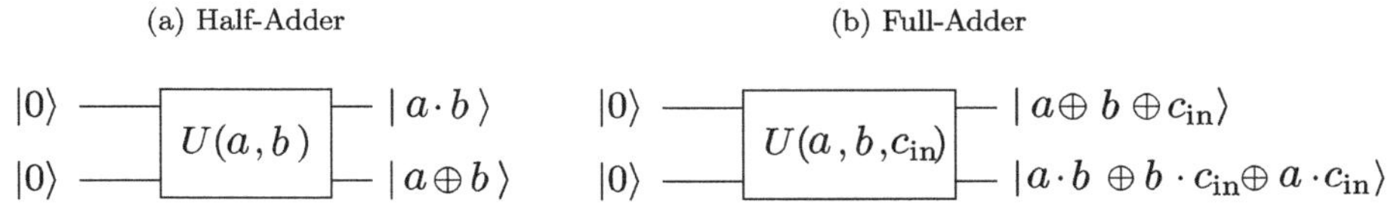

(a) Half-Adder

$$|0\rangle \quad \boxed{U(a,b)} \quad |a\cdot b\rangle$$
$$|0\rangle \quad\quad\quad\quad |a\oplus b\rangle$$

(b) Full-Adder

$$|0\rangle \quad \boxed{U(a,b,c_{\mathrm{in}})} \quad |a\oplus b\oplus c_{\mathrm{in}}\rangle$$
$$|0\rangle \quad\quad\quad\quad |a\cdot b\oplus b\cdot c_{\mathrm{in}}\oplus a\cdot c_{\mathrm{in}}\rangle$$

Fig. 1. (a) Two-Qubit QHC-Half-Adder circuit with classical binary inputs a and b implemented directly inside circuit. (b) Two-Qubit QHC-Full-Adder circuit with classical binary inputs a, b, and c_{in} implemented directly inside circuit.

Given the initial state $|00\rangle = \begin{pmatrix} 1 & 0 & 0 & 0 \end{pmatrix}^T$ we can fulfill the half-adder truth table and response as:

$$\begin{aligned}
U(0,0)|00\rangle &= |00\rangle \\
U(0,1)|00\rangle &= |01\rangle \\
U(1,0)|00\rangle &= |01\rangle \\
U(1,1)|00\rangle &= |10\rangle
\end{aligned} \tag{4}$$

We can generalize now the protocol with full-adder as depicted in Fig. 1 (b). Again the device employs two qubits prepared in $|\psi(t = 0)\rangle = |00\rangle$. A 4×4 unitary $U(a,b,c_{\mathrm{in}})$ confines dynamics to a protected sub-manifold and encodes the angles $(a,b,c_{\mathrm{in}}) \in \{0,1\}^3$ as logical inputs. The $U(a,b,c_{\mathrm{in}})$ filters and maps the resulting amplitudes to the final qubits, which are then measured in the computational basis for the SUM (XOR) and CARRY outputs.

The corresponding matrix has the following form:

$$U(a,b,c_{\mathrm{in}}) = \frac{1}{4}\begin{pmatrix}
l+2m+1 & -2n-p-iq+1 & l-2m+1 & 2n-p-iq+1 \\
2n-p-iq+1 & l+2m+1 & -2n-p-iq+1 & l-2m+1 \\
l-2m+1 & 2n-p-iq+1 & l+2m+1 & -2n-p-iq+1 \\
-2n-p-iq+1 & l-2m+1 & 2n-p-iq+1 & l+2m+1
\end{pmatrix}, \tag{5}$$

where

$$l = e^{i\pi(a+b+c_{\mathrm{in}})},$$
$$m = \cos\left(\tfrac{1}{2}\pi(a+b+c_{\mathrm{in}})\right),$$
$$n = \sin\left(\tfrac{1}{2}\pi(a+b+c_{\mathrm{in}})\right), \tag{6}$$
$$p = \cos(\pi(a+b+c_{\mathrm{in}})),$$
$$q = \sin(\pi(a+b+c_{\mathrm{in}})).$$

See Sect. A Appendix for the corresponding steps to arrive at Eq. (5).

The corresponding full-adder Hamiltonian reads

$$H_{fa} = \frac{1}{4}\pi(a + b + c_{\text{in}})h_{fa} = \frac{1}{4}\pi(a + b + c_{\text{in}})\begin{pmatrix} -1 & 1-i & -1 & 1+i \\ 1+i & -1 & 1-i & -1 \\ -1 & 1+i & -1 & 1-i \\ 1-i & -1 & 1+i & -1 \end{pmatrix}. \tag{7}$$

The time-evolution of this Hamiltonian $U(a,b) = \exp(-ih_{fa}t_{\text{eff}}(a,b,c_{\text{in}}))$ shows that the solution of the full-adder can be extracted on the timings given by $t_{\text{eff}}(a,b,c_{\text{in}}) = \frac{1}{4}\pi(a + b + c_{\text{in}})$, see Fig. 2.(b).

Applying the full-adder gate we receive the corresponding truth table

$$\begin{aligned} U(0,0,0)|00\rangle &= |00\rangle & U(0,0,1)|00\rangle &= |01\rangle \\ U(0,1,0)|00\rangle &= |01\rangle & U(0,1,1)|00\rangle &= |10\rangle \\ U(1,0,0)|00\rangle &= |01\rangle & U(1,0,1)|00\rangle &= |10\rangle \\ U(1,1,0)|00\rangle &= |10\rangle & U(1,1,1)|00\rangle &= |11\rangle. \end{aligned} \tag{8}$$

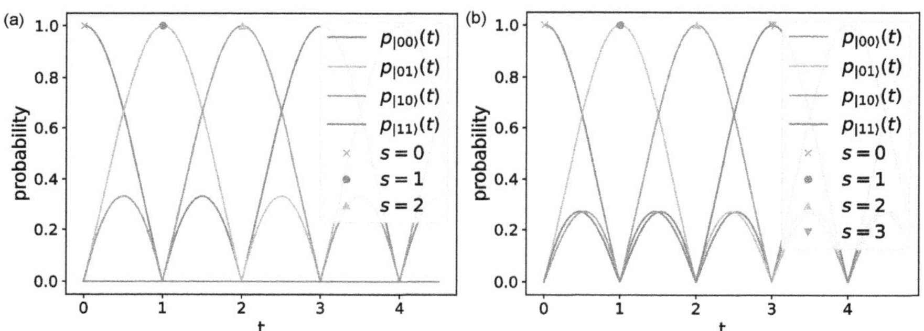

Fig. 2. Time simulation of probability density $p_{|\cdot\rangle} = |\langle\cdot|\exp(-iht_{\text{eff}})|00\rangle|^2$. (a) Simulation of half-adder Hamiltonian h_{ha} with $t_{\text{eff}} = \frac{2\pi t}{3\sqrt{3}}$. (b) Simulations of full-adder Hamiltonian h_{fa} and $t_{\text{eff}} = \frac{1}{4}\pi t$. Markers indicate the discrete steps provided by integer $s = a + b$ for the half-adder and $s = a + b + c_{\text{in}}$ for the full-adder.

Since the half-adder and full-adder unitaries correspond to reversible affine Boolean transformations on inputs (a,b) and (a,b,c_{in}), respectively, they can be implemented efficiently using Clifford operations. The optimal circuit design is $U = \left(\text{CNOT}(q_0 \to q_1);\ \text{X}(q_0)\right)^s$, which applies to both the half- and full-adder. It requires s CNOT and X gates, and must be reset after two iterations for the half-adder and after three iterations for the full-adder.

2.1 Hardware Realizations

Given the Hamiltonian formulations of h_{ha} and h_{fa}, the proposed gates can, in principle, be implemented on any hardware supporting universal quantum operations. Notably, h_{fa} has already been theoretically realized in a particle-conserving Rydberg system on a triangular grid [26], originally for quantum routing. This demonstrates how Rydberg platforms can directly realize complex-valued Hamiltonians across different lattice geometries [18,24,25].

Beyond Rydberg implementations, photonic and hybrid quantum–photonic platforms provide further motivation for Boolean logic. Besides aforementioned energy benefits of reversible logic [4,7,17] polarization-based encoding avoids wire delays and resistive losses, enabling massively parallel Boolean operations [6,15], while plasmonic optical memristive memory (OMM) allows local storage directly at the processing site, mitigating the von Neumann bottleneck [12]. Photonic hardware also offers substantial speed advantages: 10–100+ GHz electro-optic bandwidth and sub- ps all-optical switching, combined with wavelength and polarization multiplexing, yield high-throughput, low-latency control crucial for Measurement Based Quantum Computing (MBQC) and Quantum Error Correction (QEC) [27]. Since hybrid quantum algorithms demand heavy classical pre- and post-processing, implementing this functionality on the same photonic substrate eliminates costly conversion overhead and enables efficient quantum–classical co-processing [3,19]. Experimental results already support this direction: plasmonic cavities have demonstrated stable Boolean operations via hot-electron dynamics over 1.6×10^5 iterations, confirming the physical feasibility of these QHC principles [9,15].

3 Scalability, Use Cases, and Limitations

Compared to textbook ripple-carry designs, these blocks eliminate scratch qubits by moving input embedding into classical controls. In settings where the inputs are classical, these unitaries correspond to Clifford operations, see end of Sect. 2, and can be implemented exactly without T gates or approximation overhead. Our two-qubit blocks are *classically controlled* state-preparation primitives: given classical inputs $(a, b, c_{in}) \in \{0, 1\}$, they prepare $|\mathrm{CARRY}, \mathrm{SUM}\rangle$ from $|00\rangle$. While these circuits efficiently implement truth tables—which might suggest applications in Quantum Random Access Memory (QRAM) [14] or Grover oracles [23]— they are **not** coherent adders. For coherent, superposition-valued inputs promoting (a, b, c_{in}) to quantum controls reintroduces ancillas and control overhead, recovering the usual trade-offs of coherent adders with the benefit that input and output registers are nicely separated.

Since this modification reintroduces ancilla qubits and control depth, thereby recovering the standard resource requirements of coherent quantum adders [8,32] the here presented circuits are best suited for applications where classical control suffices: hybrid classical-to-quantum interfaces [19] and configurable photonic/analog control systems where minimizing logical qubit count is paramount [5].

For multi-bit *classical* addition, a two-qubit accumulator can be rippled by applying the appropriate instance of U per bit slice with classical feed-forward of the carry, preserving the low logical footprint while keeping latency linear in the word size. By design, Eqs. (1)–(8) prepare outputs from $|00\rangle$ under classical control;

Our design is thus complementary to coherent ripple-carry or QFT-based adders rather than a replacement. The same construction also applies to other 2-bit truth-tables such as comparators or multiplexers, and in programmable photonics to configurable 2×2 blocks within larger meshes, while extensions to multipliers or modular arithmetic within a coherent algorithm likely require additional qubits and fall outside the present scope. Finally, we want to mention, that promoting (a, b, c_{in}) to non-integer values allows to construct superpositions again, see Fig. 2(a-b), furthermore, it allows a quantum unique path to differentiable logic [20].

4   Conclusion

In this study, we presented Quantum Hamiltonian Computing as a method for performing quantum arithmetic operations, which reduces qubit overhead by encoding classical inputs within a single quantum gate. Our two-qubit half- and full-adder circuits in a 4×4 Hilbert space require fewer resources than traditional textbook reversible design built from a Toffoli plus clean-up gates [2,8,30,32]. Leveraging unitary evolution enables us to bypass classical CMOS energy limitations and achieve reversible logic with fewer qubits and gates, where the qubit count is only logarithmic in dependence on possible output states. Furthermore, the QHC unitary realizes the complete truth-table in one analogue pulse, as long as the gate can be implemented on the corresponding quantum hardware. While the current design targets classical logic without utilizing superposition of input and output states, it should be clear that this is possible in principle and can be as simple as converting the inputs from binary to real value inputs. Nevertheless, the found unitaries are not fully expressive but can only map a sub-manifold in the Hilbert space. Removing this limitation, would lead again to scratch qubits thus no benefit to previous implementations [32]. However, the sub-manifold can be optimized to evaluate to superposition states expected to become most prominent in the circuit without the need to introduce qubits.

We have shown that Quantum Hamiltonian Computing enables two-qubit implementations of the half- and full-adder truth tables (Eqs. (1)–(8)) with no scratch registers. Because the inputs are binary, the unitaries reduce to a finite set of two-qubit permutations that admit exact Clifford decompositions, eliminating T gates and approximation overhead. This makes the blocks attractive for hybrid classical-to-quantum interfaces and for FPGA-like quantum-configurable fabrics where the goal is to minimize logical qubits and depth.

This approach offers potential applications in quantum circuits, integrated quantum photonics, and other emerging technologies, especially as the operational speeds of integrated quantum photonic devices, for example, can exceed those of classical computers [16].

Acknowledgments. We would like to thank the Research Initiative 'Quantum Computing for Artificial Intelligence' (QC-AI) for their support.

A   Appendix: Full-Adder Unitary Matrix $U(a, c_{\text{in}}, b)$ Construction

Let the computational basis of two qubits be ordered as $\{|00\rangle, |01\rangle, |10\rangle, |11\rangle\}$. We could fix the column vector (the *control* or *eigenvector*) as:

$$V = |00\rangle, \tag{A.1}$$

and three real parameters $a, b, c_{\text{in}} \in \mathbb{R}$. The task is to build a 4×4 unitary matrix $U(a, b, c_{\text{in}})$ such that, when each parameter is restricted to the Boolean set $\{0, 1\}$, the eight input triples are mapped as

(a, c_{in}, b)	$U(a, c_{\text{in}}, b)\, V$
$(0, 0, 0)$	$|00\rangle$
$(0, 0, 1)$	$|01\rangle$
$(0, 1, 0)$	$|01\rangle$
$(0, 1, 1)$	$|10\rangle$
$(1, 0, 0)$	$|01\rangle$
$(1, 0, 1)$	$|10\rangle$
$(1, 1, 0)$	$|10\rangle$
$(1, 1, 1)$	$|11\rangle$

We define now a 4-cycle permutation matrix $\mathcal{R}$ as

$$\mathcal{R} = |01\rangle\langle 00| + |10\rangle\langle 01| + |11\rangle\langle 10| + |00\rangle\langle 11|. \tag{A.2}$$

Equation (A.2) permutes the basis cyclically $|00\rangle \to |01\rangle \to |10\rangle \to |11\rangle \to |00\rangle$; hence $\mathcal{R}^4 = I_4$. Written as a matrix,

$$\text{R} = \begin{pmatrix} 0 & 0 & 0 & 1 \\ 1 & 0 & 0 & 0 \\ 0 & 1 & 0 & 0 \\ 0 & 0 & 1 & 0 \end{pmatrix}, \qquad \mathcal{R}^\dagger \mathcal{R} = I_4.$$

Because $\mathcal{R}$ is unitary, the principal matrix logarithm

$$\mathcal{R} = e^{-iH}, \qquad H = i \log \mathcal{R}, \qquad H^\dagger = H, \tag{A.3}$$

is Hermitian. (The eigenvalues of $\mathcal{R}$ are the fourth roots of unity $\{1, i, -1, -i\}$.) the three–parameter family could be Defined as

$$U(a, b, c_{\text{in}}) = \exp\left[-i\,(a + b + c_{\text{in}})\,H\right] \tag{A.4}$$

Some algebra leads to Eq. (5).

References

1. Ample, F., Faizy, O., Kawai, H., Joachim, C.: The design of a surface atomic scale logic gate with molecular latch inputs. In: Kolmer, M., Joachim, C. (eds.) On-Surface Atomic Wires and Logic Gates, LNCS, pp. 139–155. Springer, Cham (2017). https://doi.org/10.1007/978-3-319-51847-3_9
2. Amy, M., Maslov, D., Mosca, M., Roetteler, M.: A meet-in-the-middle algorithm for fast synthesis of depth-optimal quantum circuits. IEEE Trans. Comput.-Aided Des. Integr. Circuits Syst. **32**(6), 818–830 (2013). https://doi.org/10.1109/TCAD.2013.2244643
3. Arrazola, J.M.E.A., Bergholm, V., Brádler, K., et al.: Quantum circuits with many photons on a programmable nanophotonic chip. Nature **591**(7848), 54–60 (2021). https://doi.org/10.1038/s41586-021-03202-1
4. Bennett, C.H.: Logical reversibility of computation. IBM J. Res. Dev. **17**(6), 525–532 (1973). https://doi.org/10.1147/rd.176.0525
5. Carolan, J., Harrold, C., Sparrow, C., et al.: Universal linear optics. Science **349**(6249), 711–716 (2015). https://doi.org/10.1126/science.aab3642
6. Caulfield, H.J., Dolev, S.: Why future supercomputing requires optics. Nat. Photonics **4**(5), 261–263 (2010). https://doi.org/10.1038/nphoton.2010.94
7. Chiribella, G., Yang, Y., Renner, R.: Fundamental energy requirement of reversible quantum operations. Phys. Rev. X **11**, 021014 (2021). https://doi.org/10.1103/PhysRevX.11.021014
8. Cuccaro, S.A., Draper, T.G., Kutin, S.A., Moulton, D.P.: A new quantum ripple-carry addition circuit. arXiv quant-ph/0410184 (2004). https://doi.org/10.48550/arXiv.quant-ph/0410184
9. Dell'Ova, F., Brûlé, Y., Gros, N., et al.: Compact implementation of a 1-bit adder by coherent 2-beam excitation of a single plasmonic cavity. ACS Photonics **11**(2), 752–761 (2024). https://doi.org/10.1021/acsphotonics.3c01624
10. Dridi, G., Julien, R., Hliwa, M., Joachim, C.: The mathematics of a quantum hamiltonian computing half adder boolean logic gate. Nanotechnology **26**(34), 344003 (2015). https://doi.org/10.1088/0957-4484/26/34/344003
11. Dridi, G., Namarvar, O.F., Joachim, C.: Qubits and quantum hamiltonian computing performances for operating a digital boolean 1/2-adder. Quantum Sci. Technol. **3**(2), 025005 (2018). https://doi.org/10.1088/2058-9565/aaa98b
12. Feldmann, J., Youngblood, N., Wright, C.D., Bhaskaran, H., Pernice, W.H.P.: All-optical spiking neurosynaptic networks with self-learning capabilities. Nature **569**(7755), 208–214 (2019). https://doi.org/10.1038/s41586-019-1157-8
13. Gidney, C.: Halving the cost of quantum addition. Quantum **2**, 74 (2018). https://doi.org/10.22331/q-2018-06-18-74
14. Giovannetti, V., Lloyd, S., Maccone, L.: Quantum random access memory. Phys. Rev. Lett. **100**, 160501 (2008). https://doi.org/10.1103/PhysRevLett.100.160501
15. Kumar, U., Cuche, A., Girard, C., et al.: Interconnect-free multibit arithmetic and logic unit in a single reconfigurable 3 μm^2 plasmonic cavity. ACS Nano **15**(8), 13351–13359 (2021). https://doi.org/10.1021/acsnano.1c03196
16. Labonté, L., Alibart, O., D'Auria, V., et al.: Integrated photonics for quantum communications and metrology. PRX Quantum **5**, 010101 (2024). https://doi.org/10.1103/PRXQuantum.5.010101
17. Landauer, R.: Irreversibility and heat generation in the computing process. IBM J. Res. Dev. **5**(3), 183–191 (1961). https://doi.org/10.1147/rd.53.0183

18. Lienhard, V., Scholl, P., Weber, S., et al.: Realization of a density-dependent peierls phase in a synthetic, spin-orbit coupled rydberg system. Phys. Rev. X **10**, 021031 (2020). https://doi.org/10.1103/PhysRevX.10.021031

19. McClean, J.R., Romero, J., Babbush, R., Aspuru-Guzik, A.: The theory of variational hybrid quantum-classical algorithms. New J. Phys. **18**(2), 023023 (2016). https://doi.org/10.1088/1367-2630/18/2/023023

20. Miotti, P., Niklasson, E., Randazzo, E., Mordvintsev, A.: Differentiable logic cellular automata: from game of life to pattern generation (2025). https://arxiv.org/abs/2506.04912

21. Moutinho, J.a.P., Pezzutto, M., Pratapsi, S.S., et al.: Quantum dynamics for energetic advantage in a charge-based classical full adder. PRX Energy **2**, 033002 (2023). https://doi.org/10.1103/PRXEnergy.2.033002

22. Namarvar, O.F., Giraud, O., Georgeot, B., Joachim, C.: Quantum hamiltonian computing protocols for molecular electronics boolean logic gates. Quantum Sci. Technol. **4**(3), 035009 (2019). https://doi.org/10.1088/2058-9565/ab2412

23. Nielsen, M.A., Chuang, I.L.: Quantum computation and quantum information (2010). https://doi.org/10.1017/CBO9780511976667

24. Ohler, S., Kiefer-Emmanouilidis, M., Browaeys, A., Büchler, H.P., Fleischhauer, M.: Self-generated quantum gauge fields in arrays of rydberg atoms. New J. Phys. **24**(2), 023017 (2022). https://doi.org/10.1088/1367-2630/ac4a15

25. Ohler, S., Kiefer-Emmanouilidis, M., Fleischhauer, M.: Quantum spin liquids of rydberg excitations in a honeycomb lattice induced by density-dependent peierls phases. Phys. Rev. Res. **5**, 013157 (2023). https://doi.org/10.1103/PhysRevResearch.5.013157

26. Palaiodimopoulos, N.E., Ohler, S., Fleischhauer, M., Petrosyan, D.: Chiral quantum router with rydberg atoms. Phys. Rev. A **109**, 032622 (2024). https://doi.org/10.1103/PhysRevA.109.032622

27. Pan, B., Hu, J., Huang, Y., et al.: Demonstration of high-speed thin-film lithium-niobate-on-insulator optical modulators at the $2 - \mu m$ wavelength. Opt. Express **29**(12), 17710–17717 (2021). https://doi.org/10.1364/OE.416908

28. Remaud, M.: Optimizing t and cnot gates in quantum ripple-carry adders and comparators. In: Proceedings of Recent Advances in Quantum Computing and Technology, pp. 56–61. ReAQCT '24, Association for Computing Machinery, New York, NY, USA (2024). https://doi.org/10.1145/3665870.3665875

29. Remaud, M., Vandaele, V.: Ancilla-free quantum adder with sublinear depth. In: Glück, R., Kaarsgaard, R. (eds.) Reversible Computation, LNCS, pp. 137–154. Springer, Cham (2025). https://doi.org/10.1007/978-3-031-97063-4_11

30. Selinger, P.: Quantum circuits of t-depth one. Phys. Rev. A **87**, 042302 (2013). https://doi.org/10.1103/PhysRevA.87.042302

31. Srivastava, S., Kino, H., Joachim, C.: Quantum half-adder boolean logic gate with a nano-graphene molecule and graphene nano-electrodes. Chem. Phys. Lett. **667**, 301–306 (2017). https://doi.org/10.1016/j.cplett.2016.11.009

32. Vedral, V., Barenco, A., Ekert, A.: Quantum networks for elementary arithmetic operations. Phys. Rev. A **54**, 147–153 (1996). https://doi.org/10.1103/PhysRevA.54.147

Optimizing Qubit Routing with Bridge Gates: Extending Quantum Circuit Efficiency Across Arbitrary Distances

Ward van der Schoot[1], Willem de Kok[1], and Frank Phillipson[1,2](✉)

[1] TNO, The Hague/Groningen, The Netherlands
[2] Maastricht University, Maastricht, The Netherlands
`f.phillipson@maastrichtuniversity.nl`

Abstract. Qubit routing is a critical challenge in quantum computing, essential for implementing quantum circuits on hardware with limited connectivity. This paper introduces a novel perspective on qubit routing by exploring the use of bridge gates, which enable the execution of controlled NOT (CNOT) operations over non-adjacent qubits without rerouting the qubits. The study highlights the advantages of bridge gates compared to SWAP gates, particularly their ability to preserve qubit assignments and potentially optimize subsequent routing steps. We propose an extension to the concept of bridge gates by generalizing them for arbitrary distances and provide constructions demonstrating their feasibility. By analysing their performance, we show that larger-distance bridge gates can significantly reduce the number of CNOT gates in certain circuits. Furthermore, a new qubit routing problem is defined, incorporating both SWAP and bridge gates, and we discuss how this impacts the complexity of the problem.

Keywords: Qubit Routing · Bridge Gates · Quantum Circuits · Quantum Gate Optimization

1 Introduction

Quantum computing is a revolutionary approach to computation that leverages the principles of quantum mechanics to process information in fundamentally new ways. Unlike classical computers, which rely on bits as their basic units of information, quantum computers use qubits, which can exist in multiple states simultaneously due to quantum superposition. This unique property, along with entanglement and interference, enables quantum computers to solve certain types of problems significantly faster than classical computers [1].

Its broad set of potential applications, including quantum chemistry and drug discovery [2,3], machine learning [4,5], optimization [6,7], and simulation [8,9], has fostered research efforts and corporate investment [10,11]. Though some experiments have already hinted at the possibility of quantum supremacy

F. Barbaresco and G. François (Eds.): QUEST-IS 2025, CCIS 2743, pp. 320–328, 2026.
https://doi.org/10.1007/978-3-032-13852-1_32

[12,13], substantial research efforts are needed before quantum computing can be employed in real-world settings.

Quantum algorithms are at the core of quantum computing, designed to harness these quantum properties to perform computations that would be infeasible for classical algorithms. Some well-known quantum algorithms include Shor's algorithm [14] for factoring large integers and Grover's algorithm [15] for database searching, both of which offer substantial speed-ups over their classical counterparts. To implement these algorithms on quantum hardware, they are typically expressed as quantum circuits, where qubits are manipulated through a series of quantum gates. Quantum circuits visually represent the sequence of operations that manipulate qubits to achieve the desired computational outcome, making them a crucial bridge between high-level quantum algorithms and the physical processes within a quantum computer.

When programming quantum algorithms, quantum circuits are often written in a generic way. To run a quantum circuit on specific hardware, the circuit must be translated to be compatible with the target hardware. This process is called compilation, and involves many different steps. There are different ways in which this process can be performed, but often it consists of at least the following steps, see for example [16,17]:

1. Decomposition of quantum gates: decompose the gates in the quantum circuits into one-qubit gates and a single type of two-qubit gate, usually the CNOT or CZ gate. In this work, we will assume the CNOT gate, but everything below also holds for the CZ gate.
2. Initial qubit mapping: map the qubits of the quantum circuits to the qubits on the hardware.
3. Qubit routing: explained in more detail below.
4. Native gate decomposition: decompose the current quantum gates into gates that can be natively implemented on the quantum hardware

After this process, hardware-specific instructions are produced and the required quantum circuit can be run on the target hardware. The process of compilation can be seen as an optimisation process, in which the optimal translation to hardware-specific instruction is sought. Often, this optimisation process is not easy, and it can actually be shown to be as hard as an NP-hard problem. The hardness of this optimisation process mainly results from the step of qubit routing.

Qubit routing entails the process of changing the qubit order between different quantum gates within a circuit. After the step of qubit mapping, the circuit consists of arbitrary one-qubit and a single type of two-qubit gate, often the CNOT gate. While one-qubit gates (after further decomposition in step 4 above) can directly be applied to the corresponding qubit, this is not the case for two-qubit gates. Two-qubit gates can only be applied to hardware qubits that are directly adjacent within the target hardware architecture. However, as two-qubit gates between arbitrary qubits could be present in the circuit description, this yields a problem in performing these gates on hardware.

This problem is currently tackled by changing the mapping between the circuit and hardware qubits in between two-qubit gates, so that required hardware qubits are adjacent when performing the gate. This is done by inserting SWAP gates between adjacent qubits. A SWAP gate swaps the mapping of two adjacent qubits and is built from 3 single CNOT gates, as depicted in Fig. 1. The process of remapping using SWAP gates is called qubit routing. If the quantum circuit is seen as a series of quantum gates, there is a step of qubit routing before every two-qubit gate within the circuit consisting of two non-adjacent qubits. SWAP gates are relatively expensive gates, in the sense that for current quantum hardware, the CNOT gate is the native gate with the highest fidelity error rate. For this reason, the main goal in qubit routing often is to minimise the number of SWAP gates. Formally, the following problem is considered: *Given a quantum circuit, a target qubit architecture and an initial mapping of the qubits, what is the minimal number of SWAP gates required to run the quantum circuit on the target hardware?* In literature, people worked on approaches using exact methods [18,19], heuristics approaches [20–22], machine learning approaches such as reinforcement learning [23–25] and even quantum annealing approaches [26] to solve this problem. While these can give good or even optimal results for certain configurations, they do not offer optimal solutions for general cases (at least not in reasonable time).

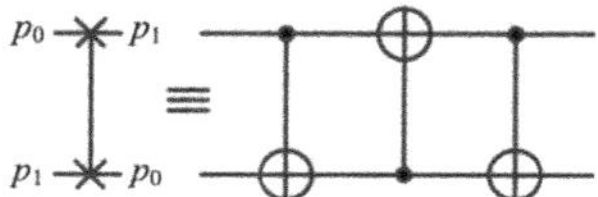

Fig. 1. The basic SWAP gate consisting of three CNOT gates.

However, the SWAP gate is not the only solution to enable the qubit routing. Also the so-called bridge gate can be used. In this paper, we investigate in what sense the qubit routing problem changes by using bridge gates and how it will influence the performance of the circuit. This work also proposes the notion of an arbitrary distance bridge gate, which has not been done before.

2 Bridge Gates

2.1 The Bridge Gate

A bridge gate is a gate that allows the operation of a CNOT gate between two qubits that have a common neighbouring qubit in the hardware lay-out. It is built out of four standard CNOT gates, as depicted in Fig. 2. These four CNOT gates together are exactly the operation of a CNOT from qubit 0 to qubit 2, without changing the qubit assignment.

Both bridge gates and SWAP gates can be used to perform a CNOT between two qubits with a common neighbouring qubit. Bridge gates do this directly,

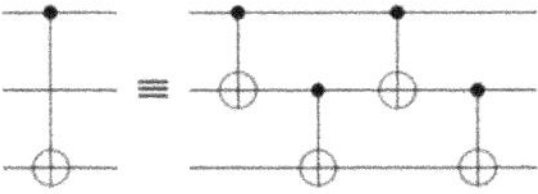

Fig. 2. A basic bridge gate consisting of four CNOT gates is equivalent to a distance 2 CNOT gate.

while SWAP gates can do this by swapping the middle qubit with one of the two outer qubits, and performing a CNOT between the two required qubits, which are now adjacent. Both methods require the same amount of four CNOT gates, and as the main cost of (currently) implementing quantum circuits comes from the number of CNOT gates to be implemented, the two gates are equally expensive. However, the difference lies in the qubit assignment after the operations: by using a SWAP gate, the qubit assignment is changed after the CNOT gate, while this is not the case when using a bridge gate. This means we have non-equivalent circuits.

As an example, consider a linear qubit assignment of three qubits numbered 0 to 2 from left to right, and a gate that needs to be executed between qubit 0 and 2. With a SWAP, this can be done by swapping 0 and 1 (costs three CNOT gates) and then performing a CNOT between 0 and 2 (total four CNOT gates), resulting in the qubit assignment 102. With a bridge gate, a CNOT can be directly executed between 0 and 2 (costs four CNOT gates), resulting in the qubit assignment 012. Using a bridge gate requires the same amount of CNOT gates, but the qubit assignment is unchanged after the full process, while this is not the case when using just SWAP gates.

As the qubit assignment influences the amount of SWAP (or bridge) gates in later qubit routing steps, both the SWAP and bridge gate options could be beneficial. In summary, the use of bridge gates in qubit routing could improve performance.

2.2 Bridge Gates over Larger Distances

In literature, such as [24], the only bridge gates that are being used, are used over distance two, i.e., with one neighbouring qubit in between the two qubits. However, in quantum algorithms, often gates need to be applied between qubits over larger distances. That is why this work introduces the notion of bridge gates over arbitrary distances. Both the construction of bridge gates over arbitrary distances, as well as their use in qubit routing, are introduced. Different constructions were considered, which are discussed below. It is important to note that this work does not claim this to be the most efficient way to perform bridge gates (in terms of number of CNOT gates). It presents the bridge gate as an alternative solution for re-routing that has advantages in some cases.

One way to construct bridge gates over larger distances, is to use the bridge gate over distance two as a building block. As an example, to construct a bridge gate over distance three, two distance-two CNOT gates and two simple CNOT gates are sufficient by following the original bridge gate building block, as shown in Fig. 3. Then, by replacing the distance-two CNOT gates by distance-two bridge gates, the distance-three bridge gate can be constructed. This can also be done for larger distance bridge gates, as seen in Fig. 4. In the end, this gives a roughly quadratic number of CNOT gates in terms of the distance.

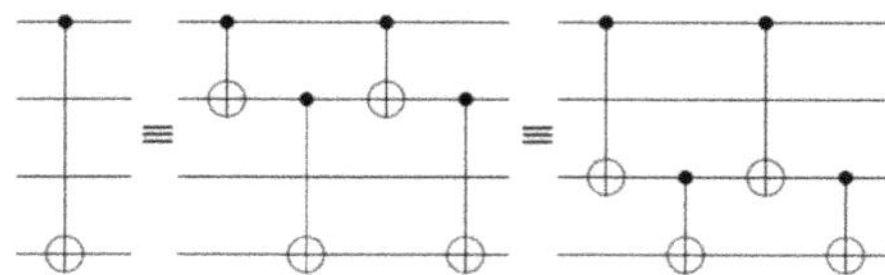

Fig. 3. Two ways of constructing a distance-three bridge gate by using a distance-two bridge gate architecture.

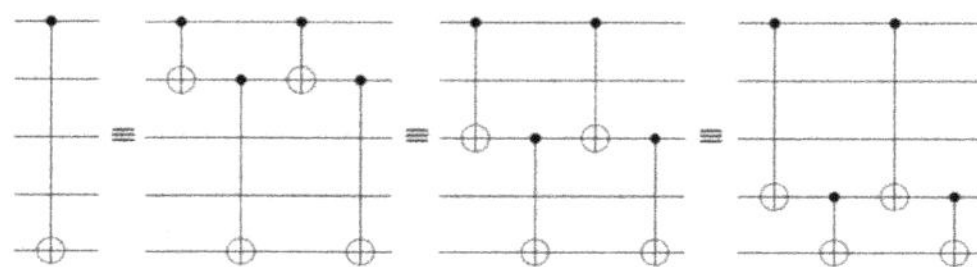

Fig. 4. Three ways of constructing a distance-four bridge gate by using a distance-two bridge gate architecture. The full bridge gate can be constructed by replacing the distance-two or -three CNOT gates by distance-two or -three bridge gates.

Next, another construction for bridge gates was constructed, as depicted in Fig. 5, specifically for a distance-five CNOT gates. This construction works for arbitrary distance CNOT gates. This yields a number of $2n + 2(n - 1) = 4n - 4$ CNOT gates for a distance n bridge gate.

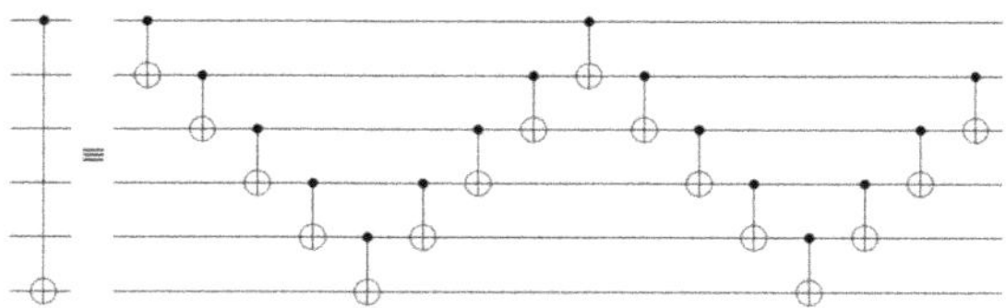

Fig. 5. A general construction for bridge gates. As an example, the construction is shown for distance-five, but the construction generalises to any distance CNOT gate.

Alternatively, a CNOT gate between two qubits at distance n can be applied by applying $n - 1$ SWAP gates and one CNOT gate, as shown in Fig. 6 This only requires $3(n - 1) + 1 = 3n - 2$ CNOT gates. While this is less, it could still be beneficial to apply the distance n bridge gate, as this qubit assignment could be beneficial for later qubit routing steps. Notice in particular that it is worse to perform SWAP gates, perform the CNOT gate, and then perform the same SWAP gates to reach the same qubit ordering as before the gate, which would yield the same effect as the distance n bridge gate.

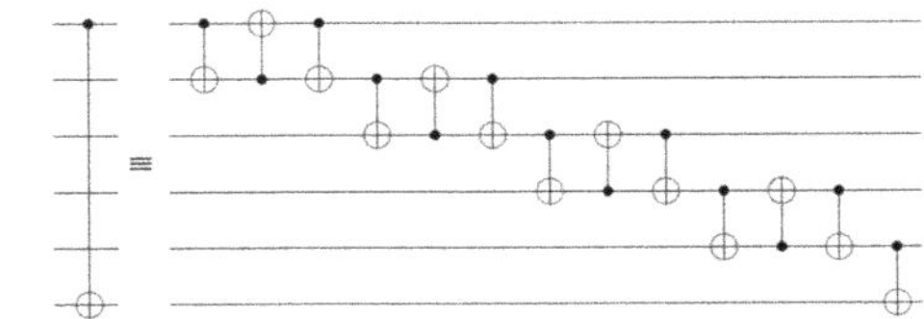

Fig. 6. A distance-five CNOT gate implemented by using only SWAP gates.

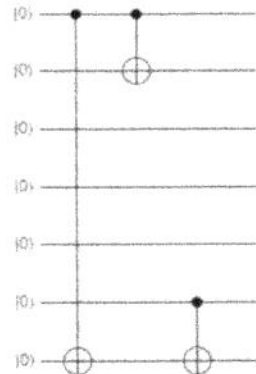

Fig. 7. A circuit showing the advantage of large distance bridge gates. Note that with large distance bridge gates, no SWAPs are required to perform this circuit.

To give an example as to why using a larger distance bridge gate could lead to less CNOT gates used, consider a linear qubit assignment of seven qubits numbered 1 to 7 from left to right. Using these qubits, a circuit is ran with three CNOT gates, the first one being between qubits 1 and 7, the second between qubits 1 and 2 and the third between qubits 6 and 7. The circuit is depicted in Fig. 7 Using large-distance bridge gates, this can be done with one distance-6 bridge gate (costs 20 CNOT gates) and then 2 CNOT gates, so 22 CNOT gates in total.

Consider now how many CNOT gates are required when using only the original distance-2 bridge gates and SWAP gates. First, the distance between qubits 1 and 7 needs to be reduced to two, which already costs at least 4 SWAPs, so $4 \times 3 = 12$ CNOT gates. After this, a bridge gate or another SWAP and a CNOT is required, bringing the total to 16 CNOT gates. After this, the qubit assignment is one of the following eleven forms: 1273456, 1723456, 2137456, 2173456, 2314756, 2317456, 2341576, 2341756, 2345167, 2345176, 2345617. Note that forms 1, 4, 9 and 10 require a single CNOT gate, and a distance-4 CNOT gate. For the latter, three SWAPs and a single CNOT gate are optimal, requiring an additional 10 qubits. This yields a total of 27 CNOT gates. Equivalently,

326   W. Schoot et al.

forms 3 and 7 require a single CNOT gate and a distance-3 CNOT gate, for which
a total of 8 extra CNOT gates are required, yielding 24 CNOT gates in total.
Form 5 requires two distance-2 CNOT gates, which also requires 8 CNOT gates,
yielding again a total of 24 CNOT gates. Forms 6 and 8 require a distance-2
CNOT gate and a distance-3 CNOT gate, which needs 11 CNOT gates in total,
yielding a total of 27 CNOT gates. Lastly, forms 2 and 11 require at least a
distance-5 CNOT gate, which needs 4 SWAPs, so 12 CNOT gates, plus a single
CNOT gate, bringing the total to 29 CNOT gates. All in all, it can be seen
that by using only SWAP gates and distance-2 CNOT gates, at least 24 CNOT
gates are required to perform the given circuit. An example of how this can
be done with 24 CNOT gates, can be seen in Fig. 8. This is worse than what
the larger distance bridge gate can achieve, which shows the advantage of using
larger distance bridge gates.

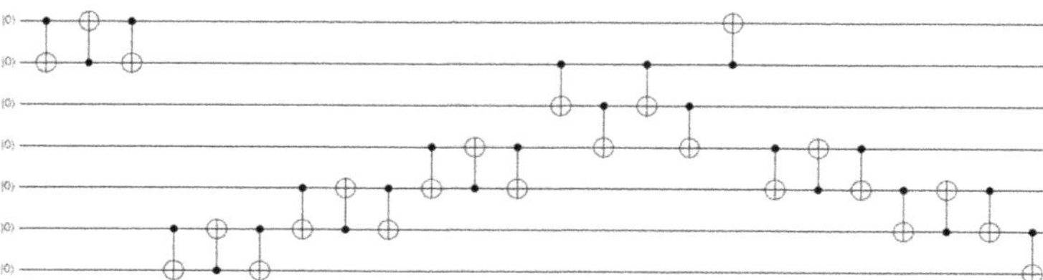

Fig. 8. The circuit in Fig. 7 implemented without using large distance bridge gates
with a minimal number of 24 CNOT gates. First, qubits 1 and 7 are swapped to reach
the qubits assignment 2137456, after which a distance-2 bridge gate is applied. Then,
qubit 7 is swapped back to neighbour qubit 6 and to apply the final 2 CNOT gates.

2.3 New Qubit Routing Problem

The addition of bridge gates introduces a different qubit routing problem, which
is the same as before, but now allowing bridge gates as well. Note that this
version of qubit routing has been proposed before, and is even used in some
state-of-the-art compilers, such as the Quantinuum's compiler, as detailed in
[27].

Problem (qubit routing with bridge gates): Consider a quantum circuit, a
target qubit architecture and an initial mapping of the qubits. For every two-
qubit gate, the gate can be implemented by using a combination of SWAP gates
to change the qubit mapping and bridge gates over arbitrary distances. What is
the minimal number of CNOT gates required to run the quantum circuit on the
target hardware?

Note that for a single two-qubit gate, also the combination of SWAP gates
and a bridge gate is allowed. Again, swapping only some of the distance between

the qubits could lead to a qubit mapping that is beneficial for later qubit routing steps. In particular, both the control and target qubit could be swapped some places before applying a bridge gate over a certain distance. This yields many different options for each single step of qubit routing.

3   Conclusion

This work considered the problem of qubit routing, and the use of bridge gates for this. These gates allow the implementation of CNOT gates over arbitrary distances, without using SWAP gates or changing the qubit mapping. This work is the first to construct bridge gates over larger distances. In addition, it is the first to define the problem of bridge gates over larger distances. Although no experiments with these bridge gates were performed in this work, it is easy to construct circuits in which bridge gates pose a considerable advantage over using just SWAP gates.

For further research, it would be particularly interesting to consider the extent of the impact of this larger distance bridge gate. Potential research questions are:

1. In which way can the qubit routing problem, including SWAP gates and bridge gates, be written as a mathematical formulation, to allow proving properties such as NP-hardness?
2. To which extent can bridge gates improve the performance of qubit routing, and which factors impact this performance increase? Potential factors are hardware architecture or circuit architecture.

References

1. Nielsen, M.A., Chuang, I.L.: Quantum Computation and Quantum Information. Cambridge University Press (2010)
2. Kumar, G., Yadav, S., Mukherjee, A., Hassija, V., Guizani, M.: Recent advances in quantum computing for drug discovery and development. IEEE Access (2024)
3. Mazzola, G.: Quantum computing for chemistry and physics applications from a monte Carlo perspective. J. Chem. Phys. **160**(1) (2024)
4. Peral-García, D., Cruz-Benito, J., García-Peñalvo, F.J.: Systematic literature review: quantum machine learning and its applications. Comput. Sci. Rev. **51**, 100619 (2024)
5. Phillipson, F.: Quantum machine learning: Benefits and practical examples. In: QANSWER, pp. 51–56 (2020)
6. Abbas, A., et al.: Challenges and opportunities in quantum optimization. Nat. Rev. Phys., 1–18 (2024)
7. Phillipson, F.: Quantum computing in logistics and supply chain management-an overview. arXiv preprint: arXiv:2402.17520 (2024)
8. Pal, S., Bhattacharya, M., Dash, S., Lee, S.-S., Chakraborty, C.: Future potential of quantum computing and simulations in biological science. Mol. Biotechnol. **66**(9), 2201–2218 (2024)

9. Fauseweh, B.: Quantum many-body simulations on digital quantum computers: state-of-the-art and future challenges. Nat. Commun. **15**(1), 2123 (2024)
10. How, M.-L., Cheah, S.-M.: Business renaissance: opportunities and challenges at the dawn of the quantum computing era. Businesses **3**(4), 585–605 (2023)
11. Ko, H., Kwon, S.: Prominence of corporate science in quantum computing research. Technol. Forecast. Soc. Chang. **212**, 123949 (2025)
12. Kalai, G., Rinott, Y., Shoham, T.: Google's 2019 quantum supremacy claims: data, documentation, and discussion. Quantum Phys.; Comput. Complex. **3**, 455–463 (2023)
13. Acharya, R., et al.: Quantum error correction below the surface code threshold. arXiv preprint: arXiv:2408.13687 (2024)
14. Shor, P.W.: Polynomial-time algorithms for prime factorization and discrete logarithms on a quantum computer. SIAM Rev. **41**(2), 303–332 (1999)
15. Grover, L.K.: A fast quantum mechanical algorithm for database search. In: Proceedings of the Twenty-eighth Annual ACM Symposium on Theory of Computing, pp. 212–219 (1996)
16. Tan, B.: Layout synthesis for quantum computing, Ph.D. dissertation, University of California, Los Angeles (2024)
17. Abbas, A., et al.: Quantum optimization: potential, challenges, and the path forward. arXiv preprint: arXiv:2312.02279 (2023)
18. Zhu, P., Cheng, X., Guan, Z.: An exact qubit allocation approach for NISQ architectures. Quantum Inf. Process. **19**(11), 391 (2020)
19. Mulderij, J., Aardal, K.I., Chiscop, I., Phillipson, F.: A polynomial size model with implicit swap gate counting for exact qubit reordering. In: International Conference on Computational Science, pp. 72–89. Springer (2023)
20. Wagner, F., Bärmann, A., Liers, F., Weissenbäck, M.: Improving quantum computation by optimized qubit routing. J. Optim. Theory Appl. **197**(3), 1161–1194 (2023)
21. Murali, P., Baker, J.M., Javadi-Abhari, A., Chong, F.T., Martonosi, M.: Noise-adaptive compiler mappings for noisy intermediate-scale quantum computers. In: Proceedings of the Twenty-Fourth International Conference on Architectural Support for Programming Languages and Operating Systems, pp. 1015–1029 (2019)
22. Steinberg, M., Bandic, M., Szkudlarek, S., Almudever, C.G., Sarkar, A., Feld, S.: Resource bounds for quantum circuit mapping via quantum circuit complexity. arXiv preprint: arXiv:2402.00478 (2024)
23. Moro, L., Paris, M.G., Restelli, M., Prati, E.: Quantum compiling by deep reinforcement learning. Commun. Phys. **4**(1), 178 (2021)
24. Pascoal, G., Fernandes, J.P., Abreu, R.: Deep reinforcement learning strategies for noise-adaptive qubit routing. In: 2024 IEEE International Conference on Quantum Software (QSW), pp. 146–156. IEEE (2024)
25. Pozzi, M.G., Herbert, S.J., Sengupta, A., Mullins, R.D.: Using reinforcement learning to perform qubit routing in quantum compilers. ACM Trans. Quantum Comput. **3**(2), 1–25 (2022)
26. Müller, S., Phillipson, F.: Quantum annealing for nearest neighbour compliance problem. Sci. Rep. **14**(1), 23340 (2024)
27. Cowtan, A., Dilkes, S., Duncan, R., Krajenbrink, A., Simmons, W., Sivarajah, S.: On the qubit routing problem. In: van Dam, W., Mančinska, L. (eds.) 14th Conference on the Theory of Quantum Computation, Communication and Cryptography (TQC 2019), ser. Leibniz International Proceedings in Informatics (LIPIcs), vol. 135. Schloss Dagstuhl – Leibniz-Zentrum für Informatik, pp. 1–32 (2019)

Resource Estimation for Matrix Inversion via QSVT with the Clifford+T Gate Set

Hiroaki Murakami[1]([✉])[iD], Kenzo Makino[1][iD], Yasunori Lee[2][iD], Keita Kanno[2][iD], and Tomonori Fukuta[1][iD]

[1] Advanced Technology R and D Center, Mitsubishi Electric Corporation, 8-1-1 Tsukaguchi Honmachi, Amagasaki-shi, Hyogo 661-8661, Japan
`Murakami.Hiroaki@aj.MitsubishiElectric.co.jp`
[2] QunaSys Inc., 1-13-7 Hakusan,, Bunkyo-ku, Tokyo 113-0001, Japan

Abstract. Solving linear systems of equations is a fundamental problem that serves as a crucial foundation in numerous fields. Initiated by the pioneering Harrow-Hassidim-Lloyd algorithm, the asymptotic performance of quantum linear system solvers has steadily improved over time. However, there remains a lack of quantitative evaluation of quantum resources based on explicit gate implementations. In this study, we particularly focus on the matrix inversion method via quantum singular value transformation and explicitly construct a quantum circuit to invert specific matrices using the standard universal gate set, Clifford+T gates. We examine the detailed methods for block-encoding, QSP angle finding, and gate decompositions, and numerically evaluate the overall resources required for the quantum circuit.

Keywords: Matrix inversion · Quantum Singular Value Transformation · Clifford+T gate set

1 Introduction

Solving linear system of equations $A\boldsymbol{x} = \boldsymbol{b}$ is one of the problems where quantum computers are expected to have an advantage over classical computers. The time and space complexity of existing classical algorithms, even for matrices with relatively good properties, is at best $\mathcal{O}(\mathrm{poly}(N))$ with respect to the matrix size $N = \max(N_{\mathrm{row}}, N_{\mathrm{col}})$ [16,17]. This limits the size of problems that can be addressed by classical algorithms. Owing to the progress in recent quantum algorithm research, initiated by the pioneering work by Harrow et al. [8], an $\mathcal{O}(\mathrm{polylog}(N))$ complexity has been achieved for certain classes of matrices. The quantum version of linear system solvers output the solution vector $\boldsymbol{x}$ as the quantum state $|x\rangle$. The entries of $\boldsymbol{x}$ cannot be accessed directly; instead, they can only be obtained by measuring the corresponding observables.

To assess the utility of quantum linear system solvers for real-world problems, a thorough quantitative evaluation of computational cost is required. Several works have done resource estimation with rigorous cost guarantees (e.g., see

F. Barbaresco and G. François (Eds.): QUEST-IS 2025, CCIS 2743, pp. 329–336, 2026.
https://doi.org/10.1007/978-3-032-13852-1_33

[3,5]). However, their primary focus is on query complexity, and they do not explore gate-level implementations in depth; consequently, the resources required for practical fault-tolerant computation are not sufficiently addressed. To bridge this gap, we explicitly constructed a quantum circuit for a specific problem using the standard universal gate set, Clifford+T gates, and numerically evaluated its accuracy and computational cost. By detailing the gate-level implementations, including exact gate decompositions, we aim to propose a novel approach for resource estimation and provide a guide for further research and devlopment.

2   Backgrounds

Given a signal $x \in [-1,1]$, there is a known method called Quantum Signal Processing (QSP) that achieves polynomial transformations of x by alternately applying a signal operator $W(x)$ that encodes the signal and a parametric signal processing operator $S(\phi)$;

$$W(x) := \begin{pmatrix} x & i\sqrt{1-x^2} \\ i\sqrt{1-x^2} & x \end{pmatrix} = e^{i\arccos(x)X}, \tag{1}$$

$$S(\phi) := \begin{pmatrix} e^{i\phi} & 0 \\ 0 & e^{-i\phi} \end{pmatrix} = e^{i\phi Z}. \tag{2}$$

For any polynomials $P(x), Q(x) \in \mathbb{C}[x]$ that satisfy certain conditions (see [7] for detail), there exists a sequence of angles $\boldsymbol{\Phi} = (\phi_0, \phi_1, \ldots, \phi_d) \in \mathbb{R}^{d+1}$ such that the following equation holds for $x \in [-1,1]$ [7,10].

$$S(\phi_0) \prod_{k=1}^{d} W(x)S(\phi_k) = \begin{pmatrix} P(x) & iQ(x)\sqrt{1-x^2} \\ iQ^*(x)\sqrt{1-x^2} & P^*(x) \end{pmatrix}. \tag{3}$$

QSP can be extended to multi-qubit system by the method called Quantum Singular Value Transformation (QSVT) [7,11]. With QSVT, when a matrix A is given in the form of a block-encoding, we can obtain a block-encoding of the matrix whose singular values $\{\sigma_i\}$ are transformed by the real part of the polynomial P.

In particular, if an appropriate polynomial P that approximates the function $1/x$ is given, it is possible to compute an approximate inverse of A by

$$P^{(\mathrm{SV})}(A)^\dagger \simeq W_\sigma \begin{pmatrix} \frac{1}{\sigma_1} & & \\ & \ddots & \\ & & \frac{1}{\sigma_N} \end{pmatrix} V_\sigma^\dagger = A^{-1}. \tag{4}$$

Here, W_σ and V_σ are two unitary matrices whose columns are left and right singular vectors of A. Now we can obtain the solution to the linear system of equations as follows:

$$\left(P^{(\mathrm{SV})}(A) \cdot \begin{smallmatrix} \cdot \\ \cdot \end{smallmatrix} \right)^\dagger \begin{pmatrix} \boldsymbol{b} \\ \boldsymbol{0} \end{pmatrix} \simeq \begin{pmatrix} A^{-1}\boldsymbol{b} \\ \cdot \end{pmatrix} = \begin{pmatrix} \boldsymbol{x} \\ \cdot \end{pmatrix}. \tag{5}$$

In the next section, we briefly summarize the methods we adopt for each step of the matrix inversion using QSVT.

3   Methods

3.1   Matrix

This study focuses on sparse and structured matrices, as the block-encoding of arbitrary dense matrices requires exponential computational cost, which undermines quantum advantage. Specifically, we consider Toeplitz matrices, whose elements along each diagonal are constant. Toeplitz matrices frequently arise in problems such as time series analysis, digital signal processing, and solving linear differential equations. Due to its structure, block-encoding circuit can be efficiently constructed using rotation gates and adders [15]. Let N_{data} be the maximal number of non-zero elements per column (or row), then the block-encoding of a Toeplitz matrix can be realized using N_{data} rotation gates and $4\log_2 N$ Toffoli gates. If N_{data} is $\mathcal{O}(\log(N))$, the total number of gates, including other Clifford gates, is $\mathcal{O}(\mathrm{polylog}(N))$.

In this study, we consider a simple matrix

$$
A = \begin{pmatrix} 2 & 1 & & \\ & 2 & \ddots & \\ & & \ddots & 1 \\ & & & 2 \end{pmatrix},
\tag{6}
$$

and explicitly implement the corresponding block-encoding circuit using the method in [15]. As an example, the block-encoding circuit for $(N = 4)$ is shown in Fig. 1. Here, we use the metod by Möttönen et al. [12] for data loading, and adopt two ripple carry adders [4] to compile the additions appearing in eq. (55) of [15].

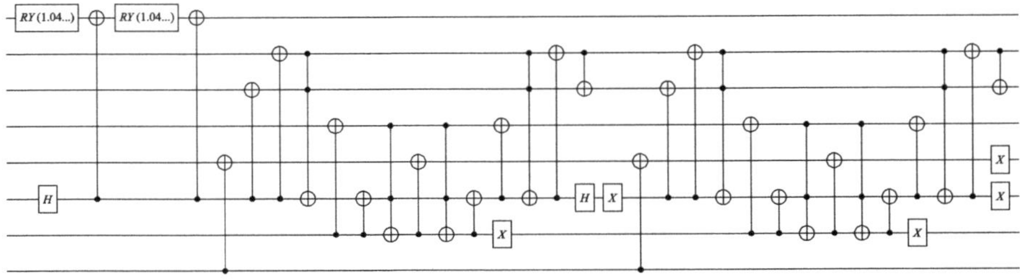

Fig. 1. A quantum circuit for block-encoding of Toeplitz matrix $(N = 4)$.

3.2 Polynomial Approximation and Angle Finding

Several methods have been proposed to obtain a polynomial that approximates the function $1/x$ over a closed interval $D_{\kappa_{\mathrm{QSVT}}} = [1/\kappa_{\mathrm{QSVT}}, 1]$, given the degree d of the polynomial. We utilize the explicit form of the optimal polynomial P and the complementary polynomial Q, proposed by [2]. Let $\epsilon_{\mathrm{QSP}} := \|P(x) - 1/x\|_\infty$. For $\kappa \gg 1$, the degree d of the polynomial scales as $\kappa \log(\kappa/\epsilon_{\mathrm{QSP}})$ (see eq. (4) of [2]). Thus, if κ increases polynomially with the size of the matrix, quantum advantage may be compromised. For the matrices considered in this study (Sect. 3.1), κ asymptotically approaches 3 as the matrix size grows, thereby preserving quantum advantage. We fix $\kappa_{\mathrm{QSVT}} = 4$ for the following experiments. We should note that for generic Toeplitz matrices, when the contributions of the off-diagonal elements are significant, the condition number may exhibit a polynomial dependence on the matrix size.

We then adopt the carving algorithm [7] to compute the corresponding QSP angle sequence $\boldsymbol{\Phi}$.

3.3 Decomposition of Non-clifford Gates

To explicitly evaluate the computational cost, we decompose the non-Clifford gates included in the constructed QSVT circuit into the Clifford$+T$ gate set. In Table 1, we summarize the non-Clifford gates appearing in our implementation. The rotation gates are approximately decomposed into a sequence of H, S and T gates using the Ross-Selinger algorithm [13]. The multi-controlled gates are decomposed using doubly-controlled $\pm iX$ gates, following [14]. Toffoli gates in ripple carry adders [4] used for block-encoding are decomposed into Clifford$+T$ gates via the method in [1].

By adding ancillary qubits, further reductions in T-count can be achieved. For example, the T-count of multi-controlled gates and ripple carry adders can approximately be halved by using "temporary logical-AND" structure (relative phase Toffoli gate) [6]. For data loading, there is a known trade-off between the number of ancillas and T-count [9]. While this paper does not delve into the specifics, these are important considerations for practical implementation.

Table 1. Non-Clifford gates.

Non-Clifford gates	# of appearance	T-count after decomp.	# of ancilla
R_Z for QSP	d	$\sim 3 \log_2 \frac{1}{\epsilon_{\mathrm{gate}}}$	0
R_Y for data loading	$N_{\mathrm{data}} d$	$\sim 3 \log_2 \frac{1}{\epsilon_{\mathrm{gate}}}$	0
Multi-controlled gate	d	$8(\log_2 N_{\mathrm{data}} + 1)$	$\log_2 N_{\mathrm{data}}$
Adder in block-encoding	$2d$	$14 \log_2 N$	0

4   Results

We numerically evaluated the implemented quantum circuit in terms of accuracy and computational cost.

4.1   Accuracy

First, we investigated the dependence of the output accuracy (evaluated by the operator norm $\|A^{-1} - P^{(\mathrm{SV})}(A)^{\dagger}\|$) on the degree d of the approximate polynomial and the matrix size $N = 2^n$. As shown in Fig. 2, we confirmed that the output accuracy improves exponentially with respect to d, and is independent of N.

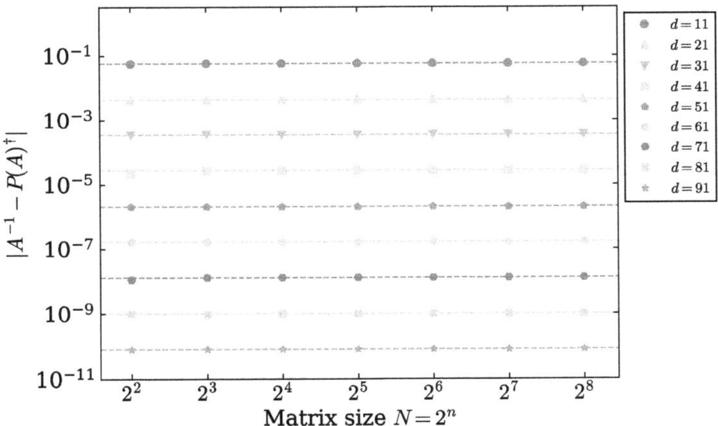

Fig. 2. Dependence of the error on the size of the matrix and the degree of the polynomial.

Next, we fixed the matrix size at $N = 8$ and decomposed the non-Clifford gates. Figure 3 shows the dependency of the output accuracy on the decomposition error ϵ_{gate} of the rotation gates for different d. In the left region of the figure, where ϵ_{gate} is large, the decomposition error of the rotation gates is dominant. On the other hand, in the right region where ϵ_{gate} is sufficiently small, the error due to polynomial approximation (indicated by the dotted lines in the figure) becomes dominant, and the output accuracy of the quantum circuit saturates.

The error due to the decomposition of the rotation gates is considered to be proportional to the product of ϵ_{gate} and the number of rotation gates $(N_{\mathrm{data}}+1)d$. Therefore, when each rotation gate is decomposed with an accuracy of ϵ_{gate}, and a final output accuracy of ϵ is required, the degree d of the approximate polynomial is determined by the minimum singular value of the scaled matrix A and some proportional constant. This, in turn, allows us to determine the computational cost discussed in the next section.

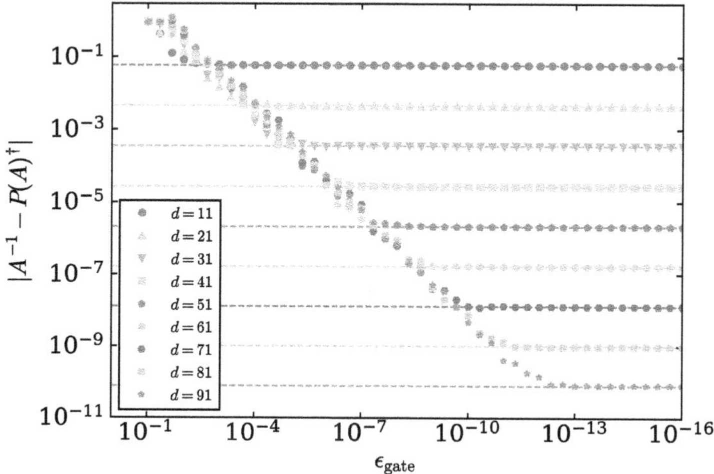

Fig. 3. Dependence of the error on the degree of the polynomial and the precision for decomposition of rotation gates.

4.2   Computational Cost

We evaluated the computational cost in regards of the number of qubits and the T-count. It was found that the number of logical qubits n_{qubits} is explicitly given as

$$n_{\text{qubits}} = 2\log_2 N + \log_2 N_{\text{data}} + 6, \tag{7}$$

which does not depend on the decomposition error ϵ_{gate} of the rotation gates or the degree d of the approximation polynomial.

Next, we examined the behavior of the T-count arises from the decomposition of non-Clifford gates. As can be seen from Fig. 4, it is confirmed that the T-count with respect to d and ϵ_{gate} matches the expectations outlined in Table 1 (indicated by the dotted lines in the figure).

In the above discussion, we treated each Clifford+T gate as error-free. However, in practical fault tolerant computations, it is necessary to take gate infidelity into account. The global error of the circuit is bounded by the sum of the (absolute) error associated with each local unitary. Therefore, to ensure that the output precision is smaller than certain value, we can establish error tolerance for each gates by dividing the required output precision by the total number of gates, including not only T gates but also all Clifford gates. Then, each gate can be prepared in a fault-tolerant manner with obtained error tolerance accordingly.

5   Summary

We explicitly constructed a quantum circuit to compute the inverse of Toeplitz matrices via QSVT with the Clifford+T gate set, and evaluated its accuracy and computational cost. As a result, we confirmed:

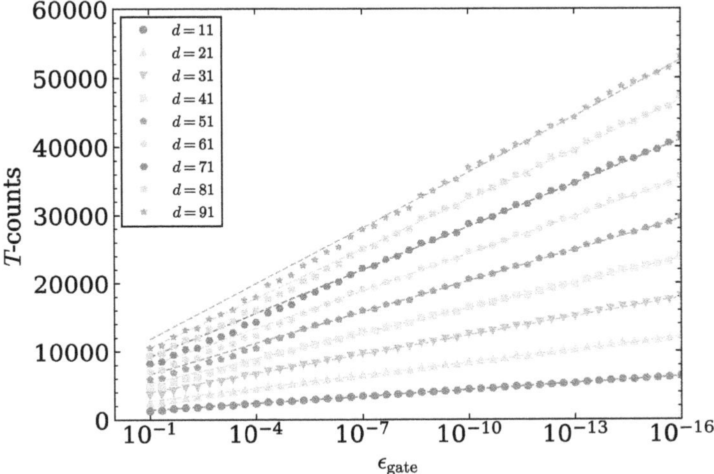

Fig. 4. Dependence of the T-count on the degree of the polynomial and the precision for decomposition of rotation gates.

(i) The output accuracy of the inverse matrix improves exponentially with respect to the degree d of the approximate polynomial.
(ii) The output accuracy does not depend on the matrix size N, since the condition number of the matrix is bounded by a constant in our specific choice of matrix.
(iii) The T-count depends on d and the decomposition error of the rotation gates ϵ_{gate} as shown in Table 1, and it depends logarithmically on N through the adder in the block-encoding circuit.

In order to construct an end-to-end quantum linear system solver, the preparation of the input quantum states and the readout of the output states must also be implemented. It is known that both, if implemented naively, may require exponential cost, which would undermine quantum advantage. Addressing this issue remains a challenge for future work.

Disclosure of Interests. The authors have no competing interests to declare that are relevant to the content of this article.

References

1. Amy, M., Maslov, D., Mosca, M., Roetteler, M.: A meet-in-the-middle algorithm for fast synthesis of depth-optimal quantum circuits. IEEE Trans. Comput. Aided Des. Integr. Circuits Syst. **32**(6), 818–830 (2013). https://doi.org/10.1109/TCAD.2013.2244643
2. Berntson, B.K., Sünderhauf, C.: Two exact quantum signal processing results. In: 2024 IEEE Computer Society Annual Symposium on VLSI (ISVLSI), pp. 625–626 (2024). https://doi.org/10.1109/ISVLSI61997.2024.00118

3. Costa, P.C., An, D., Sanders, Y.R., Su, Y., Babbush, R., Berry, D.W.: Optimal scaling quantum linear-systems solver via discrete adiabatic theorem. PRX Quantum **3**, 040303 (2022). https://doi.org/10.1103/PRXQuantum.3.040303

4. Cuccaro, S.A., Draper, T.G., Kutin, S.A., Moulton, D.P.: A new quantum ripple-carry addition circuit (2004). https://doi.org/10.48550/arXiv.quant-ph/0410184

5. Dalzell, A.M.: A shortcut to an optimal quantum linear system solver (2024). https://arxiv.org/abs/2406.12086

6. Gidney, C.: Halving the cost of quantum addition. Quantum **2**, 74 (2018). https://doi.org/10.22331/q-2018-06-18-74

7. Gilyén, A., Su, Y., Low, G.H., Wiebe, N.: Quantum singular value transformation and beyond: exponential improvements for quantum matrix arithmetics. In: Proceedings of the 51st Annual ACM SIGACT Symposium on Theory of Computing, pp. 193–204. STOC 2019, Association for Computing Machinery, New York, NY, USA (2019). https://doi.org/10.1145/3313276.3316366

8. Harrow, A.W., Hassidim, A., Lloyd, S.: Quantum algorithm for linear systems of equations. Phys. Rev. Lett. **103**, 150502 (2009). https://doi.org/10.1103/PhysRevLett.103.150502

9. Low, G.H., Kliuchnikov, V., Schaeffer, L.: Trading T gates for dirty qubits in state preparation and unitary synthesis. Quantum **8**, 1375 (2024). https://doi.org/10.22331/q-2024-06-17-1375

10. Low, G.H., Yoder, T.J., Chuang, I.L.: Methodology of resonant equiangular composite quantum gates. Phys. Rev. X **6**, 041067 (2016). https://doi.org/10.1103/PhysRevX.6.041067

11. Martyn, J.M., Rossi, Z.M., Tan, A.K., Chuang, I.L.: Grand unification of quantum algorithms. PRX Quantum **2**, 040203 (2021). https://doi.org/10.1103/PRXQuantum.2.040203

12. Möttönen, M., Vartiainen, J.J., Bergholm, V., Salomaa, M.M.: Quantum circuits for general multiqubit gates. Phys. Rev. Lett. **93**(13) (2004). https://doi.org/10.1103/physrevlett.93.130502

13. Ross, N.J., Selinger, P.: Optimal ancilla-free Clifford+T approximation of z-rotations. Quantum Inf. Comput. **16**(11&12), 0901–0953 (2016). https://doi.org/10.26421/QIC16.11-12-1

14. Selinger, P.: Quantum circuits of t-depth one. Phys. Rev. A **87**, 042302 (2013). https://doi.org/10.1103/PhysRevA.87.042302

15. Sünderhauf, C., Campbell, E., Camps, J.: Block-encoding structured matrices for data input in quantum computing. Quantum **8**, 1226 (2024). https://doi.org/10.22331/q-2024-01-11-1226

16. Trefethen, L.N., Bau, D.: Numerical linear algebra. Society for Industrial and Applied Mathematics, Philadelphia (1997). https://doi.org/10.1137/1.9780898719574

17. Vorst, H.A.v.d.: Iterative Krylov methods for large linear systems. No. 13 in Cambridge monographs on applied and computational mathematics, Cambridge University Press, Cambridge; New York (2003). https://doi.org/10.1017/CBO9780511615115

Session 4 Quantum Algorithms, Computing and Simulation – Quantum Algorithms for Finance and Industry

Q-PORT: Quantum Portfolio Optimization with Resource-Efficient Encoding and Scalability Analysis

Alberto Marchisio[1,2]($\boxtimes$), Muhammad Umair Hafeez[1,2], Nouhaila Innan[1,2], Muhammad Kashif[1,2], and Muhammad Shafique[1,2]

[1] eBRAIN Lab, Division of Engineering, New York University Abu Dhabi (NYUAD), Abu Dhabi, UAE
{alberto.marchisio,muh2008,nouhaila.innan,muhammadkashif, muhammad.shafique}@nyu.edu
[2] Center for Quantum and Topological Systems (CQTS), NYUAD Research Institute, NYUAD, Abu Dhabi, UAE

Abstract. Quantum Approximate Optimization Algorithm (QAOA) and Variational Quantum Eigensolver (VQE) have emerged as promising approaches for solving portfolio optimization tasks. However, the practical scalability of these methods remains a challenge due to the inherent noise and limitations of Noisy Intermediate-Scale Quantum (NISQ) devices. In this paper, we present *Q-PORT* (*Quantum Portfolio Optimization with Resource-efficient Encoding and Scalability Analysis*), a systematic study on the trade-offs between quantum circuit depth, stock encoding strategies, and scalability in quantum portfolio optimization.

We investigate the impact of multi-qubit representations per stock and multi-stock encodings per qubit while varying circuit repetitions and Ansatz types. Our experimental results indicate that increasing qubits per stock offers negligible precision gains compared to classical Mean-Variance Optimization (MVO), while encoding multiple stocks per qubit significantly improves efficiency with minimal precision loss. These findings provide a new pathway toward resource-efficient and scalable quantum portfolio optimization, paving the way for near-term financial applications in quantum computing.

Keywords: Quantum Computing · Portfolio Optimization · Quantum Finance

1 Introduction

Portfolio optimization remains a cornerstone of quantitative finance, seeking to balance asset allocation to maximize returns while controlling risk [1]. Classical approaches, such as Mean-Variance Optimization (MVO), model this trade-off using expected returns and covariance matrices, with numerous extensions addressing practical constraints [2–4].

Nonetheless, the scalability limitations and computational overhead of classical methods have propelled interest in quantum computing as a next-generation

F. Barbaresco and G. François (Eds.): QUEST-IS 2025, CCIS 2743, pp. 339–347, 2026.
https://doi.org/10.1007/978-3-032-13852-1_34

alternative [5]. Hybrid quantum-classical algorithms like the Quantum Approximate Optimization Algorithm (QAOA) [6] and the Variational Quantum Eigensolver (VQE) [7] have been successfully adapted to portfolio optimization [8–12]. Complementary advances, including quantum annealing for multi-objective portfolios [13], hardware-efficient ansatz design [14], and quasi-binary encoding techniques [15], have recently pushed the state of the art in this domain. However, the majority of these implementations assign a single qubit per stock, creating scalability barriers under the constraints of Noisy Intermediate-Scale Quantum (NISQ) hardware.

This work raises two fundamental research questions: (1) Can multi-qubit representations per stock improve solution fidelity, and do such gains justify the additional resource overhead? (2) Can multi-stock encodings per qubit meaningfully reduce resource requirements while preserving solution quality?

To address these questions, we propose *Q-PORT* (*Quantum Portfolio Optimization with Resource-efficient Encoding and Scalability Analysis*), a methodology to systematically investigate the impact of stock-to-qubit mappings, circuit repetitions, and Ansatz selection on portfolio optimization performance using VQE- and QAOA-based algorithms.

Our contributions are threefold: (i) we develop a comprehensive methodology for evaluating quantum portfolio scalability; (ii) we demonstrate that multi-qubit representations per stock yield negligible precision gains, indicating limited practical benefit; (iii) we show that multi-stock encodings per qubit significantly improve efficiency with minimal precision loss, enabling more scalable quantum portfolio optimization even under NISQ-era limitations.

2 Background and Related Work

2.1 Classical Portfolio Optimization

Classical portfolio optimization, rooted in Markowitz's Modern Portfolio Theory (MPT), aims to allocate assets to maximize expected return for a given level of risk, typically measured by portfolio variance [1]. Extensions to MPT have incorporated various constraints and considerations, such as transaction costs and investor preferences. Brandt et al. proposed a parametric portfolio policy that models portfolio weights as functions of asset characteristics, enhancing scalability for high-dimensional asset spaces [3]. Comprehensive treatments of these methodologies are detailed in the work of Elton et al.'s *Modern Portfolio Theory and Investment Analysis* [4].

2.2 Quantum Optimization Algorithms

Quantum computing introduces novel approaches in financial applications [16–28]. The most commonly used algorithms are the VQE [7] and the QAOA [6]. These hybrid quantum-classical algorithms utilize parameterized quantum circuits optimized via classical routines, effectively navigating complex solution spaces inherent in Quadratic Unconstrained Binary Optimization (QUBO) formulations.

Recent studies have explored the application of these algorithms to portfolio optimization. Kerenidis et al. [29] developed a quantum algorithm for constrained portfolio optimization, demonstrating potential polynomial speedups over classical counterparts. Buonaiuto et al. [9] conducted experiments using VQE on real quantum devices, analyzing the impact of various hyperparameters on performance. Zaman et al. [11] introduced PO-QA, a framework for portfolio optimization using quantum algorithms, providing insights into circuit configurations and their effects on optimization outcomes.

Despite these advancements, challenges remain regarding scalability and precision, particularly under the constraints of NISQ devices. *Our work addresses these challenges by investigating multi-qubit representations per asset and multi-asset encodings per qubit, aiming to enhance scalability while maintaining solution quality.*

3   Q-PORT Methodology

Our proposed Q-PORT methodology systematically investigates the scalability of quantum portfolio optimization by analyzing various encoding strategies and quantum circuit configurations. The overall methodology, shown in Figure 1, consists of the following steps:

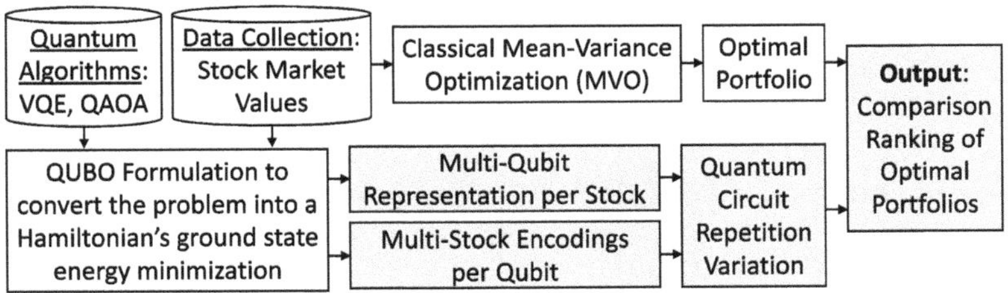

Fig. 1. Overview of our Q-PORT methodology.

- **Data Collection and Classical Benchmarking:** We begin by collecting historical stock market data, including expected returns and covariance matrices for a selected set of assets. As a classical reference, we employ the classical MVO method [1] to compute the optimal asset allocations under the classical framework. This serves as the baseline for evaluating quantum solutions.
- **QUBO Formulation:** The portfolio optimization problem is reformulated into a Quadratic Unconstrained Binary Optimization (QUBO) model, enabling its translation into the ground state energy minimization of a quantum Hamiltonian. This formulation captures both the expected returns and risk terms, as well as budget and allocation constraints, in a structure amenable to quantum algorithms.

- **Quantum Encoding Strategies:** To explore scalability, we introduce two encoding strategies:
  - **Multi-Qubit Representation per Stock:** In this approach, each stock is mapped to a group of qubits, introducing redundancy that expands the solution space and enables potentially finer-grained portfolio allocations, albeit at higher resource cost. After measurement, individual qubit outcomes are aggregated using majority voting, where the most frequent state across the qubits determines the final allocation for the stock.
  - **Multi-Stock Encoding per Qubit:** In this scheme, multiple stocks are jointly encoded into a single qubit using compressed representations, significantly reducing the total number of qubits required and directly addressing the hardware limitations of NISQ devices. To identify optimal allocations within this compressed space, we generate a pool of candidate portfolios and evaluate each based on its expected return and associated risk. The resulting portfolio candidates are then clustered using the k-means algorithm to group solutions exhibiting similar performance characteristics.
- **Quantum Optimization and Circuit Repetition Variation:** We employ hybrid quantum-classical algorithms, specifically VQE [7] and QAOA [6], to solve the QUBO-formulated portfolio optimization problem. Within this stage, we systematically vary quantum circuit configurations, such as varying the number of circuit repetitions (i.e., layers).
- **Evaluation and Comparative Analysis:** The output portfolios obtained from different encoding strategies under varying quantum circuit depths are compared.

4   Evaluation of Q-PORT Methodology

4.1   Experiment Setup

For our experiments, we collected historical stock price data from Yahoo Finance, selecting four distinct stock groups to evaluate scalability. The datasets cover the period from January 1, 2020, to January 1, 2024, and include:

- 3 stocks: Apple, Microsoft, and Google;
- 4 stocks: Apple, Microsoft, Google, and Amazon;
- 12 stocks: Apple, Microsoft, Google, Amazon, Tesla, Nvidia, Meta, Netflix, JP Morgan, Disney, Visa, and Berkshire Hathaway Inc.

We employed both VQE and QAOA algorithms, using layered ansatz circuits composed of parameterized single-qubit rotations and entangling gates arranged in a fixed connectivity pattern. Specifically, each ansatz begins with Hadamard gates to initialize qubits into superposition, followed by parameterized R_Z rotations for the cost Hamiltonian, two-qubit R_{ZZ} gates to induce entanglement reflective of the QUBO problem structure, and R_X gates implementing the mixer

Hamiltonian. By varying the number of layers (repetitions), from 1 to 5, we systematically control the circuit depth and the level of multipartite entanglement, thereby expanding the ansatz expressivity.

The risk tolerance parameter was fixed at 0.2 across all experiments, while the investment budget was defined as the total number of stocks minus one (e.g., budget of 3 for 4-stock scenarios). The classical optimizer COBYLA was employed with a maximum of 5000 iterations. All simulations were conducted using statevector-based simulators, avoiding sampling noise inherent to real quantum hardware. Experiments were executed on Google Colab instances equipped with Intel Xeon E5 CPUs.

4.2 Evaluation of Multi-Qubit Representations per Stock

Figures 2 and 3 present the ranking of the portfolios that correspond to the classical MVO-optimal solution within the output distributions of QAOA and VQE under the multi-qubit per stock encoding, across different numbers of circuit repetitions. In these plots, the value of *Ranking* on the y-axis denotes the position of the optimal solution (computed by the classical MVO) in the sorted list of quantum-generated portfolios. A lower ranking indicates that the quantum algorithm assigns a higher probability to the optimal portfolio, thus reflecting better alignment with the classical benchmark. Hence, a ranking of 1 implies that the solution found by the quantum algorithm matches the classically optimal solution.

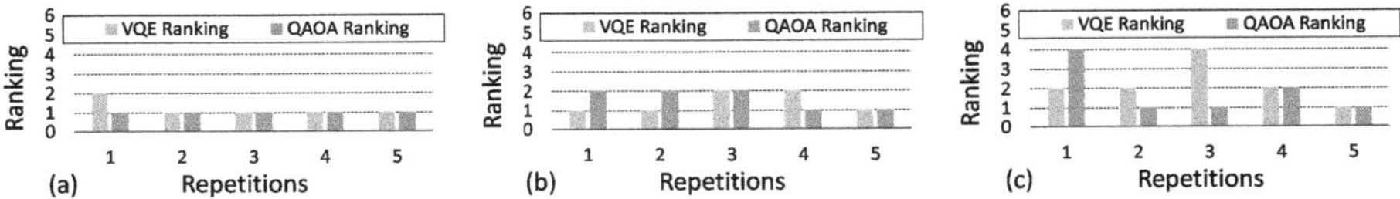

Fig. 2. Results of the multi-qubit representations per 3 stocks, as a function of circuit repetitions. (a) 1 qubit per stock, i.e., 3 qubits. (b) 2 qubits per stock, i.e., 6 qubits. (c) 3 qubits per stock, i.e., 9 qubits.

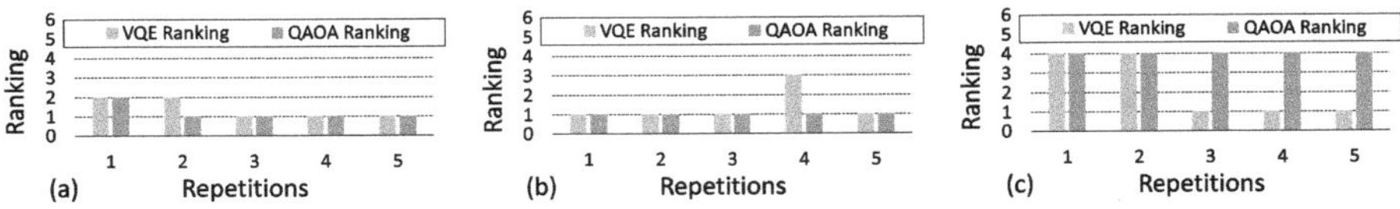

Fig. 3. Results of the multi-qubit representations per 4 stocks, as a function of circuit repetitions. (a) 1 qubit per stock, i.e., 4 qubits. (b) 2 qubits per stock, i.e., 8 qubits. (c) 3 qubits per stock, i.e., 12 qubits.

The results for 3-stock and 4-stock configurations demonstrate that increasing the number of repetitions generally improves the alignment between quantum and classical solutions, particularly for the QAOA. Notably, in the 3-stock case Fig. 2, the QAOA consistently ranks the optimal solution among the top positions even at low depth, whereas the VQE requires higher depth to converge to competitive rankings. For the 4-stock case Fig. 3, both QAOA and VQE exhibit increased variance, but higher repetitions still tend to yield lower (better) rankings overall.

Across subplots (a), (b), and (c) of Figs. 2 and 3, we analyze the effect of increasing the number of qubits assigned per stock from 1 to 3. In subplot (a), the default configuration with one qubit per stock serves as a baseline. As shown, both QAOA and VQE are capable of identifying the classically optimal solution with relatively low rankings, particularly when the number of circuit repetitions is increased.

However, in subplot (b), where each stock is represented using two qubits, we observe a modest increase in variability and, in some cases, a decline in ranking performance. This trend becomes more pronounced in subplot (c), where three qubits per stock are used. Here, the quantum algorithms exhibit greater difficulty in consistently ranking the MVO-optimal solution near the top, particularly at lower circuit depths.

This behavior can be attributed to the expansion of the solution space due to redundant encoding. While redundancy through multi-qubit representation may increase robustness in noisy environments, it also introduces additional complexity in optimization, making it harder for the quantum algorithms to converge to the classically optimal solution. Overall, these results indicate that increasing the number of qubits per stock does not necessarily improve performance and may, in fact, degrade solution quality unless accompanied by deeper and more expressive quantum circuits.

4.3   Evaluation of Multi-Stock Encoding per Qubit

Figure 4 presents the ranking performance of the classically optimal MVO solution under different multi-stock per qubit configurations for a 12-stock portfolio. The three subplots (a), (b), and (c) correspond to encoding 1, 2, and 3 stocks per qubit, respectively. As before, lower rankings (closer to 1) indicate stronger alignment between quantum-generated results and the classical MVO benchmark.

In subplot (a), where each stock is assigned to a distinct qubit, we observe baseline performance for both QAOA and VQE across varying circuit repetitions. The QAOA generally ranks the optimal portfolio higher than the VQE, particularly at lower depths. Subplot (b) shows the configuration where two stocks are compressed into each qubit. Interestingly, despite a reduction in qubit resources, the QAOA continues to perform comparably well, often preserving the optimal solution within the top few ranks. The VQE, while showing slightly more variance, maintains a generally acceptable ranking trend with increased repetitions.

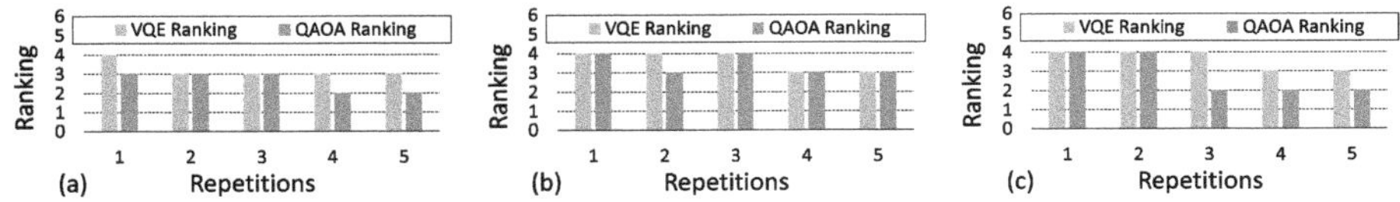

Fig. 4. Results of the multi-stock encoding for 12 stocks, as a function of circuit repetitions. (a) 1 stock per qubit, i.e., 12 qubits. (b) 2 stocks per qubit, i.e., 6 qubits. (c) 3 stocks per qubit, i.e., 4 qubits.

Subplot (c), which encodes three stocks per qubit, highlights the benefits and limitations of aggressive compression. While some degradation in ranking performance is expected due to the loss of individual stock granularity, the results demonstrate that the QAOA is still able to identify the optimal solution with competitive rankings, particularly at higher circuit depths. The VQE exhibits more fluctuation and tends to benefit from deeper circuits to stabilize performance.

Overall, these results suggest that multi-stock encoding is a viable strategy for scaling portfolio optimization to larger asset sets under limited quantum hardware resources. Although increasing the number of stocks per qubit introduces approximation errors, careful tuning of circuit depth can mitigate precision loss, particularly when using QAOA, which appears more resilient under compressed encodings.

5   Conclusion

In this work, we proposed *Q-PORT*, a systematic methodology to explore the scalability of quantum portfolio optimization under different encoding strategies and quantum circuit configurations. We introduced and analyzed two alternative encoding paradigms, the multi-qubit representation per stock and the multi-stock encoding per qubit. We then evaluated their effectiveness using VQE and QAOA across multiple portfolio sizes and circuit repetitions. Our findings demonstrate that increasing the number of qubits per stock yields negligible improvements in solution quality, while significantly increasing resource demands. In contrast, encoding multiple stocks per qubit enables substantial reductions in qubit usage with only minor losses in precision, especially when using QAOA with deeper circuits. These results highlight practical pathways for scaling quantum portfolio optimization under the constraints of NISQ hardware. As future work, we plan to extend our experimental evaluation to larger-scale portfolios and more complex financial datasets to further validate the generalizability and robustness of the proposed methodology.

Acknowledgment. This work was supported in part by the NYUAD Center for Quantum and Topological Systems (CQTS), funded by Tamkeen under the NYUAD Research Institute grant CG008.

References

1. Markowitz, H.: Portfolio selection. J. Financ. **7**(1), 77–91 (1952)
2. Fabozzi, F. J., Kolm, P. N., Pachamanova, D. A., Focardi, S. M.: Robust portfolio optimization and management. John Wiley & Sons (2007)
3. Brandt, M.W., Santa-Clara, P., Valkanov, R.: Parametric portfolio policies: exploiting characteristics in the cross-section of equity returns. Rev. Financ. Stud. **22**(9), 3411–3447 (2009)
4. Elton, E. J., Gruber, M. J., Brown, S. J., Goetzmann, W. N.: Modern portfolio theory and investment analysis. John Wiley & Sons, 9 ed., 2014
5. Preskill, J.: Quantum computing in the NISG era and beyond. Quantum **2**, 79 (2018)
6. Farhi, E., Goldstone, J., Gutmann, S.: A quantum approximate optimization algorithm. arXiv preprint arXiv:1411.4028 (2014)
7. Peruzzo, A., et al.: A variational eigenvalue solver on a quantum processor. Nat. Commun. **5**, 4213 (2014)
8. Venturelli, D., Kondratyev, A.: Quantum optimization of fully connected spin glasses. Quant. Sci. Technol. **4**(2), 025002 (2019)
9. Buonaiuto, G., Gargiulo, F., De Pietro, G., Esposito, M., Pota, M.: Best practices for portfolio optimization by quantum computing, experimented on real quantum devices. Sci. Rep. **13**(1), 19434 (2023)
10. Hodson, T., Green, T., Chakhmakhchyan, L., Grout, I.: Portfolio optimization and variational quantum algorithms. Quant. Inf. Process. **21**(2), 76 (2022)
11. Zaman, K., Marchisio, A., Kashif, M., Shafique, M.: PO-GA: a framework for portfolio optimization using quantum algorithms. In: 2024 IEEE International Conference on Quantum Computing and Engineering (QCE), vol. 1, pp. 1397–1403, IEEE (2024)
12. Innan, N., Saleem, A., Marchisio, A., Shafique, M.: Quantum portfolio optimization with expert analysis evaluation. arXiv preprint arXiv:2507.20532 (2025)
13. Aguilera, E., de Jong, J., Phillipson, F., Taamallah, S., Vos, M.: Multi-objective portfolio optimization using a quantum annealer. Mathematics **12**(9), 1291 (2024)
14. Morapakula, S. N., et al.: End-to-end portfolio optimization with quantum annealing. arXiv preprint arXiv:2504.08843 (2025)
15. Chen, B., Wu, H., Yuan, H., Wu, L., Li, X.: Quasi-binary encoding based quantum alternating operator ansatz. arXiv preprint arXiv:2304.06915 (2023)
16. Innan, N., Marchisio, A., Bennai, M., Shafique, M.: QFNN-FFD: quantum federated neural network for financial fraud detection. arXiv preprint arXiv:2404.02595 (2024)
17. Dutta, S., Innan, N., Marchisio, A., Yahia, S. B., Shafique, M.: Qadqn: quantum attention deep q-network for financial market prediction. In: 2024 IEEE International Conference on Quantum Computing and Engineering (QCE), vol. 2, pp. 341–346, IEEE (2024)
18. Innan, N., et al.: Financial fraud detection using quantum graph neural networks. Quant. Mach. Intell. **6**(1), 7 (2024)
19. Innan, N., Khan, M.A.-Z., Bennai, M.: Financial fraud detection: a comparative study of quantum machine learning models. Int. J. Quant. Inf. **22**(02), 2350044 (2024)
20. Innan, N., Marchisio, A., Bennai, M., Shafique, M.: LEP-QNN: loan eligibility prediction using quantum neural networks. arXiv preprint arXiv:2412.03158 (2024)

21. Innan, M., Kashif, M., Marchisio, A., Bennai, M., Shafique, M.:Next-generation quantum neural networks: enhancing efficiency, security, and privacy. In: 2025 IEEE 31st International Symposium on On-Line Testing and Robust System Design (IOLTS), pp. 1–4, IEEE (2025)
22. Zaman, K., Marchisio, A., Hanif, M. A., Shafique, M.: A survey on quantum machine learning: Current trends, challenges, opportunities, and the road ahead. arXiv preprint arXiv:2310.10315 (2023)
23. Alami, M. E., Innan, N., Shafique, M., Bennai, M.: Comparative performance analysis of quantum machine learning architectures for credit card fraud detection. tarXiv preprint arXiv:2412.19441 (2024)
24. Choudhary, P. K., Innan, N., Shafique, M., Singh, R: HQNN-FSP: a hybrid classical-quantum neural network for regression-based financial stock market prediction. arXiv preprint arXiv:2503.15403 (2025)
25. Sawaika, A., et al.: A privacy-preserving federated framework with hybrid quantum-enhanced learning for financial fraud detection. arXiv preprint arXiv:2507.22908 (2025)
26. Pathak, P., Oad, V., Prajapati, A., Innan, N.: Resource allocation optimization in 5g networks using variational quantum regressor. In: 2024 International Conference on Quantum Communications, Networking, and Computing (QCNC), pp. 101–105, IEEE (2024)
27. Innan, N., Marchisio, A., Bennai, M., Shafique, M.: QFNN-FFD: quantum federated neural network for financial fraud detection. In: 2025 IEEE International Conference on Quantum Software (QSW), pp. 41–47, IEEE (2025)
28. Kashif, M., Khalid, S., Innan, N., Marchisio, A., Shafique, M.: Evaluating quantum amplitude estimation for pricing multi-asset basket options. arXiv preprint arXiv:2509.09432 (2025)
29. Kerenidis, I., Prakash, A., Szilágyi, D.: Quantum algorithms for portfolio optimization. arXiv preprint arXiv:1908.08040 (2019)

Quantum Amplitude Estimation in Practice: A Case Study in Option Pricing

Nouhaila Innan[1,2]($\boxtimes$), Muhammad Kashif[1,2], Alberto Marchisio[1,2], Muhammad Moonis Usman[1,2], and Muhammad Shafique[1,2]

[1] eBRAIN Lab, Division of Engineering, New York University Abu Dhabi (NYUAD), Abu Dhabi, UAE
{nouhaila.innan,muhammadkashif,alberto.marchisio,mu2173, muhammad.shafique}@nyu.edu
[2] Center for Quantum and Topological Systems (CQTS), NYUAD Research Institute, NYUAD, Abu Dhabi, UAE

Abstract. Accurately estimating expected payoffs is central to the pricing of European call options, especially when valuation depends on low-probability events in the distribution tail. This study evaluates the performance of Quantum Amplitude Estimation (QAE), using Iterative Amplitude Estimation (IAE) and Maximum Likelihood Amplitude Estimation (MLAE), in pricing European call options based on historical Apple market data. Despite the theoretical advantages of QAE, our experiments show that both quantum estimators return an expected payoff of zero, even in scenarios where classical methods, Black-Scholes and Monte Carlo simulation yield significantly positive values. This outcome stems from the limited resolution imposed by limited uncertainty qubits, which inadequately encode small-amplitude, in-the-money price regions. While QAE circuits correctly identify the realized market outcome, they fail to capture the full expectation implied by the distribution. These results highlight the current limitations of QAE under realistic constraints and underscore the importance of enhanced encoding strategies for future quantum financial applications.

1 Introduction

Pricing European call options, contracts that give the holder the right to buy an asset at strike K on maturity T, is a cornerstone problem in quantitative finance. Classical methods such as the Black-Scholes (BS) model provide closed-form solutions under idealized assumptions [1], while Monte Carlo (MC) simulation offers a flexible simulation-based alternative at the cost of $O(1/\epsilon^2)$ sample complexity when targeting accuracy ϵ [2,3].

Recently, quantum computing-based techniques have been applied to financial applications [4–14]. Quantum Amplitude Estimation (QAE) promises an alternate route to expectation estimation, with asymptotic scaling $O(1/\epsilon)$ in the

© The Author(s), under exclusive license to Springer Nature Switzerland AG 2026
F. Barbaresco and G. François (Eds.): QUEST-IS 2025, CCIS 2743, pp. 348–358, 2026.
https://doi.org/10.1007/978-3-032-13852-1_35

ideal, fault-tolerant limit [15]. Practical variants, Iterative Amplitude Estimation (IAE) and Maximum Likelihood Amplitude Estimation (MLAE) [16,17], adapt QAE to near-term, noisy quantum devices by replacing deep phase-estimation subroutines with iterative Grover-based measurements and classical inference.

In this study, we do not assume that quantum methods will immediately outpace classical techniques [18]. Instead, our objective is to investigate whether QAE-based algorithms can faithfully replicate the structure and outcomes of classical option-pricing workflows. Specifically, we ask:

Can IAE and MLAE, when applied to a real-world European option pricing problem, produce expected-payoff estimates that align with BS and MC benchmarks, given current hardware constraints and state-preparation limitations?

To address this question, we construct a pipeline in Qiskit comprising:

1. **Distribution Encoding**: Discretize a log-normal model of asset returns into 2^n bins and prepare the corresponding quantum superposition.
2. **Payoff Circuit**: Implement the European call payoff $f(B) = \max(B - K, 0)$ as a controlled rotation on an ancilla qubit via a piecewise linear mapping.
3. **Amplitude Estimation**: Run IAE and MLAE to extract the amplitude corresponding to the expected payoff, each with carefully bounded Grover depths.
4. **Classical Benchmarks**: Compute analytic (BS) and numerical (MC) reference values using the same discretization grid for direct comparison.

By comparing quantum estimates against classical baselines on historical data, we identify the primary sources of deviation, such as finite-qubit discretization and limited iteration counts, and assess to what extent QAE algorithms can mirror classical pricing behavior. Our findings clarify the practical viability of quantum methods for financial expectation tasks and highlight concrete steps toward closing the gap between current quantum capabilities and established classical workflows.

2 Background

2.1 Black-Scholes Algorithm

The BS model is a foundational analytical framework for pricing European call options. Given an underlying asset price S, strike price K, risk-free interest rate r, time to maturity T, and volatility σ, the theoretical price C of the option is derived using the following expressions: $d_1 = \dfrac{\ln\left(\frac{S}{K}\right) + \left(r + \frac{\sigma^2}{2}\right)T}{\sigma\sqrt{T}}, d_2 = d_1 - \sigma\sqrt{T}$, and $C = S\,\Phi(d_1) - K\,e^{-rT}\,\Phi(d_2)$, where $\Phi(\cdot)$ denotes the cumulative distribution function of the standard normal distribution. While efficient, this closed-form solution assumes constant volatility, frictionless markets, and log-normal price distributions, assumptions that may not hold in practice.

2.2   Monte Carlo Simulation

MC methods are widely used to simulate asset paths under the risk-neutral measure, providing a flexible numerical approach to option pricing [19]. The terminal asset price under geometric Brownian motion is modeled as: $S(T) = S_0 \exp\left((r - \frac{1}{2}\sigma^2)T + \sigma\sqrt{T}Z\right)$, $Z \sim \mathcal{N}(0,1)$. For each simulated path, the payoff is computed as $\max(S(T) - K, 0)$, and the option value is estimated by discounting the average payoff: $V_{\mathrm{MC}} = e^{-rT} \cdot \frac{1}{N} \sum_{i=1}^{N} \max(S_i(T) - K, 0)$.

MC simulation generalizes naturally to asset options by aggregating the weighted outcomes of multiple assets. Although accuracy improves with the number of samples, convergence rate is bounded by $\mathcal{O}(1/\epsilon^2)$, where ϵ is the desired error tolerance.

2.3   Quantum Amplitude Estimation

QAE is a quantum algorithm designed to estimate the amplitude a of a quantum state with quadratic speedup over classical sampling methods [20]. Let $\mathcal{A}$ be a unitary operator that prepares the state $\mathcal{A}|0\rangle = \sqrt{1-a}\,|\psi_0\rangle + \sqrt{a}\,|\psi_1\rangle$, where $a \in [0,1]$ is the target quantity encoding the expectation of a function (e.g., payoff). The standard QAE algorithm employs phase estimation on the Grover-like operator $Q = \mathcal{A}S_0\mathcal{A}^\dagger S_\chi$ to extract an estimate $\hat{a}$, using a number of queries that scales as $\mathcal{O}(1/\epsilon)$ in the noiseless setting. However, implementing this version requires deep quantum circuits with inverse operations and the quantum Fourier transform, which are impractical on current noisy devices.

Iterative Amplitude Estimation. IAE avoids the need for quantum phase estimation and uses repeated Grover iterations combined with classical statistical inference. It performs a sequence of measurements after controlled applications of the Grover operator to refine the estimate of a iteratively (see Algorithm 1). The method preserves the asymptotic advantage while using shallower circuits, making it suitable for near-term hardware.

Algorithm 1. IAE

1: **Input:** Unitary $\mathcal{A}$, Grover operator $Q = \mathcal{A}S_0\mathcal{A}^\dagger S_\chi$, confidence level δ
2: Initialize estimate interval $[a_{\min}, a_{\max}] \leftarrow [0,1]$
3: **for** $j = 1$ to J **do**
4:     Choose number of Grover iterations m_j
5:     Apply $Q^{m_j}\mathcal{A}|0\rangle$, measure ancilla
6:     Update confidence interval for a using binomial likelihood
7: **end for**
8: **Return:** Final estimate $\hat{a}$ as midpoint of final confidence interval

---

Maximum Likelihood Amplitude Estimation. MLAE also avoids phase estimation but relies on sampling measurement outcomes for different Grover powers m and uses a classical maximum likelihood procedure to reconstruct the amplitude [21]. For each m, the probability of measuring a success outcome is $\sin^2((2\,m + 1)\theta)$, where $\theta = \arcsin(\sqrt{a})$. MLAE solves for the value $\hat{a}$ that maximizes the likelihood of the observed results (see Algorithm 2).

Algorithm 2. MLAE

1: **Input:** Unitary $\mathcal{A}$, list of powers $\{m_j\}$, number of shots N_j per m_j
2: **for** $j = 1$ to J **do**
3:      Apply $Q^{m_j}\mathcal{A}|0\rangle$, measure ancilla qubit
4:      Record number of successes k_j out of N_j trials
5: **end for**
6: Define log-likelihood function $\ell(\hat{a}) = \sum_j \log \Pr(k_j \mid m_j, \hat{a})$
7: **Return:** $\hat{a} = \arg\max_{a \in [0,1]} \ell(a)$

MLAE can be more robust than IAE under noise, especially when the likelihood function is sharply peaked and the number of samples is sufficient. Both IAE and MLAE serve as practical alternatives to QAE in noisy intermediate-scale quantum settings [22].

3   Methodology

This study explores the application of quantum amplitude estimation algorithms to the problem of pricing European call options. As shown in Fig. 1, the workflow comprises 4 main stages: (1) preparing the data, (2) modeling the uncertainty of the asset price using a log-normal distribution, (3) encoding the payoff function into a quantum circuit, and (4) estimating the expected payoff using two amplitude estimation algorithms, IAE and MLAE, and benchmarking the results against classical references.

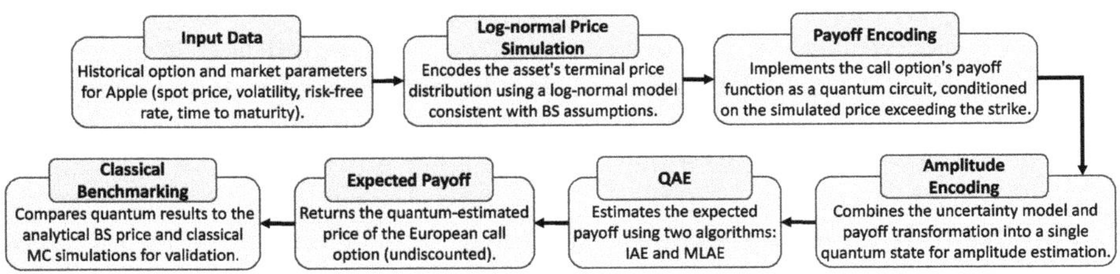

Fig. 1. Pipeline for quantum pricing of European call options. Starting from historical market data, the method proceeds through quantum encoding and amplitude estimation before comparing the quantum results with classical references (BS and MC).

3.1   Input Data and Preprocessing

We start with historical option and market data for Apple (AAPL) over the study period. The relevant features extracted include spot price S, strike price K, implied volatility σ, time to maturity T, and risk-free rate r. Only a subset of this data, filtered by data quality and maturity length, is used in quantum simulation. These features are then used to parameterize the log-normal model consistent with the BS model.

3.2   Log-Normal Price Simulation

We model the terminal asset price S_T using a log-normal distribution, consistent with the assumptions of the BS framework. The quantum state preparation uses the `LogNormalDistribution` component from Qiskit Finance[1] to encode the discretized probability distribution into a quantum circuit. The distribution parameters are derived from market inputs: initial spot price S, volatility σ, risk-free interest rate r, and time to maturity T. The support of the distribution is truncated to the interval $[S_{\min}, S_{\max}]$, spanning three standard deviations around the mean to ensure sufficient coverage. Let $\mu = \log(S) + \left(r - \frac{\sigma^2}{2}\right) T$, $\sigma_T = \sigma\sqrt{T}$, then the quantum state encodes the distribution as $|\psi\rangle = \sum_{i=0}^{2^n - 1} \sqrt{p_i}\,|i\rangle$, where $p_i \approx \mathbb{P}(S_T \in \mathrm{bin}_i)$.

3.3   Encoding the Payoff Function

The circuit is extended to represent the payoff function of a European call option, defined as $\max(S_T - K, 0)$. This is implemented as a piecewise linear transformation using Qiskit's `LinearAmplitudeFunction`. The operation applies controlled rotations on an objective qubit, conditioned on whether the simulated price exceeds the strike. A scaling factor c ensures the resulting payoff fits within the interval $[0, 1]$, satisfying amplitude estimation requirements. The quantum circuit consists of three key components:

- **State qubits**, which encode the log-normal distribution over asset prices. Each computational basis state $|i\rangle$ corresponds to a price bin s_i, with amplitude $\sqrt{p_i}$, forming the state $|\psi\rangle = \sum_i \sqrt{p_i}|i\rangle$.
- **An objective qubit**, initialized in $|0\rangle$, whose amplitude is modulated according to the payoff function. Using breakpoints and slopes, the payoff $\max(s_i - K, 0)$ is mapped linearly to amplitudes. The objective qubits are used to represent the payoff function with higher precision before it is mapped into amplitude space.
- **Ancilla qubits**, used to support the conditional logic required by the piecewise linear implementation. These ensure correct breakpoint handling and amplitude normalization.

Together, these elements form the composite state: $\mathcal{A}|0\rangle = \sqrt{1 - a}\,|\psi_0\rangle + \sqrt{a}\,|\psi_1\rangle$, where $a \in [0, 1]$ is proportional to the expected payoff.

---

[1] https://qiskit-community.github.io/qiskit-finance/.

3.4 Amplitude Estimation Algorithms

To extract the value of a, we apply two variants of amplitude estimation on the same quantum problem:

- IAE refines the estimate of a through successive rounds of Grover-like amplification and classical post-processing, avoiding the need for quantum Fourier transforms. It is initialized with a target precision ϵ and confidence level α, and proceeds with a fixed number of Grover operator applications per round. The final estimate is computed via a maximum likelihood strategy over all rounds.
- MLAE samples different Grover powers and formulates a classical optimization problem to infer the most likely amplitude $\hat{a}$. Unlike IAE, MLAE does not require pre-specifying a precision threshold and instead focuses purely on statistical likelihood maximization, making it potentially more resilient to sampling noise.

Both methods use identical quantum circuits for state preparation and payoff encoding, ensuring a fair comparison across the two estimation strategies.

3.5 Expected Payoff and Classical Benchmarking

The estimated amplitude a is rescaled to recover the expected payoff, expressed as $\mathbb{E}\left[\max(S_T - K, 0)\right] \approx a \cdot c$, where c is the payoff scaling factor. To validate the quantum estimates, we compute two classical references: The analytical price using the BS model, and the MC simulated price using standard sampling of log-normal paths. Both references are computed over the same discretized price space to ensure fair comparison. Additionally, delta estimates representing $\mathbb{P}(S_T > K)$ are derived from the quantum circuit by querying ancilla qubits associated with price thresholds.

4 Results and Analysis

4.1 Experimental Settings

As presented in Table 1, the experimental setup uses real-world financial data for Apple Inc. (AAPL), collected via the Yahoo Finance API [23], covering the period from January 2, 2022, to February 13, 2024. A 5% shift is applied to the mean log returns to simulate a bullish market scenario. The volatility is clipped to a minimum of 0.20 to ensure realistic risk estimates. The time to maturity is fixed at one year, and the strike price is set to 5% above the initial asset value. For quantum simulation using Qiskit [24], we allocate 6 uncertainty qubits to discretize the terminal asset price into 64 bins. The payoff is approximated using a piecewise linear function mapped onto 3 objective qubits, enabling finer-grained approximation of the payoff and more accurate encoding of small but significant values, with the help of 3 ancilla qubits. The amplitude estimation routines are configured as follows:

- **IAE**: Maximum of 3 Grover iterations, with precision $\epsilon = 0.01$ and confidence level $\alpha = 0.05$.
- **MLAE**: Evaluation schedule set to $\{0, 1, 3, 7\}$.

Classical benchmarks include a dot-product-based expected value from the discretized payoff distribution (BS) and a MC estimate using 100,000 samples.

Table 1. Simulation parameters for quantum and classical configurations.

Parameter	Value/Description
Assets	[''AAPL'']
Date Range	2022-01-02 to 2024-02-13
Return Shift	+5% to mean log returns
Min. Volatility	0.20 (capped)
Maturity Time T	1 year
Strike Price	5% above initial asset value
Uncertainty Qubits	6
Objective Qubits	3
Ancilla Qubits	3
IAE Parameters	$\epsilon = 0.01$, $\alpha = 0.05$, max 3 iterations
MLAE Schedule	$\{0, 1, 3, 7\}$
MC Samples	100,000
MC Passes	Single run per trial

4.2   Empirical Results and Analysis

Tables 2 and 3 report the expected payoff and final asset price estimates for two distinct market scenarios, each calibrated from historical AAPL data with a strike set 5% above the initial asset value. In both cases, the realized payoff is zero (terminal price < strike), yet classical methods predict non-zero expected payoffs ($17.60 in 2023–2024 and $17.06 in 2022–2023), by integrating the tail of the price distribution. These values are obtained analytically via a dot product of the payoff and probability vectors and confirmed by MC (100,000 samples), which show deviations below 1%. In contrast, both quantum methods (IAE and MLAE) yield an expected payoff of 0.00 in both scenarios. This underestimation does not reflect a failure of the algorithms themselves, but rather the limitations imposed by coarse amplitude encoding: with only six qubits discretizing the price range into 64 bins, the contributions from low-probability, in-the-money regions are too small to be resolved. As a result, these significant but subtle amplitudes fall below the estimation threshold, causing QAE to return zero despite the positive expected payoff predicted by classical models.

This effect also appears in the estimated asset prices: classical approaches give averages near \$198 (2023–2024) and \$192 (2022–2023), whereas quantum outputs cluster around \$185 and \$179, respectively. The disparity underscores the sensitivity of payoff estimation to distribution resolution, low-qubit implementations struggle to capture low-probability, high-payoff events that drive the true expectation.

Table 2. Expected payoff and asset price (2023–2024).

Method	Expected Payoff (USD)	Asset Price (USD)
Strike: 194.97 / Actual Payoff: 0.00		
Classical (Exact)	17.5987	198.0063
MC	17.8259	198.1746
IAE	0.0000	185.3681
MLAE	0.0000	185.3681

Table 3. Expected payoff and asset price (2022–2023).

Method	Expected Payoff (USD)	Asset Price (USD)
Strike: 193.97 / Actual Payoff: 0.00		
Classical (Exact)	17.0597	191.9414
MC	17.2799	192.1046
IAE	0.0000	179.6903
MLAE	0.0000	179.6749

4.3   Interpretation and Limitations

The disparity between quantum and classical estimates highlights the critical role of distribution resolution in QAE-based pricing. Six uncertainty qubits yield only 64 discrete bins, insufficient to capture the small but financially significant amplitudes above the strike. Consequently, low-probability events that dominate the expected payoff are effectively invisible to the amplitude estimation routines. Both IAE and MLAE correctly identify the most probable outcome (zero payoff), demonstrating stability and potential in binary classification or threshold-detection tasks. However, recovering the full expectation requires finer amplitude granularity, either via more qubits, advanced encoding schemes, or adaptive binning strategies. Our results also suggest that, even in a noiseless simulator with fully expressive circuits (e.g., IAE with 172,000 gates, MLAE with 12,600 gates), algorithmic limitations arise from the encoded probability landscape rather than circuit depth. Future work should therefore prioritize enhanced state preparation and payoff mapping, such as nonuniform bin spacing around strike regions or variational encoding, to better resolve tail events.

4.4 Discussion

The experimental results emphasize the current limitations of QAE for realistic financial estimation tasks, even when using expressive circuits. The coarse discretization from six qubits remains the dominant factor behind payoff underestimation, as bins near the strike fail to capture the relevant variations in asset prices. Though noise and hardware effects are excluded, transpilation reflects some realistic overheads, offering a practical perspective on achievable performance under ideal conditions. It is encouraging that both IAE and MLAE consistently predicted the realized market payoff, suggesting utility in binary event detection even under coarse resolution. These findings point toward promising avenues for improvement: increasing qubit counts, using nonuniform binning, or designing more targeted amplitude distributions. The presented pipeline offers a benchmark for testing such enhancements in future iterations of QAE.

4.5 Conclusion

This study evaluated the use of QAE algorithms for pricing European call options using historical asset data in a simulated quantum environment. While classical methods, dot-product integration (BS) and MC, accurately captured the expected payoff, quantum estimators returned a value of zero, underestimating the true financial expectation. The performance gap stems from resolution limitations and the inability to amplify low-amplitude but significant payoff states, not from insufficient circuit expressivity. This underscores the importance of encoding strategies and amplitude shaping in financial QAE applications.

Importantly, quantum circuits still produced stable and interpretable outputs aligned with the realized payoff, indicating near-term utility in classification or binary decision tasks. As quantum technology advances, the foundational pipeline developed here can guide progress toward accurate and scalable quantum financial analytics.

Acknowledgment. This work was supported in part by the NYUAD Center for Quantum and Topological Systems (CQTS), funded by Tamkeen under the NYUAD Research Institute grant CG008.

References

1. Black, F., Scholes, M.: The pricing of options and corporate liabilities. J. Polit. Econ. (1973)
2. Glasserman, P.: Monte Carlo Methods in Financial Engineering. Springer (2004)

3. Sharma, P., et al.: Review of research on option pricing: a bibliometric analysis. Qualit. Res. Finan. Mark. (2024)
4. Innan, N., Marchisio, A., Bennai, M., Shafique, M.: QFNN-FFD: quantum federated neural network for financial fraud detection. In: 2025 IEEE International Conference on Quantum Software (QSW), pp. 41–47. IEEE (2025)
5. Dutta, S., et al.: QADQN: quantum attention deep q-network for financial market prediction. In: 2024 IEEE International Conference on Quantum Computing and Engineering (QCE), vol. 2, pp. 341–346. IEEE (2024)
6. Innan, N., Marchisio, A., Bennai, M., Shafique, M.: LEP-QNN: loan eligibility prediction using quantum neural networks. arXiv preprint: arXiv:2412.03158 (2024)
7. Innan, N., Kashif, M., Marchisio, A., Bennai, M., Shafique, M.: Next-generation quantum neural networks: enhancing efficiency, security, and privacy. In: 2025 IEEE 31st International Symposium on On-Line Testing and Robust System Design (IOLTS), pp. 1–4. IEEE (2025)
8. Zaman, K., Marchisio, A., Hanif, M.A., Shafique, M.: A survey on quantum machine learning: Current trends, challenges, opportunities, and the road ahead. arXiv preprint: arXiv:2310.10315 (2023)
9. Zaman, K., Marchisio, A., Kashif, M., Shafique, M.: PO-QA: a framework for portfolio optimization using quantum algorithms. In: 2024 IEEE International Conference on Quantum Computing and Engineering (QCE), vol. 1, pp. 1397–1403. IEEE (2024)
10. Alami, M.E., Innan, N., Shafique, M., Bennai, M.: Comparative performance analysis of quantum machine learning architectures for credit card fraud detection. arXiv preprint: arXiv:2412.19441 (2024)
11. Choudhary, P.K., Innan, N., Shafique, M., Singh, R.: HQNN-FSP: a hybrid classical-quantum neural network for regression-based financial stock market prediction. arXiv preprint: arXiv:2503.15403 (2025)
12. Sawaika, A., et al.: A privacy-preserving federated framework with hybrid quantum-enhanced learning for financial fraud detection. arXiv preprint: arXiv:2507.22908 (2025)
13. Innan, N., Saleem, A., Marchisio, A., Shafique, M.: Quantum portfolio optimization with expert analysis evaluation. arXiv preprint: arXiv:2507.20532 (2025)
14. Kashif, M., Khalid, S., Innan, N., Marchisio, A., Shafique, M.: Evaluating quantum amplitude estimation for pricing multi-asset basket options. arXiv preprint: arXiv:2509.09432 (2025)
15. Brassard, G., et al.: Quantum amplitude amplification and estimation. Contemp. Math. (2002)
16. Grinko, D., et al.: Iterative quantum amplitude estimation. NPJ Quantum Inf. (2021)
17. Tanaka, T., Suzuki, Y., Uno, S., Raymond, R., Onodera, T., Yamamoto, N.: Amplitude estimation via maximum likelihood on noisy quantum computer. Quantum Inf. Process. **20**(9), 1–29 (2021). https://doi.org/10.1007/s11128-021-03215-9
18. Stamatopoulos, N., et al.: Option pricing using quantum computers. Quantum (2020)
19. Woerner, S., Egger, D.J.: Quantum risk analysis. NPJ Quantum Inf. (2019)
20. Vazquez, A.C., Woerner, S.: Efficient state preparation for quantum amplitude estimation. arXiv preprint: arXiv:2005.07711 (2020)
21. Suzuki, Y., Uno, S., Raymond, R., Tanaka, T., Onodera, T., Yamamoto, N.: Amplitude estimation without phase estimation. Quantum Inf. Process. **19**(2), 1–17 (2020). https://doi.org/10.1007/s11128-019-2565-2

22. Maronese, M., et al.: The quantum amplitude estimation algorithms on near-term devices: a practical guide. Quantum Rep. (2023)
23. Aroussi, R.: yfinance (2023). https://pypi.org/project/yfinance/
24. Javadi-Abhari, A., et al.: Quantum computing with Qiskit. arXiv preprint: arXiv:2405.08810 (2024)

Author Index

F. Barbaresco and G. François (Eds.): QUEST-IS 2025, CCIS 2743, pp. 359–361, 2026.
https://doi.org/10.1007/978-3-032-13852-1

GPSR Compliance
The European Union's (EU) General Product Safety Regulation (GPSR) is a set
of rules that requires consumer products to be safe and our obligations to
ensure this.

If you have any concerns about our products, you can contact us on

ProductSafety@springernature.com

In case Publisher is established outside the EU, the EU authorized
representative is:

Springer Nature Customer Service Center GmbH
Europaplatz 3
69115 Heidelberg, Germany